MW01145482

Akademie Verlag Series in
Optical Metrology

Akademie Verlag Series in Optical Metrology

Edited by W. Jüptner and W. Osten

Volume 1

Thomas Kreis

Holographic Interferometry

Principles and Methods

Akademie Verlag

Author:
Dr. Thomas Kreis,
Bremen Institute of Applied Beam Technology (BIAS), Bremen/Germany

Series Editors:
Professor Dr. Werner Jüptner, Dr. Wolfgang Osten
Bremen Institute of Applied Beam Technology (BIAS),
Klagenfurter Str. 2, D–28359 Bremen, Germany

With 163 figures and 8 tables

1st edition
Library of Congress Card Number pending

Die Deutsche Bibliothek – CIP-Einheitsaufnahme

Kreis, Thomas:
Holgraphic interferometry : principles and methods / Thomas
Kreis. – 1. ed. – Berlin : Akad. Verl., 1996
(Akademie Verlag series in optical metrology ; Vol. 1)
ISBN 3-05-501644-0
NE: GT

ISSN 1430-3965

Akademie Verlag is a member of the VCH Publishing Group.

Printed on non-acid paper.
The paper used corresponds to both the U.S. standard ANSI Z.39.48 – 1984 and the European standard ISO TC 46.

Printing: GAM Media GmbH, Berlin
Bookbinding: Verlagsbuchbinderei Mikolai GmbH, Berlin

Printed in the Federal Republic of Germany

Akademie Verlag GmbH
Postfach, D-13162 Berlin
Federal Republic of Germany

VCH Publishers, Inc.
220 East 23rd Street
New York, NY 10010-4606

Preface

Holographic interferometry is a method for measuring all changes in physical quantities that can be transformed into a variation of the phase of an optical wave. It has gained more and more interest and applications in recent decades. Holographic interferometry as a metrologic tool profits a great deal from the progress in such fields as optics, electronics, and computer technology. However, this has the effect that a textbook on holographic interferometry, which should recognize all these aspects, would explode in extent, so one has to choose one primary aspect. The particular aspect in this book is to render the principles and methods which are needed for any computer-aided evaluation of holographic interference patterns. There are excellent books on optics, laser physics, optical engineering, optical holography, interferometry, signal analysis, image processing etc., which provide valuable and up-to-date sources of information in these fields. The aim of this book is to present a self-contained treatment of the underlying principles and numerical methods intended to help the physicist or engineer in planning a holographic interferometric measurement, writing the evaluation software, performing the experiments, and interpreting the results of the computer-aided evaluation. The employment of computer power in holographic interferometry should not be restricted to, for example the determination of interference phase distributions from recorded interference patterns, but it also enables numerical feasibility studies, simulations of holographic interferograms, the optimization of a holographic setup with regard to sensitivity and accuracy, an automatic control of the measurement process or further processing of the interference phase, for example by numerical strain and stress analysis, by finite element and boundary element methods or by computerized tomography. The book should provide the fundamentals for making these attempts, where solutions cannot be given - because either they do not yet exist or they would go beyond the scope of this book - it should act as an incentive for further research and development work.

A second important reason for making a restriction to a clear-cut topic is that the book is a part of a larger book series on "Optical Metrology" being published by Akademie-Verlag Berlin. Therefore such topics as optical, electro-optical, and mechanical elements in the holographic setup, lasers, and fields of holographic nondestructive testing, speckle metrology, or digital image processing are only dealt with very briefly in this book, because it is planned that these themes should be treated in future books of the series devoted especially to these topics.

The list of references by no means constitutes a complete bibliography on the subject. Basically I have included those references I have found useful in my own research over the years. Often the use of a particular paper was dictated more by chance than by systematic search. Generally I have not made any attempts to establish historical priorities. No value judgments should be implied by my including or excluding a particular work. I extend my apologies to anyone who was inadvertently slighted or whose work has been overlooked.

It is hoped that this book will help the reader to exploit the various possibilities of holographic interferometry, to make the best choice of methods, and to use successfully the tools and algorithms presented herein. The author aims to present an account of the present state of the methods of holographic interferometric metrology; it must be emphasized that there is a considerable amount of research and development in progress and the subject needs to be kept in constant review. I hope, therfore, that the readers will be challenged to think, criticize, read further, and quickly go beyond the confines of this volume.

Of course, I must take the blame for any mistakes, omissions of significant facts, incomprehensible descriptions, residual errors or misprintings. The readers are cordially invited for providing me any corrections, comments, hints, or suggestions to improve the presentation, which may be realized in the preparation of potential further editions of this book. Therefore the full address of the author is included.

Clearly this book would not have been completed without the help of many persons. The editors of the book series, Prof. Dr. W. Jüptner and Dr. W. Osten helped greatly by critical and stimulating discussions. The book draws heavily from the research done at BIAS by my past and present colleagues and collaborators, especially J. Geldmacher, U. Schnars, Th. Bischof, and R. Biedermann which is gratefully acknowledged. S. Knoll and Chr. Kapitza produced many of the holographic interferograms used to illustrate this book, the photographs have been carefully prepared by B. Schupp. Thanks to all colleagues, especially H.-J. Hartmann, who made it possible for me to work on the book by relieving me from much of my daily routine. Also thanks go to my family for their understanding and acceptance when I devoted my spare time to the book.

Dr. Thomas Kreis
Bremer Institut für Angewandte Strahltechnik - BIAS
Klagenfurter Straße 2
D 28359 Bremen
Germany

E-mail: thkreis@rbsoft.bias.uni-bremen.de

Contents

1

Introduction

1.1 Scope of the Book

The emerging computer technology of the last decades - increasing processing speed and memory capacity, as well as CCD-camera targets having more and smaller pixels - makes the manifold applications of what can be called 'computer-aided holographic interferometry' feasible: In the planning phase of a holographic interferometric experiment the geometry of the setup can be optimized to achieve maximum sensitivity and accuracy. The load to be applied can be optimized in its type, direction, and amplitude by numerical simulation of the holographic interferograms that results from a specific load and geometry. The determination of the interference phase distribution from the recorded interference patterns by refined methods such as phase stepping or Fourier-transform evaluation is only possible with powerful computers. Further processing of the interference phase distribution by solving linear equations to obtain displacement fields or by employing computer-tomography to calculate refractive index fields can now be effectively carried out. Methods for numerical strain and stress analysis can be combined with computerized holographic interferometry to gain far-reaching knowledge about the behaviour of the tested structure with regard to the applied load. Even structural analysis methods such as finite element methods (FEM) or boundary element methods (BEM) can be efficiently associated to holographic interferometry to assist the strain and stress calculations, to optimize the component design process, or to predict the interference patterns for a given load.

The aim of this book is to present the physical principles of holography and interferometry as far as they are needed in this context, as well as the numerical methods for evaluation of the interference patterns, which constitute the fundamentals of a computer-aided holographic interferometry. The emphasis is on quantitative measurements, while the qualitative evaluation of holographic nondestructive testing (HNDT) is treated only briefly, a book devoted exclusively to this topic is planned to appear in the same book series. The present book should provide the background needed for deriving the concepts and writing the programmes to solve problems in the above mentioned fields. To fulfil these claims but not to become too extensive and to exceed the frame set by the publishers, some topics have been intentionally omitted. The description of technical components

- lasers, optics, electro-optic devices, recording media, image processing equipment and methods, computer periphery - is restricted to a very short overview, these topics are to be treated in the detail they deserve in further books to be published in this book series. The same is true for themes such as speckle metrology, digital image processing, or particle and flow-field measurements. These items are addressed here only briefly as far as it seems necessary for a comprehensive presentation. Other applications of holography than interferometric metrology, such as display holography, computer generated holograms, holographic optical elements, colour holography, holographic data storage, etc., are excluded intentionally.

The book is organized into six chapters and three appendixes. The main body of the content is contained in chapters 2, 3, 4, and 5. Chapter 2 presents the physical prerequisites of holography, starting with the wave theory of light, describing such effects as interference, diffraction, coherence, speckle, fringe localization and how these are employed in holographic recording and reconstruction of optical wave fields.

In chapter 3 the fundamentals of holographic interferometric metrology are presented. The quantitative relations for displacements or refractive index variations and the geometric and optical parameters of the holographic setup are introduced. Further discussions center on the role of the sensitivity vectors and the localization of the interference fringes.

Chapter 4 is devoted to methods for determining the interference phase distributions from recorded intensity images. Room is given to a thorough treatment of the phase-stepping and the Fourier-transform methods as well as the relatively new method of digital holography. Systematic and statistic errors are discussed, a number of approaches for interference phase demodulation are presented.

Chapter 5 describes the further processing of the interference phase distribution. Displacement vector fields, strain and stress distributions, vibration modes, three-dimensional object contours or refractive index fields are determined. Computerized defect detection of holographic nondestructive testing is briefly addressed.

Speckle methods for deformation measurement are closely related to holographic interferometry. It can be anticipated, that when the number of pixels in CCD-targets increases and the pixel size decreases more and more, the distinction between speckle methods and holographic methods will become obsolete. The main speckle methods are discussed shortly in chapter 6, a detailed treatment is left for another book in the series.

The appendices provide the reader with the essentials of Fourier transforms, methods for computerized tomography, and Bessel functions, as far as this seems necessary to understand and implement the related methods of the main chapters.

1.2 Historical Developments

Holography got its name from the Greek words 'holos' meaning whole or entire and 'graphein' meaning to write. It is a means for recording and reconstructing the whole information contained in an optical wavefront, namely amplitude and phase, and not just intensity as ordinary photography. Holography essentially is a clever combination of interference and diffraction, two phenomena based on the wave nature of light.

Diffraction was first noted by F. M. Grimaldi (1618 - 1663) as the deviation from rectilinear propagation, the interference generated by thin films was observed and described by R. Hooke (1635 - 1703). I. Newton (1642 - 1727) discovered the composition of white light from independent colours. The mathematical basis for the wave theory describing these effects was founded by Chr. Huygens (1629 - 1695), who further discovered the polarization of light. The interference principle introduced by Th. Young (1773 - 1829) and Huygens principle were used by A. J. Fresnel (1788 - 1827) to calculate the diffraction patterns of different objects. Since about 1850 the view of light as a transversal wave won against the corpuscular theory. The relations between light, electricity, and magnetism were recognized by M. Faraday (1791 - 1867). These phenomena were summarized by J. C. Maxwell (1831 - 1879) in his well known equations. A medium supporting the waves was postulated as the all pervading ether. The experiments of A. A. Michelson (1852 - 1931), published in 1881 and the work of A. Einstein (1879 - 1955) led to the recognition that there is no ether.

In 1948 D. Gabor (1900 - 1979) presented holography as a lensless process for image formation by reconstructed wavefronts [1, 2, 3]. He primarily aimed at an improvement of electron microscopy by avoiding the aberration with this approach. However, a successful application of the technique to electron microscopy has not materialized so far because of several practical problems. The validity of Gabor's ideas in the optical field was recognized and confirmed by, for example, Rogers [4], El-Sum and Kirkpatrick [5], and Lohmann [6], but the interest in holography declined after a few years, mainly because of the poor quality of the holographic images obtained in these days. The breakthrough of holography was initiated by the development of the laser which made available a powerful source of coherent light. This was accompanied by the solution of the twin-image problem encountered in Gabor's in-line arrangement. Leith and Upatnieks [7, 8, 9] recognized the similarity of Gabor's holography to the synthetic aperture antenna problem of radar technology and introduced the off-axis reference beam technique. These advances initiated an explosive growth of activity in optical holography, which among other achievements led to the discovery of holographic interferometry by Stetson and Powell in 1965 [10].

1.3 Holographic Interferometry as a Measurement Tool

In holographic interferometry, two or more wave fields are compared interferometrically, at least one of them must be holographically recorded and reconstructed. The method gives rise to interference patterns whose fringes are determined by the geometry of the holographic setup via the sensitivity vectors and by the optical path length differences. Thus holographic interference patterns can be produced by keeping the optical path length difference constant and changing the sensitivity vectors, by holding the sensitivity vectors constant and varying the optical path length differences, or by altering both of them between the object states to be compared. Especially the path lengths can be modified by a number of physical parameters. The flexibility and the precision gained by comparing the optical path length changes with the wavelength of the laser light used makes holographic interferometry an ideal means for measuring a manifold of physical quantities [11, 12, 13]. The main advantages are:

- The measurements are contactless and noninvasive. In addition to an eventual loading for inducing the optical pathlength changes, the object is only impinged by light waves. The intensities of these waves are well below the level for causing any damage, even for the most delicate of biological objects.
- A reliable analysis can be performed at low loading intensities: the testing remains non-destructive.
- Not only may two states separated by a long time be compared, but furthermore the generation and evaluation of the holographic information can be separated both temporally and locally.
- Measurements can be made through transparent windows. We can therefore make measurements in pressure or vacuum chambers or protect against hostile environments. Due to the measurement of differences of the optical path lengths instead of absolute values, low quality windows do not disturb the results.
- Holographic interferometric measurements can be accomplished at moving surfaces: Short pulse illumination makes the method insensitive to a disturbing motion, vibrations can be investigated, the holographic setup can be made insensitive to specific motion components, the rotation of spinning objects can be cancelled optically by using an image derotator.
- Deformation measurements can be performed at rough, diffusely reflecting surfaces, which occur frequently in engineering. No specular reflection of the object is required.
- The objects to be examined holographically may be of almost arbitrary shape. Using multiple illumination and observation directions or fiber optics, barely inaccessible areas can be studied.
- Holographic interferometry is nearly independent of the state of matter: Deformations of hard and soft materials can be measured. Refractive-index variations in solids, fluids, gases and even plasmas can be determined.
- Lateral dimensions of the examined subjects may range from a few mm to several m.
- The measurement range extends roughly speaking from a hundredth to several hundreds of a wavelength, for example displacements can be measured from about 0.005 μm to 500 μm.
- The achievable resolution and accuracy of a holographic interferometric displacement measurement permit subsequent numerical strain and stress calculations.
- Two-dimensional spatially continuous information is obtained: local singularities, for example local deformation extrema, cannot go undetected.
- Multiple viewing directions using a single hologram are possible, enabling the application of computerized tomography to obtain three-dimensional fields.

2

Optical Foundations of Holography

This chapter discusses the physical basis of holography and holographic interferometry. The primary phenomena constituting holography are interference and diffraction, which take place because of the wave nature of light. So this chapter begins with a description of the wave theory of light as far as it is required to understand the recording and reconstruction of holograms and the effect of holographic interferometry. In holographic interferometry the variation of a physical parameter is measured by its influence on the phase of an optical wave field. Therefore the dependence of the phase upon the geometry of the optical setup and the different parameters to be measured is outlined.

2.1 Light Waves

2.1.1 Solutions of the Wave Equation

Light is a transverse, electromagnetic wave characterized by time-varying electric and magnetic fields. Since electromagnetic waves obey the Maxwell equations, the propagation of light is described by the wave equation which follows from the Maxwell equations. The *wave equation* for propagation of light in vacuum is

$$\nabla^2 \boldsymbol{E} - \frac{1}{c^2}\frac{\partial^2 \boldsymbol{E}}{\partial t^2} = 0 \tag{2.1}$$

where $\boldsymbol{E}$ is the *electric field strength*, ∇^2 is the *Laplace operator*

$$\nabla^2 = \frac{\partial^2}{\partial x^2} + \frac{\partial^2}{\partial y^2} + \frac{\partial^2}{\partial z^2} \tag{2.2}$$

(x, y, z) are the Cartesian spatial coordinates, t denotes the temporal coordinate, the time, and c is the propagation speed of the wave. The *speed of light* in vacuum c_0 is a constant of nature

$$c_0 = 299\,792\,458 \text{ m/s} \qquad \text{or almost exactly} \quad c_0 = 3 * 10^8 \text{ m/s} \tag{2.3}$$

Transverse waves vibrate at right angles to the direction of propagation and so they must be described in vector notation. The wave may vibrate horizontally, vertically, or in any direction combined of these. Such effects are called *polarization* effects. Fortunately for the most applications it is not necessary to use the full vectorial description of the fields, so we can assume a wave vibrating in a single plane. Such a wave is called *plane polarized.* For a plane polarized wave field propagating in the z-direction the *scalar wave equation* is sufficient

$$\frac{\partial^2 E}{\partial z^2} - \frac{1}{c^2}\frac{\partial^2 E}{\partial t^2} = 0 \tag{2.4}$$

It is easily verified that

$$E(z,t) = f(z - ct) \qquad \text{or} \qquad E(z,t) = g(z + ct) \tag{2.5}$$

are also solutions of this equation, which means that the wavefield retains its form during propagation. Due to the linearity of (2.4)

$$E(z,t) = a\, f(z - ct) + b\, g(z + ct) \tag{2.6}$$

is likewise a solution to the wave equation. This *superposition principle* is valid for linear differential equations in general and thus for (2.1) also.

The most important solution of (2.4) is the *harmonic wave*, which in real notation is

$$E(z,t) = E_0 \cos(kz - \omega t) \tag{2.7}$$

E_0 is the *real amplitude* of the wave, the term $(kz - \omega t)$ gives the *phase* of the wave. The *wave number* k is associated to the *wavelength* λ by

$$k = \frac{2\pi}{\lambda} \tag{2.8}$$

Typical figures of λ for visible light are 514.5 nm (green line of argon-ion Laser) or 632.8 nm (red light of helium-neon laser). The *angular frequency* ω is related to the *frequency* ν of the wave by

$$\omega = 2\pi\nu \tag{2.9}$$

ν is the number of periods per second, that means

$$\nu = \frac{c}{\lambda} \qquad \text{or} \qquad \nu\lambda = c \tag{2.10}$$

If we have not the maximum amplitude at $x = 0$ and $t = 0$, we have to introduce the *relative phase* ϕ

$$E(z,t) = E_0 \cos(kz - \omega t + \phi) \tag{2.11}$$

With the *period* T, the time for a full 2π-cycle, we can write

$$E(z,t) = E_0 \cos(\frac{2\pi}{\lambda} z - \frac{2\pi}{T} t + \phi) \tag{2.12}$$

Fig. 2.1 displays two aspects of this wave. Fig. 2.1a shows the temporal distribution of the field at two points $z = 0$ and $z = z_1 > 0$, Fig. 2.1b gives the spatial distribution of

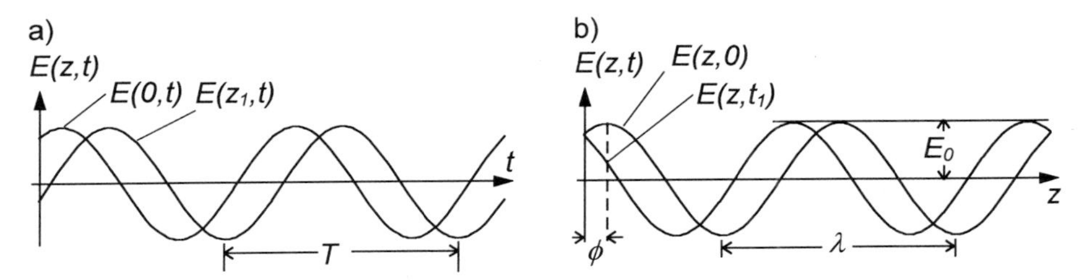

Figure 2.1: Spatial and temporal distribution of a scalar harmonic wave

two periods for time instants $t = 0$ and $t = t_1 > 0$. We see that a point of constant phase moves with the so called *phase velocity*, the speed c.

The use of trigonometric functions leads to cumbersome calculations, which can be circumvented by using the complex exponential which is related to the trigonometric functions by *Euler's formula*

$$\mathrm{e}^{\mathrm{i}\alpha} = \cos\,\alpha + \mathrm{i}\,\sin\,\alpha \tag{2.13}$$

where $i = \sqrt{-1}$ is the imaginary unit. Since the cosine now is

$$\cos\,\alpha = \frac{1}{2}(\mathrm{e}^{\mathrm{i}\alpha} + \mathrm{e}^{-\mathrm{i}\alpha}) \tag{2.14}$$

the harmonic wave (2.11) is

$$E(z,t) = \frac{1}{2}E_0\;\mathrm{e}^{\mathrm{i}(kz-\omega t+\phi)} + \frac{1}{2}E_0\;\mathrm{e}^{-\mathrm{i}(kz-\omega t+\phi)} \tag{2.15}$$

The second term on the right hand side is the complex conjugate of the first term and can be omitted as long as it is understood that only the real part of $E(z,t)$ represents the physical wave. Thus the harmonic wave in complex notation is

$$E(z,t) = \frac{1}{2}E_0\;\mathrm{e}^{\mathrm{i}(kz-\omega t+\phi)} \tag{2.16}$$

A *wavefront* refers to the spatial distribution of the maxima of the wave, or other surfaces of constant phase, as these surfaces propagate. The wavefronts are normal to the direction of propagation. A *plane wave* is a wave which has constant phase in all planes orthogonal to the propagation direction for a given time t. For describing the spatial distribution of the wave, we can assume $t = 0$ in an arbitrary time scale. Since

$$\boldsymbol{k}\cdot\boldsymbol{r} = \text{const.} \tag{2.17}$$

is the equation for a plane in three-dimensional space, with the *wave vector* $\boldsymbol{k} = (k_x, k_y, k_z)$ and the spatial vector $\boldsymbol{r} = (x, y, z)$, a plane harmonic wave at time $t = 0$ is

$$E(\boldsymbol{r}) = E_0\;\mathrm{e}^{\mathrm{i}\boldsymbol{k}\cdot\boldsymbol{r}+\phi} \tag{2.18}$$

This wave repeats after the wavelength λ in direction $\boldsymbol{k}$, which can easily be proved using $|\boldsymbol{k}| = k = 2\pi/\lambda$ by

$$E(\boldsymbol{r} + \lambda\frac{\boldsymbol{k}}{k}) = E(\boldsymbol{r}) \tag{2.19}$$

The expression

$$E(\boldsymbol{r}, t) = E_0 \, \mathrm{e}^{\mathrm{i}(\boldsymbol{k} \cdot \boldsymbol{r} - \omega t + \phi)} \tag{2.20}$$

describes the temporal dependence of a plane harmonic wave propagating in the direction of the wavevector or

$$E(\boldsymbol{r}, t) = E_0 \, \mathrm{e}^{\mathrm{i}(\boldsymbol{k} \cdot \boldsymbol{r} + \omega t + \phi)} \tag{2.21}$$

if the wave propagates contrary to the direction of $\boldsymbol{k}$.

Another waveform often used is the *spherical wave* where the phase is constant on each spherical surface. The importance of spherical waves comes from the *Huygens' principle* which states that each point on a propagating wavefront can be considered as radiating itself a spherical wavelet.

For a mathematical treatment the wave equation now has to be described in polar coordinates (r, θ, ψ), transformed by $x = r \sin\theta \cos\psi$, $y = r \sin\theta \sin\psi$, $z = r\cos\theta$. Due to the spherical symmetry, a spherical wave is not dependent on θ and ψ. Then the scalar wave equation is

$$\frac{1}{r}\frac{\partial^2}{\partial r^2}(rE) - \frac{1}{c^2}\frac{\partial^2 E}{\partial t^2} = 0 \tag{2.22}$$

The solutions of main interest are the harmonic spherical waves

$$E(r, t) = \frac{E_0}{r} \, \mathrm{e}^{\mathrm{i}(kr - \omega t + \phi)} \tag{2.23}$$

One observes that the amplitude E_0/r decreases proportionally to $1/r$. Furthermore in a long distance from the origin the spherical wave locally approximates a plane wave.

The complex amplitudes of wavefronts scattered by a surface are generally very complicated, but due to the superposition principle (2.6) they can be treated as the sum of plane waves or spherical waves. There are still other solutions to the wave equation. An example are the *Bessel waves* of the class of *nondiffracting beams* [14]. But up to now they have not found applications in holographic interferometry, so here we restrict on the plane and on the spherical waves.

2.1.2 Intensity

The only parameter of light which is directly amenable to sensors, - eye, photodiode, CCD-target, etc. - is the *intensity* (and in a rough scale the frequency as colour). Intensity is defined by the energy flux through an area per time. From the Maxwell equations we get

$$I = \varepsilon_0 c E^2 \tag{2.24}$$

where we only use the proportionality of the intensity I to E^2

$$I \sim E^2 \tag{2.25}$$

It has to be recognized that the intensity has a nonlinear dependence on the electric field strength. Since there is no sensor which can follow the frequency of light, we have to integrate over a *measuring time* T_m, the momentary intensity is not measurable. So if $T_m \gg T = 2\pi/\omega$, omitting proportionality constants we define

$$I = E_0 \, E_0^* = |E_0|^2 \tag{2.26}$$

where * denotes the complex conjugate. The intensity of a general stationary wave field is

$$I(\boldsymbol{r}) = \langle E \, E^* \rangle = \lim_{T_m \to \infty} \frac{1}{T_m} \int_{-T_m/2}^{T_m/2} E(\boldsymbol{r}, t') E^*(\boldsymbol{r}, t') dt' \tag{2.27}$$

This intensity is the limit of the *short time intensity*

$$I(\boldsymbol{r}, t, T_m) = \frac{1}{T_m} \int_{t-T_m/2}^{t+T_m/2} E(\boldsymbol{r}, t') E^*(\boldsymbol{r}, t') dt' \tag{2.28}$$

which is a sliding average of a temporal window centered around t with width T_m. The measuring time T_m always is large compared with the period of the light wave but has to be short in the time scale of the investigated process.

2.2 Interference of Light

2.2.1 Interference of Two Waves with Equal Frequency

The *interference* effect which occurs if two or more coherent light waves are superposed, is the basis of holography and holographic interferometry. So we first consider two waves, emitted by the same source, which differ in the directions $\boldsymbol{k_1}$ and $\boldsymbol{k_2}$, and the phases ϕ_1 and ϕ_2, but for convenience have the same amplitude E_0 and frequency ω and are linearly polarized in the same direction. Then in scalar notation

$$\begin{aligned} E_1(\boldsymbol{r}, t) &= E_0 \, \mathrm{e}^{\mathrm{i}(\boldsymbol{k_1} \cdot \boldsymbol{r} - \omega t + \phi_1)} \\ E_2(\boldsymbol{r}, t) &= E_0 \, \mathrm{e}^{\mathrm{i}(\boldsymbol{k_2} \cdot \boldsymbol{r} - \omega t + \phi_2)} \end{aligned} \tag{2.29}$$

For determination of the superposition of these waves we decompose the vectors $\boldsymbol{k_1}$ and $\boldsymbol{k_2}$ into components of equal and opposite directions, Fig. 2.2, $\boldsymbol{k}' = (\boldsymbol{k_1} + \boldsymbol{k_2})/2$ and

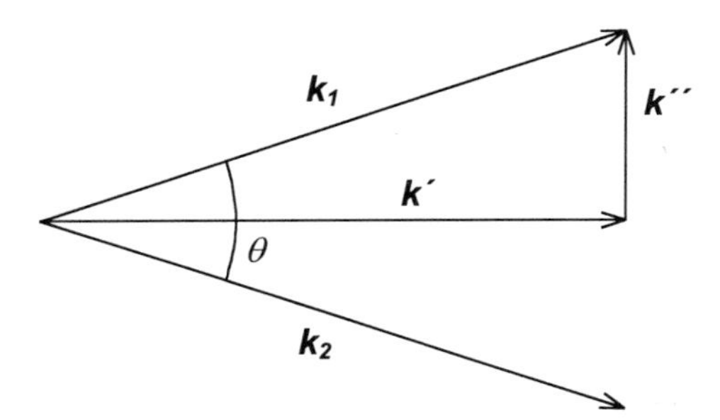

Figure 2.2: Decomposition of wave vectors

$\boldsymbol{k}'' = (\boldsymbol{k_1} - \boldsymbol{k_2})/2$. If θ is the angle between $\boldsymbol{k_1}$ and $\boldsymbol{k_2}$ then

$$|\boldsymbol{k}''| = \frac{2\pi}{\lambda} \sin \frac{\theta}{2} \tag{2.30}$$

In the same way we define the mean phase $\phi = (\phi_1 + \phi_2)/2$ and the half phase difference $\Delta\phi = (\phi_1 - \phi_2)/2$. Now the superposition gives the field

$$\begin{aligned}(E_1 + E_2)(\boldsymbol{r}, t) &= E_0\,\mathrm{e}^{\mathrm{i}(\boldsymbol{k_1}\cdot\boldsymbol{r}-\omega t+\phi_1)} + E_0\,\mathrm{e}^{\mathrm{i}(\boldsymbol{k_2}\cdot\boldsymbol{r}-\omega t+\phi_2)} \\ &= E_0\{\mathrm{e}^{\mathrm{i}(\boldsymbol{k}'\cdot\boldsymbol{r}+\boldsymbol{k}''\cdot\boldsymbol{r}-\omega t+\phi+\Delta\phi)} + \mathrm{e}^{\mathrm{i}(\boldsymbol{k}'\cdot\boldsymbol{r}-\boldsymbol{k}''\cdot\boldsymbol{r}-\omega t+\phi-\Delta\phi)}\} \\ &= E_0\,\mathrm{e}^{\mathrm{i}(\boldsymbol{k}'\cdot\boldsymbol{r}-\omega t+\phi)}\{\mathrm{e}^{\mathrm{i}(\boldsymbol{k}''\cdot\boldsymbol{r}+\Delta\phi)} + \mathrm{e}^{\mathrm{i}(-\boldsymbol{k}''\cdot\boldsymbol{r}-\Delta\phi)}\} \\ &= 2\,E_0\,\mathrm{e}^{\mathrm{i}(\boldsymbol{k}'\cdot\boldsymbol{r}-\omega t+\phi)}\cos(\boldsymbol{k}''\cdot\boldsymbol{r}+\Delta\phi) \end{aligned} \tag{2.31}$$

In this field the exponential term is a temporally varying phase but the cosine term is independent from time. Thus we get the temporally constant intensity

$$\begin{aligned} I(\boldsymbol{r}) &= (E_1 + E_2)(E_1 + E_2)^* \\ &= 4\,E_0^2\,\cos^2(\boldsymbol{k}''\cdot\boldsymbol{r}+\Delta\phi) \end{aligned} \tag{2.32}$$

This means the intensity is minimal where $\cos^2(\boldsymbol{k}''\cdot\boldsymbol{r}+\Delta\phi) = 0$. These are the loci where

$$\boldsymbol{k}''\cdot\boldsymbol{r}+\Delta\phi = (2n+1)\frac{\pi}{2} \qquad n \in \boldsymbol{Z} \tag{2.33}$$

Here the wavefronts are said to be *anti-phase*, we speak of destructive interference. The intensity is maximal where

$$\boldsymbol{k}''\cdot\boldsymbol{r}+\Delta\phi = n\,\pi \qquad n \in \boldsymbol{Z} \tag{2.34}$$

Here the wavefronts are *in-phase*, we have constructive interference.

The resulting time independent pattern is called *interference pattern*, the fringes are called *interference fringes*. For plane waves they are oriented parallel to $\boldsymbol{k}'$ and have a distance of $\pi/|\boldsymbol{k}''|$ in the direction $\boldsymbol{k}''$. This is shown in moiré analogy in Fig. 2.3.

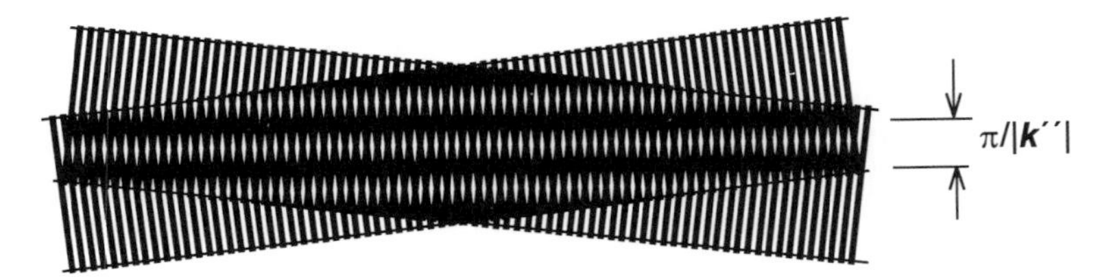

Figure 2.3: Interference fringes constant in time

2.2.2 Interference of Two Waves with Different Frequencies

In the following we investigate the interference of two waves where not only the propagation directions and the phases, but additionally the frequencies $\nu_i = \omega_i/(2\pi)$ are different.

$$\begin{aligned} E_1(\boldsymbol{r}, t) &= E_0\,\mathrm{e}^{\mathrm{i}(\boldsymbol{k_1}\cdot\boldsymbol{r}-2\pi\nu_1 t+\phi_1)} \\ E_2(\boldsymbol{r}, t) &= E_0\,\mathrm{e}^{\mathrm{i}(\boldsymbol{k_2}\cdot\boldsymbol{r}-2\pi\nu_2 t+\phi_2)} \end{aligned} \tag{2.35}$$

Besides the definitions of $\boldsymbol{k}'$, $\boldsymbol{k}''$, ϕ and $\Delta\phi$ now let $\nu = (\nu_1 + \nu_2)/2$ and $\Delta\nu = (\nu_1 - \nu_2)/2$. Then we have

$$\begin{aligned}
(E_1 + E_2)(\boldsymbol{r}, t) &= E_0\{\mathrm{e}^{\mathrm{i}(\boldsymbol{k}'\cdot\boldsymbol{r}+\boldsymbol{k}''\cdot\boldsymbol{r}-2\pi\nu t-2\pi\Delta\nu t+\phi+\Delta\phi)} + \mathrm{e}^{\mathrm{i}(\boldsymbol{k}'\cdot\boldsymbol{r}-\boldsymbol{k}''\cdot\boldsymbol{r}-2\pi\nu t+2\pi\Delta\nu t+\phi-\Delta\phi)}\} \\
&= E_0\,\mathrm{e}^{\mathrm{i}(\boldsymbol{k}'\cdot\boldsymbol{r}-2\pi\nu t+\phi)}\{\mathrm{e}^{\mathrm{i}(\boldsymbol{k}''\cdot\boldsymbol{r}-2\pi\Delta\nu t+\Delta\phi)} + \mathrm{e}^{\mathrm{i}(-\boldsymbol{k}''\cdot\boldsymbol{r}+2\pi\Delta\nu t-\Delta\phi)}\} \\
&= 2\,E_0\,\mathrm{e}^{\mathrm{i}(\boldsymbol{k}'\cdot\boldsymbol{r}-2\pi\nu t+\phi)}\cos(\boldsymbol{k}''\cdot\boldsymbol{r} - 2\pi\Delta\nu t + \Delta\phi)
\end{aligned} \tag{2.36}$$

and the intensity is

$$\begin{aligned}
I(\boldsymbol{r}, t) &= 4\,E_0^2\,\cos^2(\boldsymbol{k}''\cdot\boldsymbol{r} + \Delta\phi - 2\pi\Delta\nu t) \\
&= 2\,E_0^2[1 + \cos(2\boldsymbol{k}''\cdot\boldsymbol{r} + 2\Delta\phi - 4\pi\Delta\nu t)]
\end{aligned} \tag{2.37}$$

If the frequency difference is small enough, $\nu_1 \approx \nu_2$, a detector can register an intensity at $\boldsymbol{r}$ oscillating with the *beat frequency* $2\Delta\nu = \nu_1 - \nu_2$. The phase of this modulation is the phase difference $2\Delta\phi = \phi_1 - \phi_2$ of the superposed waves. Contrary to the frequencies of the optical waves the beat frequency can be measured electronically and further evaluated as long as it remains in the kHz- or MHz-range. The measurement of the beat frequency $\Delta\nu$ enables one to calculate the motion of a reflector via the *Doppler shift* or to determine the phase difference $\Delta\phi$ between different points of an object where the intensity oscillates with the same constant beat frequency.

2.2.3 Interference of Two Waves with Different Amplitudes

If we have plane linearly polarized waves of the same frequency, but different direction and phase and moreover different amplitudes

$$\begin{aligned}
E_1(\boldsymbol{r}, t) &= E_{01}\,\mathrm{e}^{\mathrm{i}(\boldsymbol{k_1}\cdot\boldsymbol{r}-\omega t+\phi_1)} \\
E_2(\boldsymbol{r}, t) &= E_{02}\,\mathrm{e}^{\mathrm{i}(\boldsymbol{k_2}\cdot\boldsymbol{r}-\omega t+\phi_2)}
\end{aligned} \tag{2.38}$$

we get the intensity

$$\begin{aligned}
&I(\boldsymbol{r}, t) \\
&= \left(E_{01}\,\mathrm{e}^{\mathrm{i}(\boldsymbol{k_1}\cdot\boldsymbol{r}-\omega t+\phi_1)} + E_{02}\,\mathrm{e}^{\mathrm{i}(\boldsymbol{k_2}\cdot\boldsymbol{r}-\omega t+\phi_2)}\right)\left(E_{01}\,\mathrm{e}^{-\mathrm{i}(\boldsymbol{k_1}\cdot\boldsymbol{r}-\omega t+\phi_1)} + E_{02}\,\mathrm{e}^{-\mathrm{i}(\boldsymbol{k_2}\cdot\boldsymbol{r}-\omega t+\phi_2)}\right) \\
&= E_{01}^2 + E_{02}^2 + E_{01}E_{02}\{\mathrm{e}^{\mathrm{i}(\boldsymbol{k_1}\cdot\boldsymbol{r}+\boldsymbol{k_2}\cdot\boldsymbol{r}+\phi_1-\phi_2)} + \mathrm{e}^{\mathrm{i}(\boldsymbol{k_2}\cdot\boldsymbol{r}-\boldsymbol{k_1}\cdot\boldsymbol{r}+\phi_2-\phi_1)}\} \\
&= E_{01}^2 + E_{02}^2 + 2E_{01}E_{02}\cos(2\boldsymbol{k}''\cdot\boldsymbol{r} + 2\Delta\phi)
\end{aligned} \tag{2.39}$$

This result can be written as

$$I = I_1 + I_2 + 2\sqrt{I_1 I_2}\cos(2\boldsymbol{k}''\cdot\boldsymbol{r} + 2\Delta\phi) \tag{2.40}$$

or using the identity $\cos\alpha = 2\cos^2(\alpha/2) - 1$ for comparison with (2.32) as

$$I = E_{01}^2 + E_{02}^2 + 4E_{01}E_{02}\cos^2(\boldsymbol{k}''\cdot\boldsymbol{r} + \Delta\phi) - 2E_{01}E_{02} \tag{2.41}$$

The special case $E_{01} = E_{02} = E_0$ gives (2.32).

In general the result of superposing two waves consists of one part that is the addition of the intensities and another part, the interference term, (2.40). Up to now we only have investigated *parallel polarized waves*. The other extreme are *orthogonally polarized waves*. These waves do not interfere, their superposition only consists of the addition of the intensities

$$I = I_1 + I_2 \tag{2.42}$$

For other angles between the polarization directions the field vector has to be decomposed into components of parallel and orthogonal polarizations, the result contains interference parts as well as an addition of intensities.

Reasons for the additive intensity term not only may be mutually oblique polarization directions or different intensities, but also an insufficient coherence of the interfering waves. Because in the superposition of incoherent light we always observe a pure addition of the intensities but no interference, the additive term often is called the *incoherent part*, or we speak of *incoherent superposition*.

The *visibility* or *contrast* of the interference pattern is defined by

$$V = \frac{I_{max} - I_{min}}{I_{max} + I_{min}} \tag{2.43}$$

If two parallel polarized waves of the same intensity interfere, we have the maximal contrast of $V = 1$, we have minimal contrast $V = 0$ for incoherent superposition. For example, if the ratio of the intensities of interfering waves is 5:1, the contrast is 0.745.

2.3 Coherence

With sunlight or lamp-light we rarely observe interference. Only light of sufficient coherence will exhibit this effect. Roughly speaking coherence means the ability of light waves to interfere. Precisely, coherence describes the correlation between individual light waves. The two aspects of the general spatio-temporal coherence are the temporal and the spatial coherence.

2.3.1 Temporal Coherence

Temporal coherence describes the correlation of a wave with itself as it behaves at different time instants [15, 16]. It is best explained with the help of a *Michelson interferometer*, Fig. 2.4. At a beam splitter the incoming wave field is divided into two parts, one being reflected into the orthogonal direction, one passing the splitter and maintaining the original direction. This type of wavefront division is called *amplitude division*.

To keep the mathematics easy we assume that the beam splitter reflects 50 percent of the incident light and transmits the other 50 percent. The reflected wave travels to the fixed mirror, is reflected again and part of it hits the detector or screen for observation, where it meets the other part which was reflected at the movable mirror. At the detector mutually time shifted parts of the wave are superimposed, the time shift can be varied

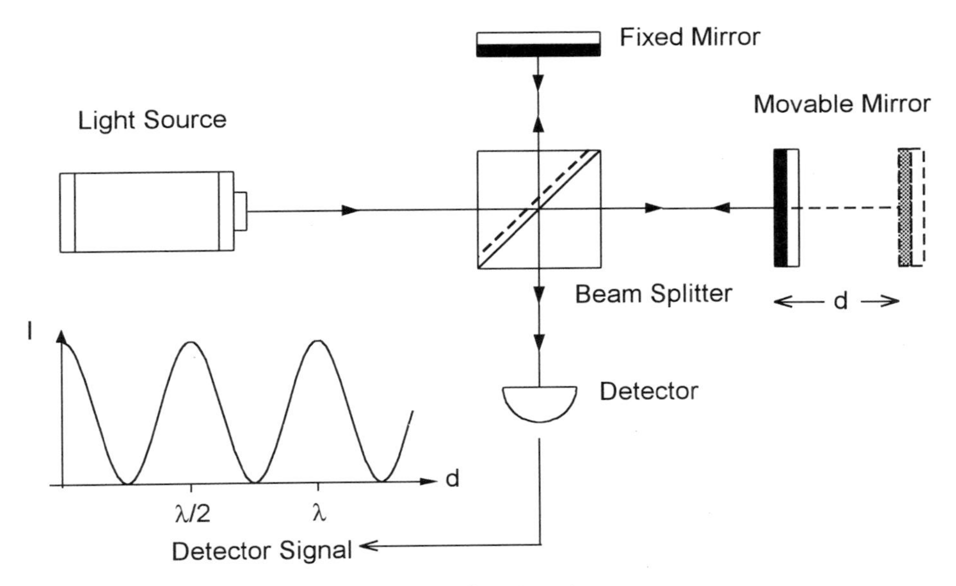

Figure 2.4: Michelson interferometer

by changing the mirror shift d. If the mirrors are perfectly orthogonal we see a constant intensity over the extended screen, but with a minute tilt of one mirror around, say, the x-axis, we observe fringes parallel to the x-axis, Fig. 2.4.

Let the waves be E_1 and E_2. For a fixed point on the screen, the position of the detector, we have

$$E_2(t) = E_1(t+\tau) \qquad \text{or} \qquad E_1(t) = E_2(t-\tau) \tag{2.44}$$

where

$$\tau = \frac{2d}{c} \tag{2.45}$$

We have to recognize that the distance d is travelled forward and backward, therefore the factor 2 in (2.45).

At the observation point we have the superposition

$$E(t) = E_1(t) + E_2(t) = E_1(t) + E_1(t+\tau) \tag{2.46}$$

and see the intensity

$$\begin{aligned} I &= \langle E\, E^* \rangle \\ &= \langle E_1\, E_1^* \rangle + \langle E_2\, E_2^* \rangle + \langle E_2\, E_1^* \rangle + \langle E_1\, E_2^* \rangle \\ &= 2I_1 + 2\mathrm{Re}(\langle E_1\, E_2^* \rangle) \end{aligned} \tag{2.47}$$

due to our assumption of equal amplitudes.

According to (2.27) we define the complex *self coherence* $\Gamma(\tau)$ as

$$\Gamma(\tau) = \langle E_1^* E_1(t+\tau)\rangle = \lim_{T_m\to\infty}\frac{1}{T_m}\int_{-T_m/2}^{T_m/2} E_1^*(t)\, E_1(t+\tau)dt \quad (2.48)$$

which is the autocorrelation of $E_1(t)$. The normalized quantity

$$\gamma(\tau) = \frac{\Gamma(\tau)}{\Gamma(0)} \quad (2.49)$$

defines the *degree of coherence*. Since $\Gamma(0) = I_1$ is always real and the maximal value of $|\Gamma(\tau)|$, we have

$$|\gamma(\tau)| \leq 1 \quad (2.50)$$

and

$$I(\tau) = 2I_1(1 + \mathrm{Re}\gamma(\tau)) \quad (2.51)$$

The degree of coherence or the self coherence are not directly measurable, but it can be shown [16] that for the contrast V, which is easily measurable, we have

$$V(\tau) = |\gamma(\tau)| \quad (2.52)$$

for waves of equal intensity, otherwise we get additional factors. Now we can discriminate perfectly coherent light with $|\gamma(\tau)| = 1$, which nearly is emitted by a stabilized single mode laser, incoherent light with $|\gamma(\tau)| = 0$ for all $\tau \neq 0$ where we have a statistically fluctuating phase, e. g. in sunlight, and partially coherent light, $0 \leq |\gamma(\tau)| \leq 1$. Often the contrast $p(\tau)$ decreases monotonically in τ. So we can introduce the *coherence time* τ_c as the time shift at which the contrast is fallen down to 1/e. The time shift is realized in interferometers by different optical pathlengths, so one can define the *coherence length*

$$l_c = c\tau_c \quad (2.53)$$

If we have a periodic instead of a monotonically decreasing contrast function, e. g. from a two-mode laser, we take the time shift corresponding to the first minimum as the coherence time.

2.3.2 Spatial Coherence

Spatial coherence describes the mutual correlation of different parts of the same wavefront [15, 16] and is explained at *Young's double aperture interferometer*, Fig. 2.5. This type of interferometer picks two geometrically different parts of the wavefront and brings them to interference, therefore it is called a *division of wavefront* interferometer. let an opaque screen contain two small holes or parallel slits with a mutual distance d. For the moment we assume one point $S_1 = (0, y_{S1}, -R - L)$ of an extended source placed at the distance R behind that illuminates the opaque screen. Only the light passing through the holes forms an interference pattern on the observation screen placed some distance L in front of them. The distances of S_1 to the holes are r_1, r_2, those of the holes to the observation

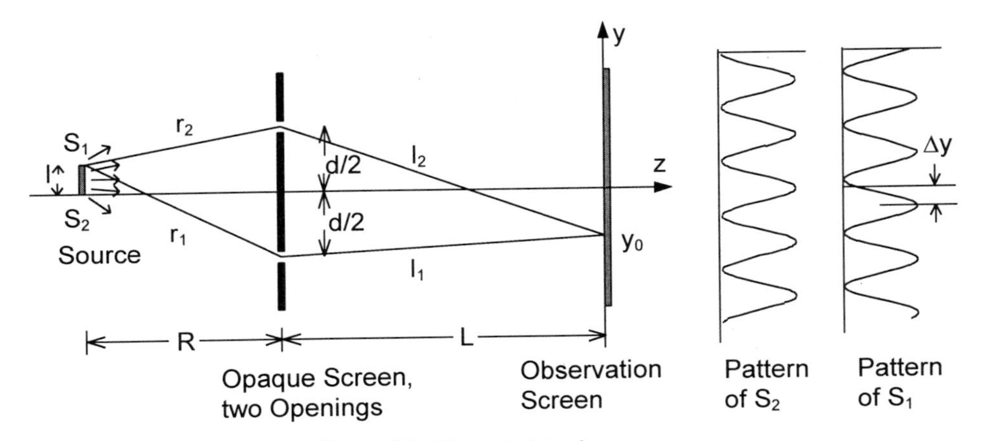

Figure 2.5: Young's interferometer

point are l_1, l_2. We can assume that the intensities of the two spherical waves are equal, therefore the intensity at the observation screen is (2.32)

$$I(x, y) = 4I_0(x, y) \cos^2 \Delta\phi(x, y) \tag{2.54}$$

The half phase difference $\Delta\phi$ is

$$\Delta\phi = (\frac{2\pi}{\lambda}\Delta l)/2 \tag{2.55}$$

where Δl is the difference in the pathlength of the light from the source S_1 to the observation point $B = (0, y_0, 0)$

$$\begin{aligned} \Delta l &= r_2 + l_2 - r_1 - l_1 \qquad (2.56) \\ &= \sqrt{R^2 + (\frac{d}{2} - y_{S1})^2} + \sqrt{L^2 + (\frac{d}{2} - y_0)^2} - \sqrt{R^2 + (\frac{d}{2} + y_{S1})^2} - \sqrt{L^2 + (\frac{d}{2} + y_0)^2} \end{aligned}$$

Since y_0, y_{S1} and d are small compared with R and L, the square roots can be approximated by

$$\Delta l = -d(\frac{y_{S1}}{R} + \frac{y_0}{L}) \tag{2.57}$$

The resulting irradiance now is proportional to

$$I = I_0 \cos^2[\frac{\pi d}{\lambda}(\frac{y_{S1}}{R} + \frac{y_0}{L})] \tag{2.58}$$

and describes a pattern of fringes parallel to the x-axis with a spacing of $\lambda L/d$ in the y-direction.

Next we investigate an extended source of perimeter l, Fig. 2.5. Source point S_2 on the optical axis emits a spherical wave which reaches the holes with equal phase, so we get an intensity maximum where the optical axis hits the observation screen. The point S_1 gives rise to a fringe system which is shifted laterally because here r_1 and r_2 do not have equal length. Therefore the phase difference between the two spherical waves originating at the two holes is

$$\phi_1 - \phi_2 = \frac{2\pi}{\lambda}(r_1 - r_2) \tag{2.59}$$

which results in a lateral shift of the interference pattern by the amount

$$\Delta y = \frac{L}{d}(r_1 - r_2) \tag{2.60}$$

If there is a fixed phase relation between S_1 and S_2, simultaneous emission from S_1 and S_2 will produce an interference pattern similar to the one of one point source alone. If on the other hand we have a randomly fluctuating phase between S_1 and S_2, in the mean we get the sum of the intensities. As a condition for visibility of the fringes therefore we have to demand a lateral shift Δy less than a half fringe spacing

$$|\Delta y| < (\frac{\lambda L}{d})/2 \tag{2.61}$$

or equivalently

$$|r_1 - r_2| < \lambda/2 \tag{2.62}$$

To express this in terms of the optical setup using the same arguments which led to (2.57) we get $|r_1 - r_2| = dl/R$ and thus

$$\frac{dl}{R} < \lambda/2 \tag{2.63}$$

The derivation was carried out for the two points S_1 and S_2. But since these are the farest points of the extended source, the condition (2.62), if fulfilled, is valid for all points between.

For points near the optical axis the path lengths from the holes to the observation screen are nearly the same. There the fringe pattern gives us information about the similarity of the wavefront $E(\boldsymbol{r_1}, t)$ and $E(\boldsymbol{r_2}, t)$ at the apertures at $\boldsymbol{r_1}$ and $\boldsymbol{r_2}$ without time shift. This similarity can be expressed by the *spatial coherence function*

$$\Gamma(\boldsymbol{r_1}, \boldsymbol{r_2}, 0) = \Gamma_{12}(0) = \langle E(\boldsymbol{r_1}, t)E^*(\boldsymbol{r_2}, t)\rangle \tag{2.64}$$

The general *spatio-temporal coherence function* now is

$$\begin{aligned} \Gamma(\boldsymbol{r_1}, \boldsymbol{r_2}, t_1, t_2) &= \Gamma_{12}(t_2 - t_1) = \Gamma(\tau) \\ &= \langle E(\boldsymbol{r_1}, t + \tau)E^*(\boldsymbol{r_2}, t)\rangle \end{aligned} \tag{2.65}$$

which can be normalized to give the *mutual degree of coherence*

$$\gamma_{12}(\tau) = \frac{\Gamma_{12}(\tau)}{\sqrt{\Gamma_{11}(0)\Gamma_{22}(0)}} \tag{2.66}$$

where $\Gamma_{11}(0)$ is the intensity at $\boldsymbol{r_1}$, $\Gamma_{22}(0)$ is the intensity at $\boldsymbol{r_2}$.

Spatial coherence of thermal or gas discharge sources is associated primarily with the spatial extent of the source. If we are far enough from the source, R very large, the condition (2.63) is fulfilled. So we receive coherent light from stars, big thermal sources, at the earth, which is used in the field of stellar interferometry. This shows that coherence is not a property of the source, but of the light wave. The spatial coherence of laser light, as used in holographic interferometry, is associated with the transverse mode structure of the resonance cavity. For lasers resonating in the TEM_{00} mode, all points on the wavefront essentially have the same phase, therefore they have extremely good spatial coherence.

2.4 Diffraction Theory

Holography and holographic interferometry strongly build upon the wave nature of light. A description of the propagation of light waves must recognize diffraction effects. Only a few aspects of scalar diffraction theory are summarized in the sequel, as they are necessary for the description of digital holography, Sec. 4.7, or the optical Fourier transform, Sec. 6.1 and Sec. 6.3.

2.4.1 The Fresnel-Kirchhoff Diffraction Formula

A detailed discussion of the preliminaries of a scalar diffraction theory and a rigorous derivation of the diffraction formulas based on Green's functions are given by Goodman [15]. He considers the problem of diffraction by an aperture Σ in an infinite opaque screen. Fig. 2.6 gives the geometry of the aperture in the $(z = 0)$-plane of a Cartesian coordinate

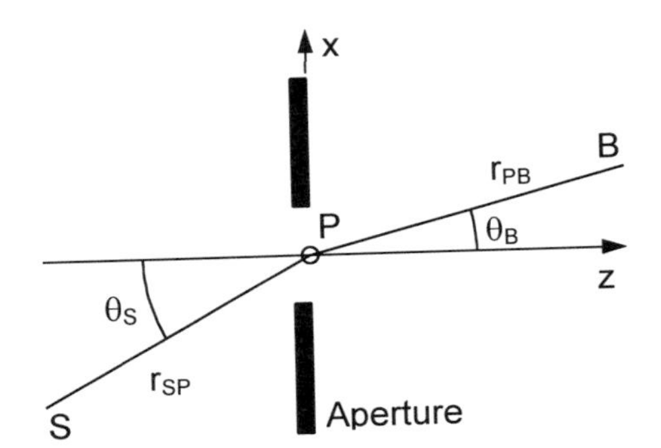

Figure 2.6: Geometry for the Fresnel-Kirchhoff diffraction formula

system. The aperture is illuminated by a spherical wave $E(r) = (E_0/r)\exp\{ikr\}$ emitted from point S and observed at B. For an arbitrary point P of the aperture the distance from S to P is r_{SP} and from P to B is r_{PB}, the corresponding vectors form the angles θ_S and θ_B with the z-axis. Then the field at the observation point is given by the *Fresnel-Kirchhoff diffraction formula*

$$E(B) = \frac{\mathrm{i}E_0}{\lambda} \int\int_\Sigma \frac{\mathrm{e}^{-\mathrm{i}k(r_{SP} + r_{PB})}}{r_{SP}r_{PB}} \left(\frac{\cos\theta_B - \cos\theta_S}{2} \right) dxdy \tag{2.67}$$

Here it has been assumed, that $k \gg 1/r_{PB}$, meaning an observation point many optical wavelengths apart from the aperture.

2.4.2 Diffraction at a Transmittance

The more interesting case is the illumination of an amplitude transmittance $\tau(x, y)$ which is conveniently placed in the $(z = 0)$-plane, Fig. 2.7. Let the transmittance be illumina-

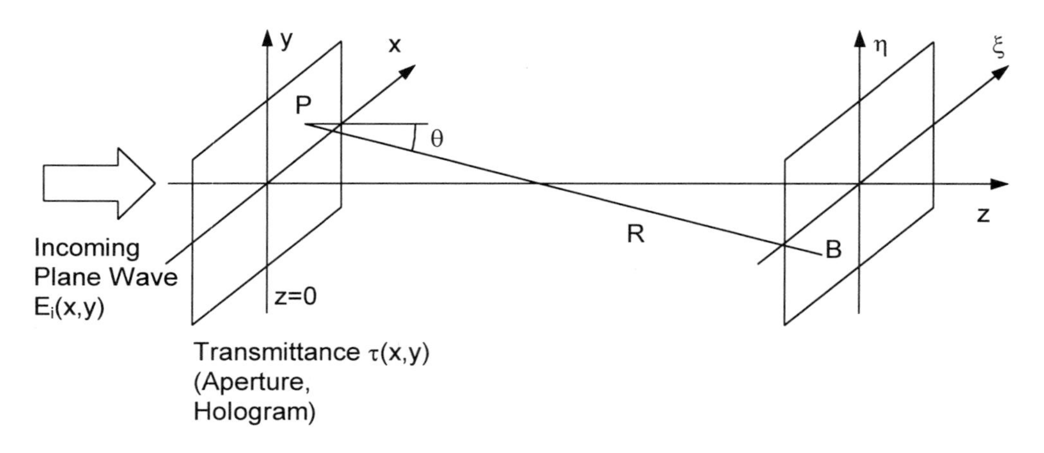

Figure 2.7: Plane wave illumination of a transmittance

ted by a plane wave $E_i(x, y) = E_0 \exp\{\mathrm{i}\boldsymbol{k} \cdot \boldsymbol{r}\}$ propagating parallel to the z-axis. This corresponds to a hologram illuminated by a plane reference wave. Directly behind the transmittance $(z = 0)$ the field is

$$E(x, y) = \tau(x, y) E_i(x, y) \tag{2.68}$$

The field at points farer from the $z = 0$-plane is given by integration over all spherical waves emitted from the $(x, y, 0)$

$$E(\xi, \eta, z) = \frac{\mathrm{i}E_0}{\lambda} \int_{-\infty}^{\infty} \int_{-\infty}^{\infty} \tau(x, y) \frac{\mathrm{e}^{-\mathrm{i}kR}}{R} \left(\frac{1}{2} + \frac{1}{2} \cos\theta \right) dx dy \tag{2.69}$$

which is the so called *Fresnel-Kirchhoff integral.*

For x- and y-values as well as for ξ- and η-values which are small compared to the distance z of the observed pattern from the diffracting transmittance, $|x| \ll z$, $|y| \ll z$, $|\xi| \ll z$, $|\eta| \ll z$, the angle θ becomes small, its cosine can be set to 1. In this *paraxial approximation* furthermore the inverse length $1/R$ can be replaced by $1/z$ and the R in the argument of the exponential is replaced by the first terms of the Taylor-series

$$\begin{aligned} R &= \sqrt{(\xi - x)^2 + (\eta - y)^2 + z^2} = z\sqrt{1 + \frac{(\xi - x)^2}{z^2} + \frac{(\eta - y)^2}{z^2}} \\ &\approx z + \frac{(\xi - x)^2}{2z} + \frac{(\eta - y)^2}{2z} \end{aligned} \tag{2.70}$$

This gives the *Fresnel approximation* of the Fresnel-Kirchhoff integral

$$E(\xi, \eta, z) = \frac{\mathrm{i}E_0\, \mathrm{e}^{-\mathrm{i}kz}}{\lambda z} \int_{-\infty}^{\infty} \int_{-\infty}^{\infty} \tau(x, y) \mathrm{e}^{-\frac{\mathrm{i}k}{2z}[(\xi - x)^2 + (\eta - y)^2]} dx dy \tag{2.71}$$

The factor $\mathrm{e}^{-\mathrm{i}kz} = \mathrm{e}^{-\mathrm{i}(2\pi/\lambda)z}$ can be omitted, since it only affects the overall phase independently of (ξ, η). If we carry out the multiplications in the argument of the exponential under the integrals we get

$$E(\xi,\eta,z) = \frac{\mathrm{i}E_0}{\lambda z}\mathrm{e}^{-\frac{\mathrm{i}\pi}{\lambda z}(\xi^2+\eta^2)} \int_{-\infty}^{\infty}\int_{-\infty}^{\infty} \tau(x,y)\mathrm{e}^{-\frac{\mathrm{i}\pi}{\lambda z}(x^2+y^2)}\mathrm{e}^{\frac{\mathrm{i}2\pi}{\lambda z}(x\xi+y\eta)}dxdy \quad (2.72)$$

A comparison of this formula with the definition of the *two-dimensional Fourier transform* (A.60) shows, that the Fresnel-approximation is, up to a spherical phase factor, the inverse Fourier transform of the transmittance $\tau(x, y)$, which is multiplied by the spatially varying phase factor $\exp\{-\mathrm{i}\pi(x^2 + y^2)/(\lambda z)\}$.

The Fresnel approximation can be used for distances greater than about ten wavelengths. For really large distances of the observation plane from the diffracting plane, $z \gg \frac{\pi}{\lambda}(x^2 + y^2)$, the quadratic phase factor to the transmittance approaches 1 and the resulting *Fraunhofer approximation* is the Fourier transform of $\tau(x, y)$ up to the multiplication of the result with a spherical phase factor.

2.4.3 The Spherical Lens

A thin convex spherical lens is characterized by its *focal length* f. A collimated beam directed along the lens axis which impinges on the lens is brought to focus at the distance f from the lens. Assuming no absorption, the effect of the lens is merely to introduce a phase delay $\Delta\phi(x, y)$ which can be written

$$\Delta\phi(x,y) = \frac{\pi}{\lambda f}(x^2+y^2) \quad (2.73)$$

With this phase delay the thin lens can be considered as a transparency with the complex amplitude transmittance

$$T_L(x,y) = \mathrm{e}^{\frac{\mathrm{i}\pi}{\lambda f}(x^2+y^2)} \quad (2.74)$$

If now a transparency with amplitude transmittance $\tau(x, y)$ is placed directly in front of a lens, the net transmitted amplitude is $\tau(x, y)T_L(x, y)$. The complex amplitude in the back focal plane $z = f$ of the lens is calculated from (2.72) to

$$E(\xi,\eta,z=f) = \frac{\mathrm{i}E_0}{\lambda f}\mathrm{e}^{-\frac{\mathrm{i}\pi}{\lambda f}(\xi^2+\eta^2)} \int_{-\infty}^{\infty}\int_{-\infty}^{\infty} \tau(x,y)\mathrm{e}^{\frac{\mathrm{i}2\pi}{\lambda f}(x\xi+y\eta)}dxdy \quad (2.75)$$

Now consider the transparency to be placed in the front focal plane of the lens, Fig. 2.8. According to (2.71) the complex amplitude $E_P(x', y')$ in the pupil of the lens is

$$E_P(x',y') = \frac{\mathrm{i}E_0}{\lambda f}\int_{-\infty}^{\infty}\int_{-\infty}^{\infty} \tau(x,y)\mathrm{e}^{-\frac{\mathrm{i}\pi}{\lambda f}[(x'-x)^2+(y'-y)^2]}dxdy \quad (2.76)$$

The complex amplitude in the back focal plane we get by replacing $E_0\tau(x, y)$ by $E_P(x', y')$ in (2.75)

$$E(\xi,\eta,z=2f) = \frac{\mathrm{i}}{\lambda f}\mathrm{e}^{-\frac{\mathrm{i}\pi}{\lambda f}(\xi^2+\eta^2)} \int_{-\infty}^{\infty}\int_{-\infty}^{\infty} E_P(x',y')\mathrm{e}^{\frac{\mathrm{i}2\pi}{\lambda f}(x'\xi+y'\eta)}dx'dy' \quad (2.77)$$

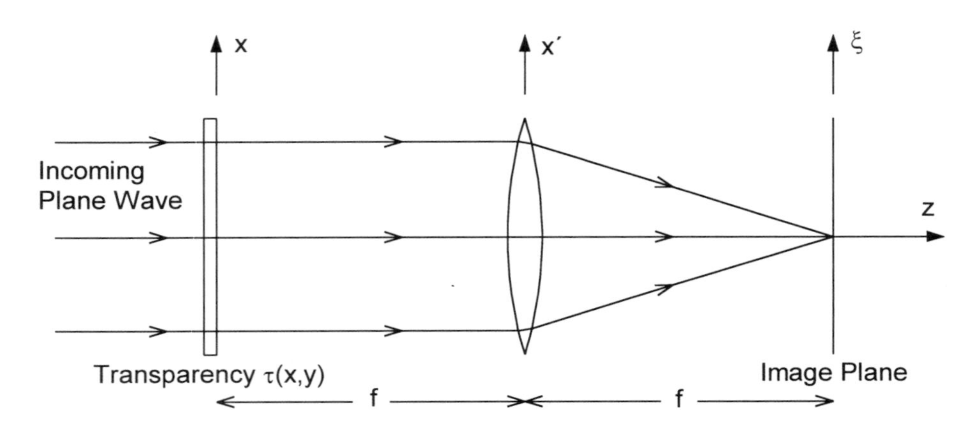

Figure 2.8: Arrangement for optical Fourier transformation

With $\xi' = \xi/(\lambda f)$ and $\eta' = \eta/(\lambda f)$ this can be written as, see (A.2) and (A.60)

$$E(\xi', \eta') = \frac{\mathrm{i}}{\lambda f} \mathrm{e}^{\mathrm{i}\pi\lambda f(\xi'^2 + \eta'^2)} \mathcal{F}^{-1}\{E_P\}(\xi', \eta') \tag{2.78}$$

$E_P(x', y')$ in (2.76) can be regarded as a convolution

$$E_P(x', y') = \frac{\mathrm{i}E_0}{\lambda f} \tau(x', y') \star \mathrm{e}^{-\frac{\mathrm{i}\pi}{\lambda f}(x'^2 + y'^2)} \tag{2.79}$$

so that using the convolution theorem, Table A.3, we obtain

$$\mathcal{F}^{-1}\{E_P\}(\xi', \eta') = \frac{E_0}{\mathrm{i}\lambda f} T(\xi', \eta') \mathrm{e}^{\mathrm{i}\pi\lambda f(\xi'^2 + \eta'^2)} \tag{2.80}$$

with $T(\xi', \eta')$ being the Fourier transform of $\tau(x', y')$. If this result is substituted in (2.78), we get

$$E(\xi', \eta') = \frac{E_0}{\lambda^2 f^2} T(\xi', \eta') \tag{2.81}$$

This means, the lens performs a two-dimensional Fourier transform from the front to the back focal plane. $T(\xi', \eta')$ is the *complex amplitude spectrum* of the transparency $\tau(x, y)$. Also we see that the *Fraunhofer diffraction pattern* is taken from infinity to the back focal plane by the lens. But only the intensity can be directly observed or recorded

$$I(\xi', \eta') = |T(\xi', \eta')|^2 = |\mathcal{F}\{\tau(x, y)\}|^2 \tag{2.82}$$

This is called the (Fraunhofer) diffraction pattern or *spectrum* of $\tau(x, y)$.

The placement of the transparency into the front focal plane was necessary to compensate the phase term. Since the phase information is canceled by taking the intensity, the original can be placed into an arbitrary plane before the lens. To avoid vignetting effects a placement directly before the lens is recommended.

2.4.4 Discrete Finite Fresnel-Transform

The Fresnel approximation to the Fresnel-Kirchhoff integral, (2.72), often called shortly the *Fresnel transform*, is the inverse Fourier transform of the input function $\tau(x, y)$ multiplied with the factor $\exp\{-i\pi(x^2 + y^2)/(\lambda z)\}$. For practical numerical calculations we have to sample the function $\tau(x, y)$ at $N \times M$ points in the (x, y)-plane, the samples being separated by Δx in the x- and by Δy in the y-direction. For abbreviation we introduce

$$\nu = \frac{\xi}{\lambda z} \quad \text{and} \quad \mu = \frac{\eta}{\lambda z} \tag{2.83}$$

Herewith (2.72) becomes

$$E(\nu, \mu, z) = \frac{iE_0}{\lambda z} e^{-i\pi\lambda z(\nu^2 + \mu^2)} \int_{-\infty}^{\infty} \int_{-\infty}^{\infty} \tau(x, y) e^{-\frac{i\pi}{\lambda z}(x^2 + y^2)} e^{i2\pi(x\nu + y\mu)} dx dy \tag{2.84}$$

The resulting sample points in the spatial frequency domain have the spacing $\Delta\xi$ and $\Delta\eta$ or in the normalized form $\Delta\nu$ and $\Delta\mu$ given by

$$\Delta x = \frac{1}{N\Delta\nu} = \frac{\lambda z}{N\Delta\xi} \quad \text{and} \quad \Delta y = \frac{1}{N\Delta\mu} = \frac{\lambda z}{N\Delta\eta} \tag{2.85}$$

Now the *discrete finite Fresnel transform* is

$$\begin{aligned} E(n, m, z) \\ = \frac{iE_0}{\lambda z} e^{-i\pi\lambda z(\frac{n^2}{N^2\Delta x^2} + \frac{m^2}{M^2\Delta y^2})} \sum_{k=0}^{N-1} \sum_{l=0}^{M-1} \tau(k, l) e^{-\frac{i\pi}{\lambda z}(k^2\Delta x^2 + l^2\Delta y^2)} e^{i2\pi(\frac{kn}{N} + \frac{lm}{M})} \end{aligned} \tag{2.86}$$

In all applications where only the intensity is of interest, the exponential before the double sum does not need to be calculated. For an effective determination of the discrete finite Fresnel transform, first $\tau(k, l)$ is multiplied pointwisely by $\exp\{-i\pi(k^2\Delta x^2 + l^2\Delta y^2)/(\lambda z)\}$, then the FFT algorithm, Sec. A.6, is used for a fast calculation of the discrete finite inverse Fourier transform.

2.5 Speckles

If a randomly scattering object like a diffusely reflecting surface is illuminated with coherent light, all illuminated object points emit spherical waves which can interfere. The resulting wave field in space is called the *speckle field* or *speckle pattern*. The *speckles* form a random pattern in space which is stationary in time but highly fluctuating from point to point. Their appearance is almost independent of the object characteristics, but strongly depends on the optical properties of the viewing system. In holographic interferometry the speckles normally are disturbing, they influence the achievable resolution and accuracy of the measurement. On the other hand a number of related methods employ the speckle effect for solving measurement problems. Some of these methods will be discussed in Chap. 6. In the following a brief summary of the statistics of speckle fields will be given [17].

2.5.1 Statistics of Speckle Intensity and Phase

Let us consider a surface which is rough on the scale of the optical wavelength, Fig. 2.9. According to the Huygens principle each point of the coherently illuminated rough surface

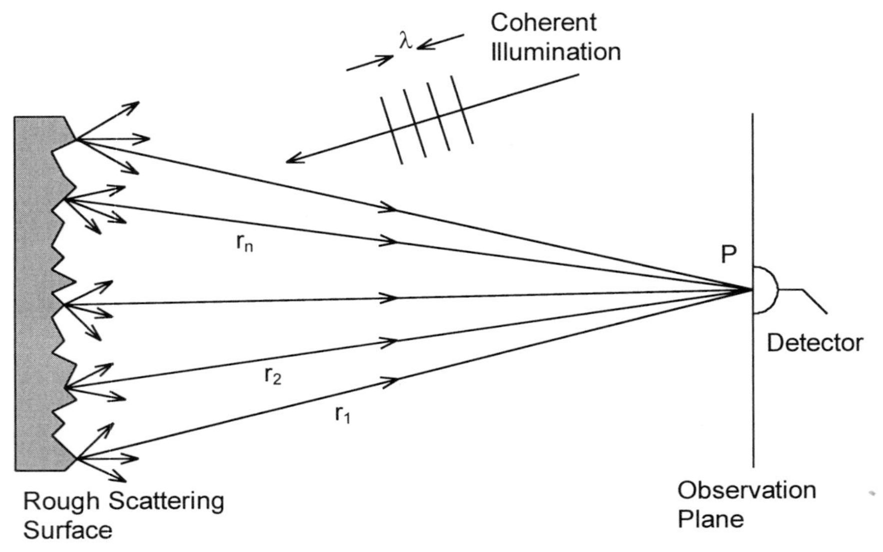

Figure 2.9: Physical origin of speckles, free space propagation

can be treated as emitting a spherical secondary wavelet (2.23)

$$E_n(r_n) = \frac{E'_{0n}}{r_n} e^{i(kr_n + \phi'_n)} \tag{2.87}$$

The mutual phases are temporally constant but strongly varying with the emission point. Thus at an observation point B in space all the individual field strengths sum up to

$$E(B) = \sum_n \frac{E'_{0n}}{r_n} e^{i(kr_n + \phi'_n)} \tag{2.88}$$

where now the r_n are the distances from the surface points to B. This summation is analogous to the *random walk problem* in two dimensions of probability theory. Each spherical wave can be represented as a vector in the complex plane, the resulting field strength is the vector sum, Fig. 2.10. For applying the central limit theorem of probability theory we write

$$E_n(r_n) = \frac{1}{\sqrt{N}} |E_{0n}| e^{i\phi_n} \tag{2.89}$$

and take the assumptions that (i) the amplitude $E_{0n}/\sqrt{N}$ and the phase ϕ_n of each wave are statistically independent from each other and from the amplitudes and phases of all other waves and (ii) the phases ϕ_n are uniformly distributed in the interval $[-\pi, \pi]$.

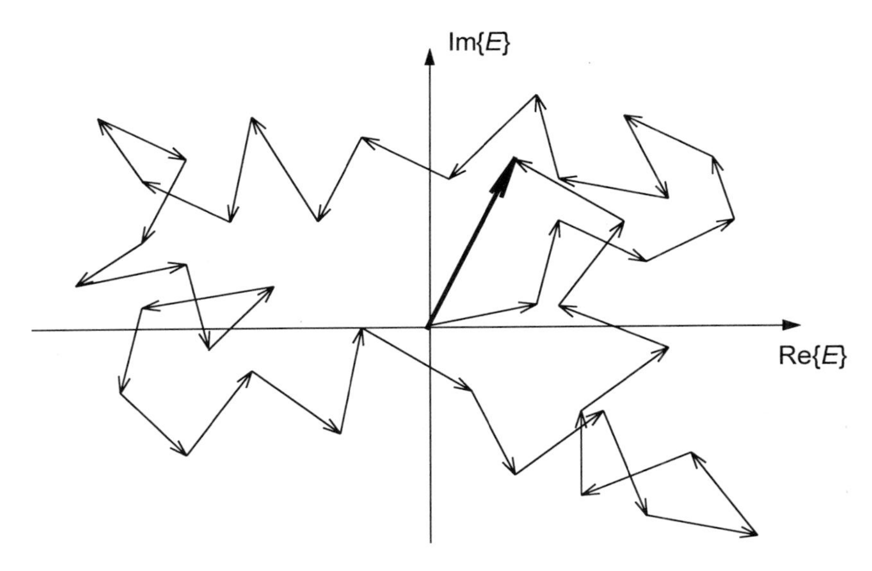

Figure 2.10: Random walk in the complex plane

These assumptions are physically justified by the facts that the elementary scattering areas of the surface are unrelated and the strength of a given scattered component bears no relation to its phase (i) and that the surface is rough compared to the wavelength (ii). Phase excursions more than 2π radians have the same effect as those with the same value modulo 2π.

Let E_r and E_i be the real and imaginary parts of E

$$\begin{aligned} E_r &= \text{Re}\{E\} = \frac{1}{\sqrt{N}} \sum_{n=1}^{N} |E_{0n}| \cos \phi_n \\ E_i &= \text{Im}\{E\} = \frac{1}{\sqrt{N}} \sum_{n=1}^{N} |E_{0n}| \sin \phi_n \end{aligned} \tag{2.90}$$

The average values over an ensemble of macroscopically similar but microscopically different surfaces are

$$\begin{aligned} \langle E_r \rangle &= \frac{1}{\sqrt{N}} \sum_{n=1}^{N} \langle |E_{0n}| \cos \phi_n \rangle = \frac{1}{\sqrt{N}} \sum_{n=1}^{N} \langle |E_{0n}| \rangle \langle \cos \phi_n \rangle = 0 \\ \langle E_i \rangle &= \frac{1}{\sqrt{N}} \sum_{n=1}^{N} \langle |E_{0n}| \sin \phi_n \rangle = \frac{1}{\sqrt{N}} \sum_{n=1}^{N} \langle |E_{0n}| \rangle \langle \sin \phi_n \rangle = 0 \end{aligned} \tag{2.91}$$

Here the independence of the individual amplitudes from the phases, assumption (i), allows to take the average over the factors separately. The uniform distribution of the phases in $[-\pi, \pi]$, assumption (ii), gives the zero values for $\langle \cos \phi_n \rangle$ and $\langle \sin \phi_n \rangle$. To calculate the variance and correlation we use

$$\begin{aligned}\langle\cos\phi_n\cos\phi_m\rangle &= \langle\sin\phi_n\sin\phi_m\rangle = \begin{cases}\frac{1}{2} & : \quad n = m\\ 0 & : \quad n \neq m\end{cases}\\ \langle\cos\phi_n\sin\phi_m\rangle &= 0\end{aligned} \tag{2.92}$$

This gives

$$\begin{aligned}\langle E_r^2\rangle &= \frac{1}{N}\sum_{n=1}^{N}\sum_{m=1}^{N}\langle|E_{0n}||E_{m0}|\rangle\langle\cos\phi_n\cos\phi_m\rangle = \frac{1}{N}\sum_{n=1}^{N}\frac{1}{2}\langle|E_{0n}|^2\rangle\\ \langle E_i^2\rangle &= \frac{1}{N}\sum_{n=1}^{N}\sum_{m=1}^{N}\langle|E_{0n}||E_{m0}|\rangle\langle\sin\phi_n\sin\phi_m\rangle = \frac{1}{N}\sum_{n=1}^{N}\frac{1}{2}\langle|E_{0n}|^2\rangle\\ \langle E_r E_i\rangle &= \frac{1}{N}\sum_{n=1}^{N}\sum_{m=1}^{N}\langle|E_{0n}||E_{m0}|\rangle\langle\cos\phi_n\sin\phi_m\rangle = 0\end{aligned} \tag{2.93}$$

Altogether we have a complex field with uncorrelated real and imaginary parts of zero mean and identical variances. So we can apply the central limit theorem which states that for $N \to \infty$, E_r and E_i are asymptotically Gaussian. The joint probability density function is

$$p_{r,i}(E_r, E_i) = \frac{1}{2\pi\sigma^2}\mathrm{e}^{-\frac{E_r^2+E_i^2}{2\sigma^2}} \tag{2.94}$$

with the variance

$$\sigma^2 = \lim_{N\to\infty}\frac{1}{N}\sum_{n=1}^{N}\frac{1}{2}\langle|E_{0n}|^2\rangle \tag{2.95}$$

Since the intensity I and the phase ϕ of the resultant field are related to the real and imaginary parts by

$$E_r = \sqrt{I}\cos\phi \qquad \text{and} \qquad E_i = \sqrt{I}\sin\phi \tag{2.96}$$

the joint probability density function $p_{I,\phi}(I,\phi)$ of intensity and phase is

$$p_{I,\phi}(I,\phi) = p_{r,i}(\sqrt{I}\cos\phi, \sqrt{I}\sin\phi)||J|| \tag{2.97}$$

where $||J||$ is the modulus of the determinant of the Jacobian matrix of the transformation (2.96). Substituting (2.96) into (2.94) gives

$$p_{I,\phi}(I,\phi) = \begin{cases}\frac{1}{4\pi\sigma^2}\mathrm{e}^{-\frac{I}{2\sigma^2}} & \text{for} \quad I > 0 \quad \text{and} \quad -\pi \leq \phi < \pi\\ 0 & \text{otherwise}\end{cases} \tag{2.98}$$

The one-dimensional marginal distributions for the intensity and the phase alone are

$$p_I(I) = \int_{-\pi}^{\pi} p_{I,\phi}(I,\phi)d\phi = \begin{cases}\frac{1}{2\sigma^2}\mathrm{e}^{-\frac{I}{2\sigma^2}} & \text{for} \quad I > 0\\ 0 & \text{otherwise}\end{cases} \tag{2.99}$$

$$p_\phi(\phi) = \int_0^{\infty} p_{I,\phi}(I,\phi)dI = \begin{cases}\frac{1}{2\pi} & \text{for} \quad -\pi \leq \phi < \pi\\ 0 & \text{otherwise}\end{cases} \tag{2.100}$$

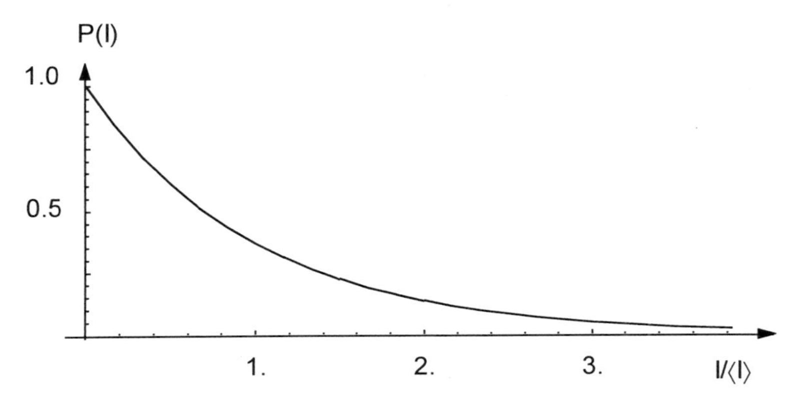

Figure 2.11: Normalized probability density function versus normalized intensity

Thus the *intensity in a speckle pattern* obeys a negative exponential probability distribution and the *phase* is uniformly distributed. Furthermore we see that the intensity and phase are statistically independent

$$p_{I,\phi}(I,\phi) = p_I(I)p_\phi(\phi) \tag{2.101}$$

The n-th moment $\langle I^n \rangle$ of the intensity is

$$\langle I^n \rangle = n!(2\sigma^2)^n = n!\langle I \rangle^n \tag{2.102}$$

especially the mean value is $\langle I \rangle = 2\sigma^2$. By calculating the variance σ_I^2 of the intensity

$$\sigma_I^2 = \langle I^2 \rangle - \langle I \rangle^2 = \langle I \rangle^2 \tag{2.103}$$

we get the result that the standard deviation σ_I of a speckle pattern equals the mean intensity. Since a common measure for the contrast V is

$$V = \sigma_I / \langle I \rangle \tag{2.104}$$

we see that the *contrast of a speckle* pattern is always unity.

Sometimes we are interested in the probability $P(I)$ that the intensity exceeds a threshold I. This probability is

$$P(I) = \int_I^\infty p_I(I')dI' = \int_I^\infty \frac{1}{\langle I \rangle} \mathrm{e}^{-\frac{I'}{\langle I \rangle}} dI' = \mathrm{e}^{-\frac{I}{\langle I \rangle}} \tag{2.105}$$

which for this special distribution equals the probability density function (2.99) normalized by $\langle I \rangle$. Fig. 2.11 shows $\langle I \rangle p_I(I)$ or $P(I)$ against the normalized intensity $I/\langle I \rangle$.

Higher order statistics and the statistics of coherently and incoherently summed speckle patterns can be found in [17].

2.5.2 Speckle Size

In the context of holographic interferometry we are more interested in the average size of the individual speckles which we can observe. The *speckle size* must be related to the pixel size of CCD-targets recording the patterns and will determine the resolution of measurement methods.

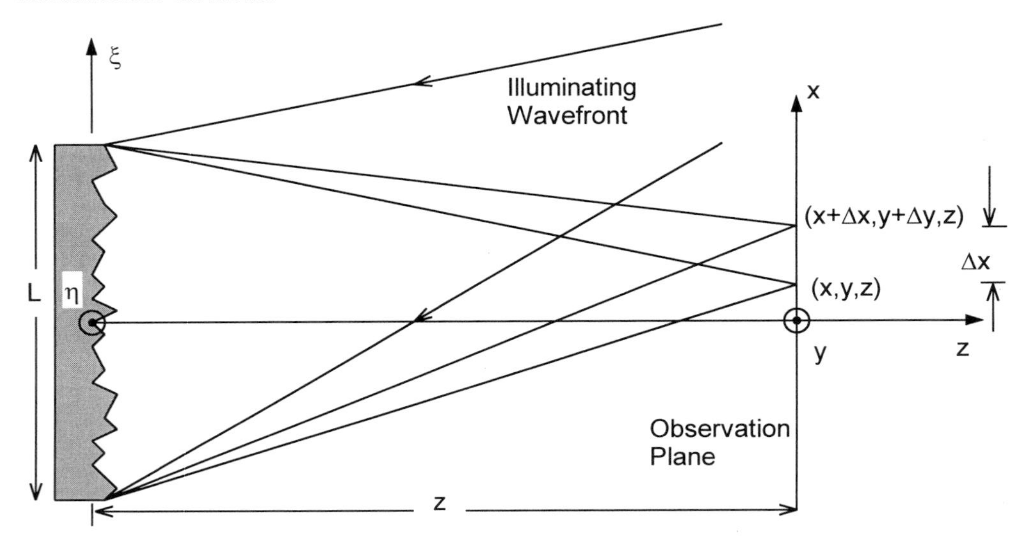

Figure 2.12: Free space propagation speckle formation

First we consider the free space propagation of the scattered field, Fig. 2.12, without imaging system. The mean speckle size can be derived over the autocorrelation function $R(x_1, y_1; x_2, y_2)$ of the intensity in the observation plane. The autocorrelation is defined as

$$R(x_1, y_1; x_2, y_2) = R(\Delta x, \Delta y) = \langle I(x_1, y_1) I(x_2, y_2) \rangle \tag{2.106}$$

In lengthy calculations involving the circular Gaussian distributions and the Huygens-Fresnel principle in [17] the expression

$$R(\Delta x, \Delta y) = \langle I \rangle^2 \left\{ 1 + \left| \frac{\int_{-\infty}^{\infty} \int_{-\infty}^{\infty} |I(\xi, \eta)|^2 e^{i \frac{2\pi}{\lambda z} (\xi \Delta x + \eta \Delta y)} d\xi d\eta}{\int_{-\infty}^{\infty} \int_{-\infty}^{\infty} |I(\xi, \eta)|^2 d\xi d\eta} \right|^2 \right\} \tag{2.107}$$

is derived with $|I(\xi, \eta)|$ denoting the intensity distribution incident on the scattering spot (ξ, η).

For the special case of a uniformly scattering square area with dimensions $L \times L$, we have

$$|I(\xi, \eta)| = \begin{cases} 1 & \text{for} \quad \left|\frac{\xi}{L}\right| \quad \text{and} \quad \left|\frac{\eta}{L}\right| \\ 0 & \text{otherwise} \end{cases} \tag{2.108}$$

and the resulting autocorrelation is

$$R(\Delta x, \Delta y) = \langle I \rangle^2 \left\{ 1 + \mathrm{sinc}^2 \left(\frac{L\Delta x}{\lambda z} \right) \mathrm{sinc}^2 \left(\frac{L\Delta y}{\lambda z} \right) \right\} \tag{2.109}$$

The average size of a speckle can be taken to be the value of Δx where $\mathrm{sinc}^2(L\Delta x/\lambda z)$ first falls to zero, given by

$$\Delta x_s = \frac{\lambda z \pi}{L} \tag{2.110}$$

The same result we could derive by taking the two extreme points a distance L apart as the two apertures in the Young's double aperture interferometer, (2.58).

We see that the speckle size at a fixed distance z from the scattering surface increases as the size of the illuminated area decreases. Because the size of these speckles only depends on the scattering surface and the plane where it is viewed, but not on any imaging system, they are called *objective speckles*.

Contrary we speak of *subjective speckles* if they are formed by an imaging system, Fig. 2.13. Now the spatial distribution of the speckles is additionally determined by the

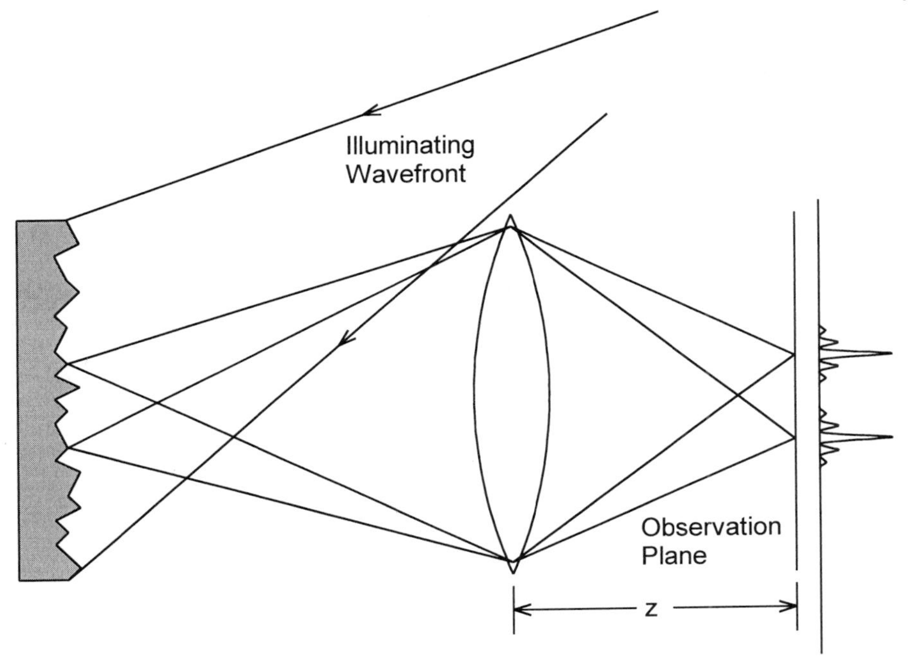

Figure 2.13: Imaging geometry for speckle formation

diffraction limit of the imaging system. For this case the autocorrelation function of the intensity in the image plane was found [17] to be

$$R(r) = \langle I \rangle^2 \left\{ 1 + \left| \frac{2\mathrm{J}_1\left(\frac{\pi D r}{\lambda z}\right)}{\frac{\pi D r}{\lambda z}} \right|^2 \right\} \tag{2.111}$$

where now D is the diameter of the circular lens pupil, J_1 is the Bessel function of first order, z is the distance of the image plane from the lens pupil plane, and $r = \sqrt{\Delta x^2 + \Delta y^2}$. The speckle size can be taken as the separation between the first two minima of J_1. Since $J_1(x) = 0$ at $x = 1.22\pi$ we get

$$d_{sp} = \frac{2.44\lambda z}{D} \tag{2.112}$$

The size of image plane speckles thus depends on the imaging system. If the aperture of the viewing lens is decreased, the speckle size will increase. It should be mentioned that the objective speckles exist more or less only theoretically, because we always have an imaging system, at least our own eyes.

The maximum spatial frequency f_{max} in the speckle pattern is given by the size of the lens aperture and the distance of the lens from the observation plane by [18]

$$\frac{1}{f_{max}} = \frac{\lambda z}{D} \tag{2.113}$$

2.6 Holographic Recording and Reconstruction

2.6.1 Hologram Recording

Each optical wave field consists of an amplitude distribution as well as a phase distribution (2.16), but all detectors or recording material like photographic film only register intensities: The phase is lost in the registration process. Now we have seen in (2.32) that if two waves of the same frequency interfere, the resulting intensity distribution is temporally stable and depends on the phase difference $\Delta\phi$. This is used in *holography* where the phase information is coded by interference into a recordable intensity. Clearly, to get a temporally stable intensity distribution, at least as long as the recording process, $\Delta\phi$ must be stationary, which means the wave fields must be mutually coherent.

It was D. Gabor who has shown that by illuminating the recorded interference pattern by one of the two interfering wave fields we can reconstruct the other one: this reconstructed wave field then consists of amplitude and phase distributions, not only the intensity. Figs. 2.14 and 2.15 show schematically two basic holographic setups, used for recording the wave field reflected from the object. This field is called the object field or object wave, while the other field, necessary for producing the interference, is called the reference field or reference wave. To be mutually coherent, both must stem from the same source of coherent light, the laser. The division into object and reference wave can be performed by wavefront division, Fig. 2.14, or by amplitude division, Fig. 2.15.

The following description uses a point source which does not restrict the generality, because by the superposition principle (2.6) the results can be extended to all points of the object surface. Let the wave reflected by an object surface point P be the spherical wave (2.23), called the *object wave*

$$E_P = \frac{E_{0P}}{p} e^{i(kp + \phi)} \tag{2.114}$$

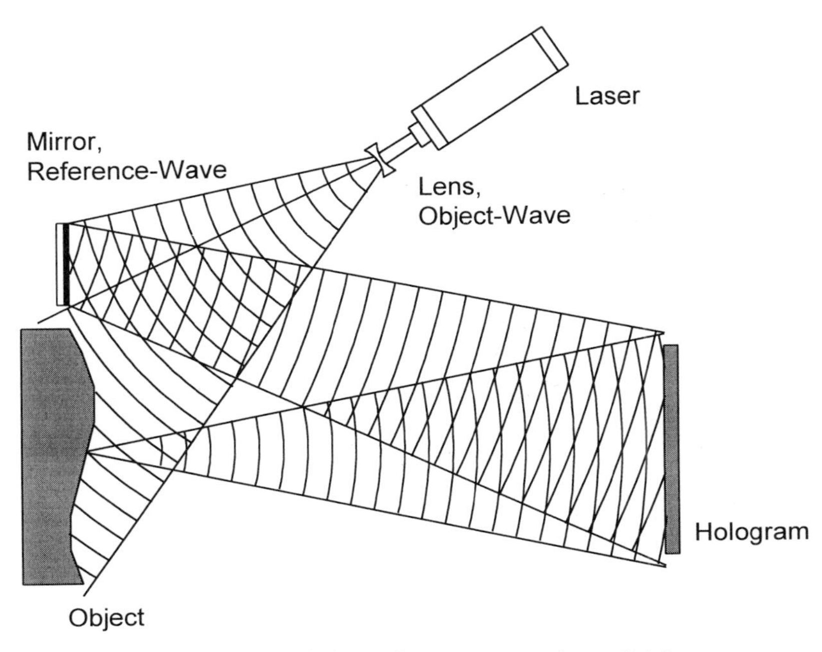

Figure 2.14: Basic holographic setup, wavefront division

where p is the distance between the point P and the point $H = (x, y, 0)$ on the photographic plate. The temporal factor ωt of (2.23) can be omitted. The *reference wave* is assumed to be a spherical wave emitted at R

$$E_R = \frac{E_{0R}}{r} \mathrm{e}^{\mathrm{i}(kr + \psi)} \tag{2.115}$$

with the distance r between R and H. The photographic plate registers the intensity

$$\begin{aligned} I(x, y) & \\ &= |E_P + E_R|^2 = E_P E_P^* + E_R E_R^* + E_P^* E_R + E_P E_R^* \\ &= \frac{E_{0P}^2}{p^2} + \frac{E_{0R}^2}{r^2} + \frac{E_{0P}}{p} \mathrm{e}^{-\mathrm{i}(kp + \phi)} \frac{E_{0R}}{r} \mathrm{e}^{\mathrm{i}(kr + \psi)} + \frac{E_{0P}}{p} \mathrm{e}^{\mathrm{i}(kp + \phi)} \frac{E_{0R}}{r} \mathrm{e}^{-\mathrm{i}(kr + \psi)} \\ &= \frac{E_{0P}^2}{p^2} + \frac{E_{0R}^2}{r^2} + \frac{2E_{0P}E_{0R}}{pr} \cos(k(r - p) + \psi - \phi) \end{aligned} \tag{2.116}$$

This intensity distribution, spatially varying because $p = p(x, y)$ and $r = r(x, y)$, is the *hologram* of a point source, the phase ϕ of the object wave relative to the phase ψ of the reference wave is coded into the intensity variation.

During the *time of exposure* t_B, the photographic plate receives the energy

$$B(x, y) = \int_0^{t_B} I(x, y, t) dt \tag{2.117}$$

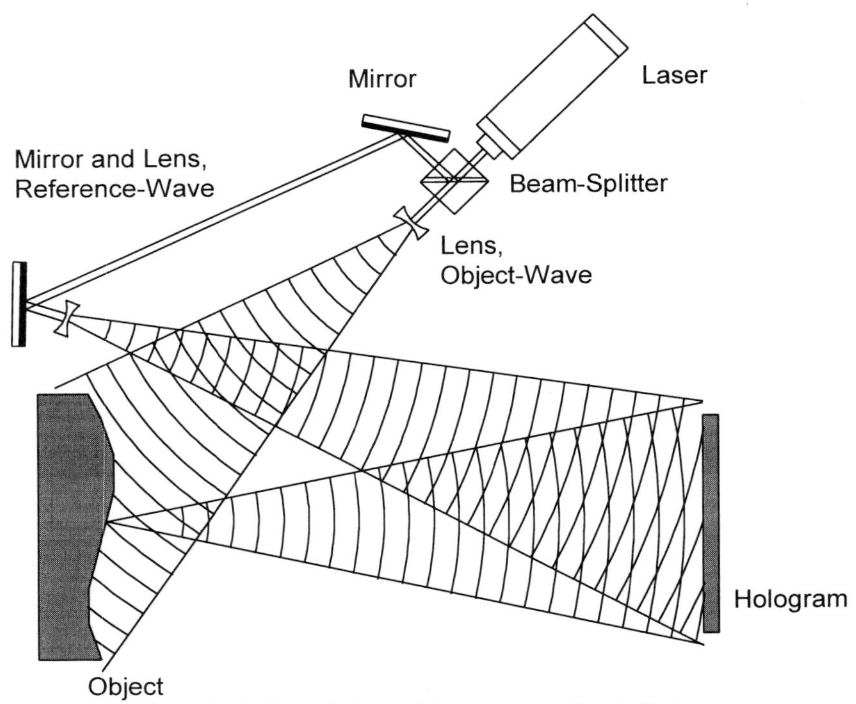

Figure 2.15: Basic holographic setup, amplitude division

By processing, this energy is translated into a blackening and a change of the refractive index. This is summarized in the complex *degree of transmission* τ which is generally a spatially varying function

$$\tau = \tau(x, y) = T(x, y)e^{i\theta(x, y)} \tag{2.118}$$

This contains the cases of the *amplitude hologram*, where $\theta = \text{const}$, or the *phase hologram* with $T = \text{const}$.

If the exposed plate is processed to produce an amplitude hologram, the real transmission T depends on the received energy B as shown in Fig. 2.16. One has to work in the linear range, where the curve is approximated by the line

$$\begin{aligned} T &= \alpha - \beta B \\ &= \alpha - \beta t_B I \end{aligned} \tag{2.119}$$

for a temporally constant intensity. α represents a uniform background transmittance, the positive value β is the slope of the amplitude transmittance. The working point B_0 is reached by adjusting the exposure time t_B. To keep the variation around B_0 small, the two wavefronts are given different amplitudes (2.40). The resulting real amplitude transmittance after processing is

$$T = \alpha - \beta t_B(E_P E_P^* + E_R E_R^* + E_P^* E_R + E_P E_R^*)$$

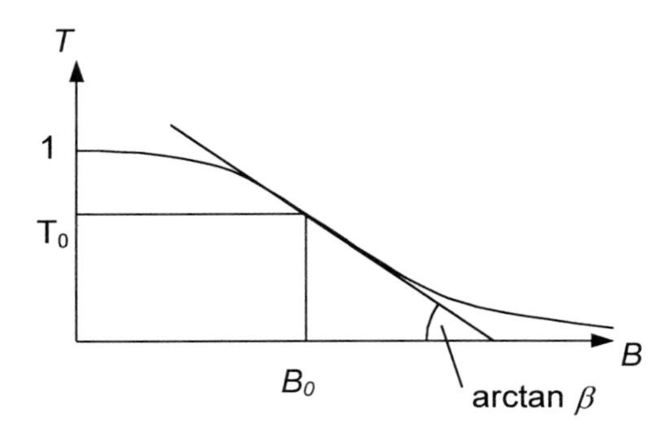

Figure 2.16: Amplitude transmittance versus received energy

$$\begin{aligned} &= \alpha - \beta t_B \left(\frac{E_{0P}^2}{p^2} + \frac{E_{0R}^2}{r^2} \right. \\ &\quad \left. + \frac{E_{0P}E_{0R}}{pr} \mathrm{e}^{\mathrm{i}(k(r-p)+\psi-\phi)} + \frac{E_{0P}E_{0R}}{pr} \mathrm{e}^{-\mathrm{i}(k(r-p)+\psi-\phi)} \right) \\ &= T_0 - \frac{E_{0P}E_{0R}}{pr} \left(\mathrm{e}^{\mathrm{i}(k(r-p)+\psi-\phi)} + \mathrm{e}^{-\mathrm{i}(k(r-p)+\psi-\phi)} \right) \end{aligned} \quad (2.120)$$

where T_0 is the mean transmittance $T_0 = \alpha - \beta t_B (E_{0P}^2/p^2 + E_{0R}^2/r^2)$.

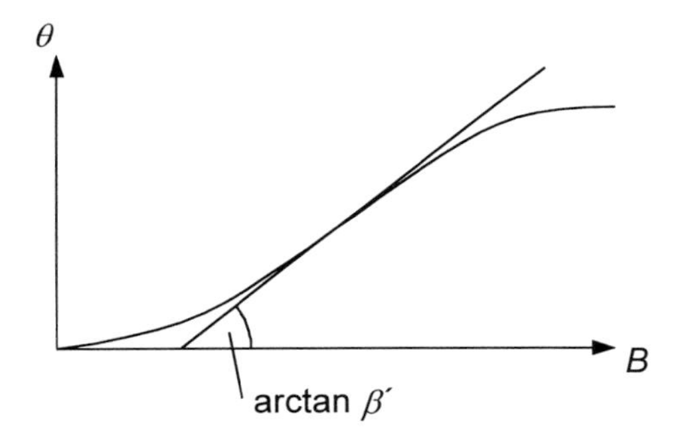

Figure 2.17: Phase shift versus received energy

If we produce a phase hologram, we must remain in the linear range of the curve describing the effective phase shift θ against exposure, Fig. 2.17. Again we have a range where we approximate by a line

$$\theta = \alpha' + \beta' t_B I \quad (2.121)$$

The complex transmission after a series expansion of the exponential and neglecting higher than linear terms is

$$\tau = \mathrm{e}^{\mathrm{i}\theta(I)} \approx 1 + \mathrm{i}\theta(I) \quad (2.122)$$

The resulting phase transmittance is

$$\tau = \mathrm{e}^{\mathrm{i}\theta(I)} \approx (1 + \mathrm{i}\alpha') + \mathrm{i}\beta t_B (E_P E_P^* + E_R E_R^* + E_P^* E_R + E_P E_R^*) \quad (2.123)$$

analogous to (2.120).

The necessary *spatial resolution* of the recording media can be estimated by (2.120) and (2.123). If the angle between reference and object wave is θ, both waves for the moment being assumed as plane waves, the fringe distance is

$$\frac{\pi}{|k''|} = \frac{\lambda}{2\sin\frac{\theta}{2}} \tag{2.124}$$

If we assume a wavelength $\lambda = 0.5\mu m$, for an angle of $\theta = 1^o$ we need a spatial resolution of better than 35 LP/mm (line-pairs per millimeter), for $\theta = 10^o$ already more than 350 LP/mm and for 30° more than 1035 LP/mm. As a consequence the spatial resolution puts an upper bound to the angular separation of the object points and thus to the object size.

2.6.2 Holographic Reconstruction of a Wave Field

For *reconstruction* of the object wave we illuminate the processed photographic plate, also called *hologram*, with the reference wave E_R, Fig. 2.18. This results in a modulation of

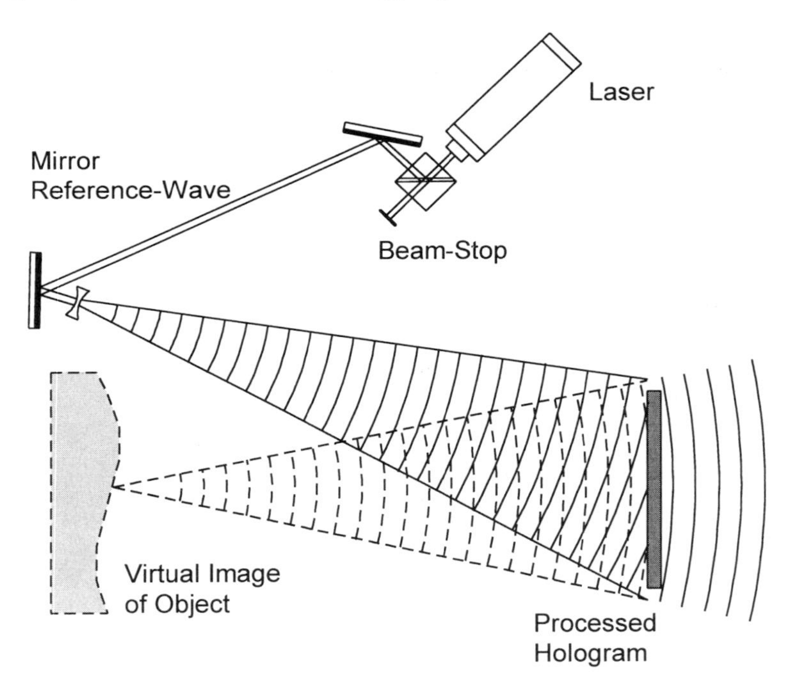

Figure 2.18: Holographic reconstruction

the reference wave by the transmission $\tau(x, y)$.

For an amplitude hologram, directly behind the hologram (2.120) we get the complex amplitude

$$
\begin{aligned}
E_{reconstr} &= TE_R \\
&= \alpha E_R - \beta t_B E_R (E_P E_P^* + E_R E_R^* + E_P^* E_R + E_P E_R^*) \\
&= \left(\alpha - \beta t_B (|E_R|^2 + |E_P|^2)\right) E_R \\
&\quad -\beta t_B E_R^2 E_P^* \\
&\quad -\beta t_B |E_R|^2 E_P
\end{aligned} \tag{2.125}
$$

The first term after the last equals sign

$$
\left(\alpha - \beta t_B (|E_R|^2 + |E_P|^2)\right) E_R = T_0 E_R \tag{2.126}
$$

is the zeroth diffraction order, only the reference wave multiplied with the mean transmittance. The second term

$$
\beta t_B E_R^2 E_P^* = \beta t_B \frac{E_{0R}^2}{r^2} \mathrm{e}^{\mathrm{i}2(kr+\psi)} \frac{E_{0P}}{p} \mathrm{e}^{-\mathrm{i}(kp+\phi)} \tag{2.127}
$$

is a *conjugated image* of the object wave. While the original wavefront was a diverging one from P, the conjugate is a converging wavefront, which converges to a focus. The *image* is *real* and *pseudoscopic*, the latter describes the property of inverted depths: Parts of the object in the background now appear in the foreground and vice versa, which gives curious parallax effects, difficult to observe. The contrary to pseudoscopic images are called *orthoscopic images*, to them we are used in the real world. The third term

$$
\beta t_B |E_R|^2 E_P = \beta t_B \frac{E_{0R}^2}{r^2} \frac{E_{0P}}{p} \mathrm{e}^{\mathrm{i}(kp+\phi)} \tag{2.128}
$$

gives the original wavefront E_P multiplied by a pure intensity term which for a plane reference wave is constant. The reconstructed wavefront produces a *virtual image* of the object which appears to stand at the place it occupied during recording of the hologram. Since the full object wave with amplitude and phase is reconstructed, the holographically recorded and reconstructed scene can be observed three-dimensionally. That means we can observe with varying depth of focus and with varying parallax.

If we reconstruct from a phase hologram (2.123), we get

$$
\begin{aligned}
E_{reconstr} &= \tau E_R \\
&\approx \left((1+\mathrm{i}\alpha') + \mathrm{i}\beta' t_B (|E_R|^2 + |E_P|^2)\right) E_R \\
&\quad +\mathrm{i}\beta' t_B E_R^2 E_P^* \\
&\quad +\mathrm{i}\beta' t_B |E_R|^2 E_P
\end{aligned} \tag{2.129}
$$

The three terms consist of the same waves as in (2.125), only the factors before differ.

The *phase conjugated wave* $E_{pc}(z,t)$ belonging to $E(z,t) = E_0 \mathrm{e}^{\mathrm{i}kz} \mathrm{e}^{-\mathrm{i}\omega t}$ is

$$
E_{pc}(z,t) = E_0 \mathrm{e}^{-\mathrm{i}kz} \mathrm{e}^{-\mathrm{i}\omega t} \tag{2.130}
$$

Contrary to $E^*(z,t)$ in $E_{pc}(z,t)$ only the spatial part is conjugated. Physically a phase conjugated wave is the original wave travelling in the opposite direction: instead of being emitted and diverging from a point the phase conjugated wave is converging to this point.

If we have used a plane wave as reference wave, it is easy to produce the phase conjugated one. We only have to turn the hologram by 180° and in the case of an amplitude hologram get

$$\begin{aligned} E_{reconstr} &= TE_R^* \\ &= \left(\alpha - \beta t_B(|E_R|^2 + |E_P|^2)\right) E_R^* \\ &\quad -\beta t_B |E_R|^2 E_P^* \\ &\quad -\beta t_B E_R^{*2} E_P \end{aligned} \tag{2.131}$$

Now the second term represents a real orthoscopic image in exactly the position relative to the hologram the object had during the recording.

To get a reconstructed wavefront undistorted by the other reconstructed waves, reference and object wave have to be separated in space. While Gabor had no sources giving light with sufficient coherence, he had to produce *in-line-holograms* where object and reference waves were travelling the same direction orthogonally to the hologram. Leith and Upatnieks [7, 8, 9] were the first who took advantage of the coherence of laser light and who gave the object and the reference beams different directions. In their *off-axis-holograms* all reconstructed waves are well separated.

To find the directions of the reconstructed images, for the moment we assume the object and reference waves as plane waves and calculate the wave in the hologram plane. Without restriction of generality we assume $y = 0$. Furthermore we obey the convention that angles are counted positive in counterclockwise direction and negative in clockwise direction.

Let the reference wave E_R illuminate the hologram plate under the angle $-\gamma_1$ and let the object wave E_P impinge under $+\gamma_2$, Fig. 2.19. Then the broadened reference wave E_R, the first term in (2.125) and (2.129), leaves the hologram plate in direction $-\gamma_1$. To find the second term, we observe that in the hologram plane ($z = 0$) with $E_P(x,0,0) = \exp(ikx \sin\gamma_2)$ the conjugate is $E_P^*(x,0,0) = \exp(-ikx\sin\gamma_2) = \exp[ikx\sin(-\gamma_2)]$. A conjugate wave would leave the hologram under $-\gamma_2$. The square of $E_R = \exp[ikx\sin(-\gamma_1)]$ is

$$E_R^2(x,0,0) = \mathrm{e}^{-ikx2\sin\gamma_1} \tag{2.132}$$

Thus the second term in (2.125) and (2.129) is

$$E_R^2 E_P^* = \mathrm{e}^{-ikx2\sin\gamma_1}\mathrm{e}^{-ikx\sin\gamma_2} = \mathrm{e}^{ikx\sin(-\gamma_3)} \tag{2.133}$$

For small γ_1 we observe that $2\sin\gamma_1 \approx \sin 2\gamma_1$, but always $2\sin\gamma_1 > \sin 2\gamma_1$. As long as γ_1, γ_2 remain small, we can set $\gamma_3 = 2\gamma_1 + \gamma_2$, but for larger γ_1, γ_2 the image shifts and becomes more and more distorted until it vanishes for

$$2\sin\gamma_1 + \sin\gamma_2 \geq 0 \tag{2.134}$$

The third term of (2.125) and (2.129) is E_P which leaves the hologram under γ_2 and constitutes the virtual image of the object.

The arrangement simplifies with a reference wave impinging orthogonally onto the hologram plane, then we have $\gamma_1 = 0$.

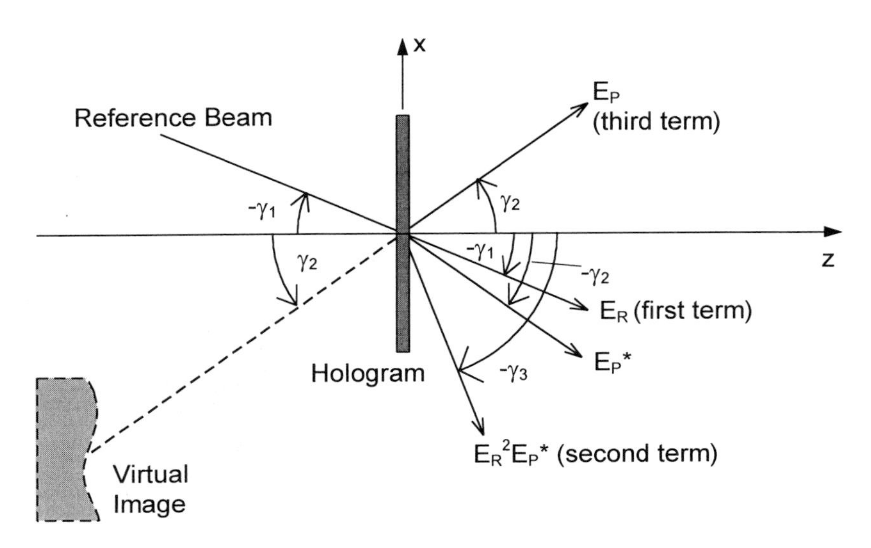

Figure 2.19: Locations of reconstructed images

2.6.3 Holographic Imaging Equations

In Sec. 2.6.2 we have assumed to reconstruct with exactly the same reference wave of the same wavelength as used for the recording of the hologram. But, for example, a hologram recorded with a pulsed ruby laser of wavelength $\lambda = 0.694$ μm has to be reconstructed for observation with a continuous laser, normally a He-Ne laser of wavelength $\lambda = 0.633$ μm. Furthermore we are interested whether the reconstructed image vanishes, if the reconstructing wave is slightly shifted from the recording wave, or what else is happening to the reconstruction [19, 20, 21, 22, 23].

The investigations are performed for a single object point $P = (x_P, y_P, z_P)$ and a hologram point $Q = (x, y, 0)$. The hologram is placed in the (x, y)-plane of the Cartesian coordinate system, Fig. 2.20. The reconstructed image of an extended object then can be determined by calculating the positions of several of its surface points. The spherical reference wave of wavelength λ during the recording diverges from point R, during reconstruction it may diverge from R' having wavelength λ'.

Reconstruction with $E_{R'}$ and λ' instead of E_R and λ gives a third term in (2.125) proportional to

$$\begin{aligned} E_P E_{R'} E_R^* &= \frac{E_{0P}}{p}\frac{E_{0R'}}{r'}\frac{E_{0R}}{r} \mathrm{e}^{\mathrm{i}(kp+\phi)} \mathrm{e}^{\mathrm{i}(k'r'+\xi)} \mathrm{e}^{-\mathrm{i}(kr+\psi)} \\ &\sim \mathrm{e}^{\mathrm{i}(kp+k'r'-kr+\phi+\xi-\phi)} \\ &= \mathrm{e}^{\mathrm{i}(k'p'+\eta)} \end{aligned} \quad (2.135)$$

which describes a wave with wave number $k' = 2\pi/\lambda'$ coming from P' to Q. p is the distance from P to Q, r is the distance from R to Q, r' is that from R' to Q and p' is the distance from P' to Q, resp.

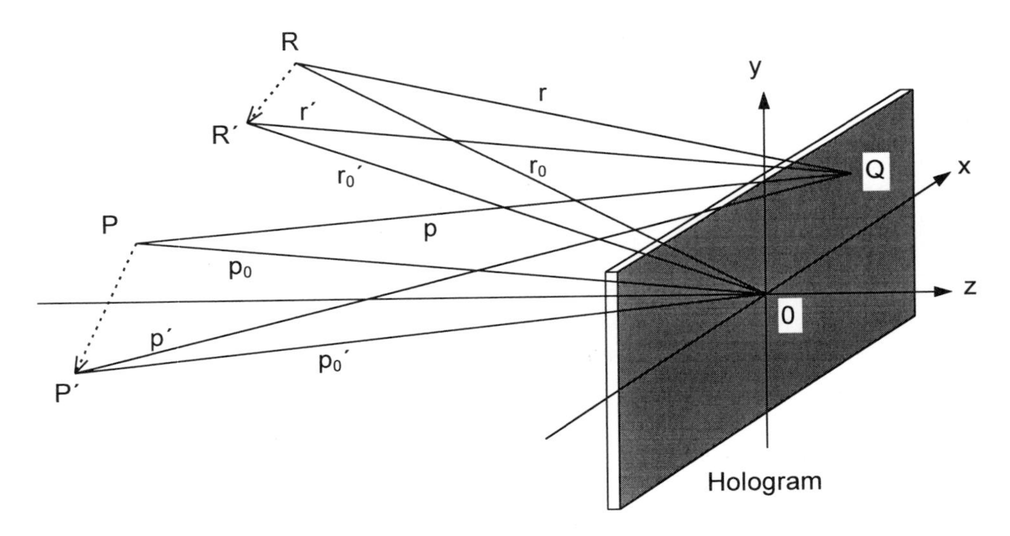

Figure 2.20: Coordinate system for the holographic imaging equations

We first investigate the distance p

$$\begin{aligned} p &= \sqrt{(x - x_P)^2 + (y - y_P)^2 + z_P^2} \\ &= \sqrt{x^2 + y^2 - 2xx_P - 2yy_P + p_0^2} \end{aligned} \tag{2.136}$$

where p_0 is the distance of P to the origin. We observe that p_0 is large compared to x, y, x_P and y_P, so we can expand p into a power series around p_0 [24]

$$p = p_0 + \frac{x^2 + y^2}{2p_0} - \frac{2xx_P + 2yy_P}{2p_0} + - \ldots \tag{2.137}$$

In the same way we expand r, r' and p'

$$r = r_0 + \frac{x^2 + y^2}{2r_0} - \frac{2xx_R + 2yy_R}{2r_0} + - \ldots \tag{2.138}$$

$$r' = r'_0 + \frac{x^2 + y^2}{2r'_0} - \frac{2xx'_R + 2yy'_R}{2r'_0} + - \ldots \tag{2.139}$$

$$p' = p'_0 + \frac{x^2 + y^2}{2p'_0} - \frac{2xx'_P + 2yy'_P}{2p'_0} + - \ldots \tag{2.140}$$

From the definition of p' (2.135) we have

$$k'p' = k(p - r) + k'r' \tag{2.141}$$

Now we insert the approximations of p, r, r' and p' and observe that (2.141) is valid for all (x, y), so we equate the coefficients of similar terms and with the definition $\mu = \lambda'/\lambda$

we get

$$\frac{1}{p_0'} = \frac{\mu}{p_0} - \frac{\mu}{r_0} + \frac{1}{r_0'} \tag{2.142}$$

$$\frac{x_P'}{p_0'} = \frac{\mu x_P}{p_0} - \frac{\mu x_R}{r_0} + \frac{x_R'}{r_0'} \tag{2.143}$$

$$\frac{y_P'}{p_0'} = \frac{\mu y_P}{p_0} - \frac{\mu y_R}{r_0} + \frac{y_R'}{r_0'} \tag{2.144}$$

These are the *holographic imaging equations* for the direct image to a first order approximation. In a similar way one can derive the imaging equations of the conjugate image, the second term in (2.125). They have the same form, only the signs preceding the μ are interchanged.

In the preceding derivation we have expanded about p_0, p_0', r_0 and r_0' [24], which gives a more accurate approximation than the expansion about the z-component z_P, z_P', z_R , z_R' [25]. This may be performed, if the x- and y-components are negligible compared to the z-components. In this case we would get the imaging equations

$$\frac{1}{z_P'} = \frac{\mu}{z_P} - \frac{\mu}{z_R} + \frac{1}{z_R'} \tag{2.145}$$

$$\frac{x_P'}{z_P'} = \frac{\mu x_P}{z_P} - \frac{\mu x_R}{z_R} + \frac{x_R'}{z_R'} \tag{2.146}$$

$$\frac{y_P'}{z_P'} = \frac{\mu y_P}{z_P} - \frac{\mu y_R}{z_R} + \frac{y_R'}{z_R'} \tag{2.147}$$

which can be rearranged to yield the better to handle and well known form [25, 26]

$$x_P' = \frac{\mu x_P z_R z_R' - \mu x_R z_P z_R' + x_R' z_P z_R}{\mu z_R z_R' - \mu z_P z_R' + z_P z_R} \tag{2.148}$$

$$y_P' = \frac{\mu y_P z_R z_R' - \mu y_R z_P z_R' + y_R' z_P z_R}{\mu z_R z_R' - \mu z_P z_R' + z_P z_R} \tag{2.149}$$

$$z_P' = \frac{z_R' z_P z_R}{\mu z_R z_R' - \mu z_P z_R' + z_P z_R} \tag{2.150}$$

We now consider some special cases. The easiest is the reconstruction with a reference wave identical to the one used for recording, even with the same wavelength. Then we have $p_0' = p_0$ and $x_P' = x_P$, $y_P' = y_P$ and $z_P' = z_P$, as expected. Next we examine a plane reference wave impinging orthogonally onto the hologram, but during reconstruction with λ' instead of λ. Then we have $r_0 = z_R \to \infty$, $r_0' = z_R' \to \infty$, $x_R/r_0 = 0$, $y_R/r_0 = 0$, $x_R'/r_0' = 0$, $y_R'/r_0' = 0$. From (2.142) we get $\mu p_0'/p_0 = 1$ and together with (2.143) and (2.144) resp.

$$x_P' = x_P \tag{2.151}$$

$$y_P' = y_P \tag{2.152}$$

$$z_P' = \frac{1}{\mu} z_P \tag{2.153}$$

This means the object is shifted, stretched or compressed in the z-direction, but the lateral dimensions remain unaffected.

The *lateral magnification* M_{lat} in the direct image can be defined [25]

$$M_{lat} = \frac{dx'_P}{dx_P} = \frac{dy'_P}{dy_P} \tag{2.154}$$

It is

$$M_{lat} = \frac{1}{1 + z_P\left(\frac{1}{\mu z'_R} - \frac{1}{z_R}\right)} \tag{2.155}$$

For the plane wave of the case above due to $z_R \to \infty$ and $z'_R \to \infty$ we get $M_{lat} = 1$.

Furthermore we may define the *angular magnification* by [25]

$$M_{ang} = \frac{d(x'_P/z'_P)}{d(x_P/z_P)} = \frac{d(y'_P/z'_P)}{d(y_P/z_P)} \tag{2.156}$$

and calculate $|M_{ang}| = \mu$. For the *longitudinal magnification* M_{long} of the primary image we obtain

$$M_{long} = \frac{dz'_P}{dz_P} = \frac{1}{\mu} M_{lat}^2 \tag{2.157}$$

The different magnifications for the conjugate image are derived analogously.

The holographic imaging equations can be used to compensate the effect of CW-reconstruction with a wavelength differing from that of the pulsed recording by a proper reference beam adjustment [27]. If the hologram is shifted during the real-time reconstruction, shearing fringes are produced which modify the deformation fringes in a controlled way [28]. If in the expansion (2.137) we retain higher order terms, we get the *aberrations* like spherical aberration, coma, astigmatism, field curvature and others [19, 20, 25, 29, 30, 31, 32]. They give information about the quality of the imaged object, which means how the image of a point is washed out. The knowledge about these aberrations will not be considered further in this book.

2.6.4 Types of Holograms

In Sec. 2.6.2 we already mentioned the distinction into Gabor's in-line holograms and the Leith-Upatnieks off-axis holograms. The *in-line arrangement* is still used for the analysis of transparent objects, or small particles like droplets, Fig. 2.21a. The reference wave is the light passing unaffected by the particles and the object wave is the wave field scattered by the particles. We may use divergent or collimated light, Fig. 2.21b. Reconstruction by the illuminating wave alone without the object gives the virtual image of the particles or the transparent object at its original position as well as the real image on the opposite side of the hologram. There is no beam splitting into reference and object beam, so the method sometimes is called *single beam holography.*

The main disadvantages of in-line holograms are the disturbed reconstruction due to the bright reference beam and the twin images: Virtual and real image are along the same

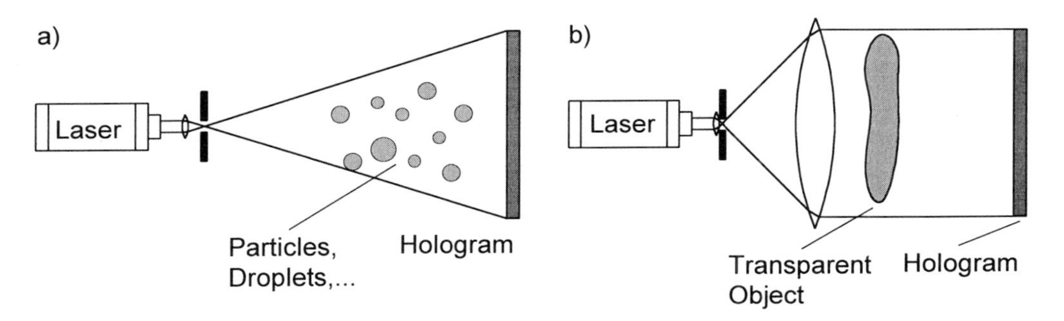

Figure 2.21: In-line holography with divergent (a) and collimated (b) light

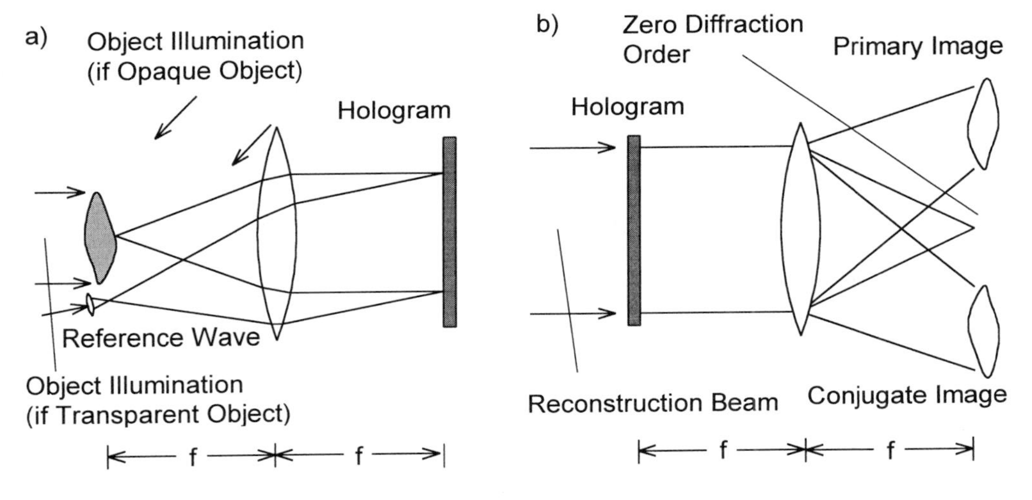

Figure 2.22: Recording (a) and reconstruction (b) of a Fourier transform hologram

line of sight, so while focusing on one of them we see it superposed by an out of focus image of the other one. These drawbacks are avoided by the *off-axis arrangement.* Here the laser beam is splitted by wavefront division, Fig. 2.14, or by amplitude division, Fig. 2.15, into reference and object wave. This approach has been called *split beam holography* or *two beam holography.* If the offset angle between reference and object wave is large enough, we have no overlap between the virtual and the real reconstructed images nor do we stare into the directly transmitted reference beam.

If we have flat or nearly flat objects we may record the Fourier transform of the object and reference waves. Then we speak about *Fourier transform holography.* Imagine an object located in the front focal plane of a lens, Fig. 2.22a, illuminated by coherent light. Let the complex amplitude leaving the object plane be $E_P(x, y)$, then the complex amplitude of the wave field at the holographic plate located in the back focal plane of the lens is the Fourier transform of $E_P(x,y)$

$$\mathcal{E}_P(u,v) = \mathcal{F}\{E_P(x,y)\} \tag{2.158}$$

The reference wave is a spherical wave emitted from a point source at (x_0, y_0) in the front focal plane. Without loss of generality we can assume unit amplitude, so $E_R(x_0, y_0) = \delta(x_0, y_0)$ is the complex amplitude at (x_0, y_0). $\delta(x,y)$ describes the *Dirac delta impulse*, see (A.6). The complex amplitude of this reference wave in the hologram plane is

$$\mathcal{E}_R(u,v) = \mathcal{F}\{\delta(x_0, y_0)\} = \mathrm{e}^{-\mathrm{i}2\pi(ux_0 + vy_0)} \tag{2.159}$$

The interference pattern formed by these two waves is recorded. This so called *Fourier transform hologram* is

$$\begin{aligned} \mathcal{I}(u,v) &= |\mathcal{E}_P + \mathcal{E}_R|^2 = (\mathcal{E}_P + \mathcal{E}_R)(\mathcal{E}_P + \mathcal{E}_R)^* \\ &= |\mathcal{E}_P|^2 + \mathcal{E}_P^* \mathrm{e}^{-\mathrm{i}2\pi(ux_0 + vy_0)} + \mathcal{E}_P \mathrm{e}^{\mathrm{i}2\pi(ux_0 + vy_0)} + 1 \end{aligned} \tag{2.160}$$

For reconstruction the processed hologram is illuminated by a parallel beam of coherent light, Fig. 2.22b. Again we assume unit amplitude for the reconstruction wave and a linear dependence of the amplitude transmittance of the processed hologram from the intensity $\mathcal{I}(u,v)$. The complex amplitude immediately behind the hologram now (2.119) is proportional to $\alpha + \beta t_B \mathcal{I}(u,v)$. In the back focal plane of the lens we get the Fourier transform of this amplitude distribution

$$\begin{aligned} \mathcal{F}\{\alpha + \beta t_B \mathcal{I}(u,v)\} &= \mathcal{F}\{\alpha + \beta t_B |\mathcal{E}_P|^2 + \beta t_B \mathcal{E}_P^* \mathrm{e}^{-\mathrm{i}2\pi(ux_0 + vy_0)} \\ &\quad + \beta t_B \mathcal{E}_P \mathrm{e}^{\mathrm{i}2\pi(ux_0 + vy_0)} + \beta t_B\} \\ &= \alpha\delta(x,y) + \beta t_B E_P(x,y) \star E_P(x,y) + \beta t_B E_P^*(x_0 - x, y_0 - y) \\ &\quad + \beta t_B E_P(x - x_0, y - y_0) + \beta t_B \delta(x,y) \end{aligned} \tag{2.161}$$

The first and last term of this reconstructed wave field is a focus on the optical axis while the second term constitutes a halo around this focus. The forth term is proportional to the original object wave field but shifted by (x_0, y_0) out of the origin. The third term is the inverted conjugate of the original wave shifted by $(-x_0, -y_0)$. Both images are real and can be registered by film or TV-camera in the back focal plane. The conjugated image in the registered intensity distribution is identified only by its geometric inversion. The main advantage of such a Fourier hologram is the stationary reconstructed image even when the hologram is translated in its own plane, due to the shift invariance of the Fourier transform intensity, see Appendix A. Furthermore the resolution requirements on the recording medium are less.

Collimation by the lens in Fourier transform holography means that the object points and the reference source are at infinity. *Lensless Fourier transform holography* is possible if object point and reference source are at a finite but the same distance from the holographic plate.

Let the wave leaving the object be $E_P(x,y)$, Fig. 2.23. The Fresnel-Kirchhoff integral

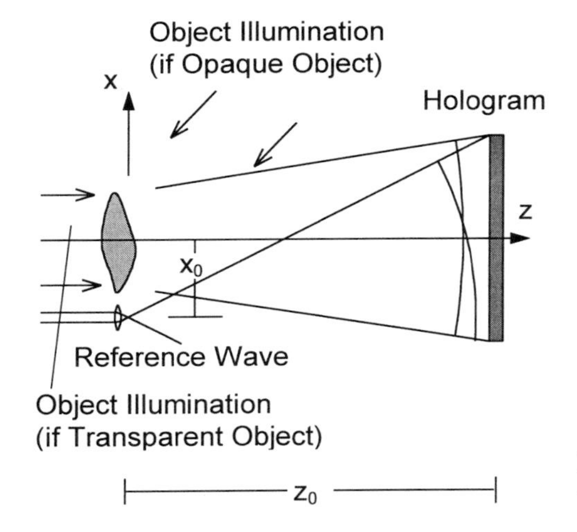

Figure 2.23: Recording of a lensless Fourier transform hologram

(2.71) implies that the complex amplitude in the hologram plane is

$$E_P(x, y, z_0) = \frac{\mathrm{i}}{\lambda z_0} \mathrm{e}^{-\frac{\mathrm{i}\pi}{\lambda z_0}(x^2 + y^2)} \mathcal{E}_P(u, v) \tag{2.162}$$

with z_0 the distance between object plane and hologram plane, $u = \frac{x}{\lambda z_0}$, $v = \frac{y}{\lambda z_0}$, and

$$\mathcal{E}_P(u, v) = \mathcal{F}\{E_P(x, y, 0)\mathrm{e}^{-\frac{\mathrm{i}\pi}{\lambda z_0}(x^2 + y^2)}\} \tag{2.163}$$

This is the Fourier transform of the object wave multiplied by a spherical phase factor which depends on the distance z_0 between object and hologram. The same formalism holds for the reference wave

$$E_R(x, y, z_0) = \mathrm{e}^{-\frac{\mathrm{i}\pi}{\lambda z_0}(x^2 + y^2)} \mathrm{e}^{-\mathrm{i}2\pi(ux_0 + vy_0)} \tag{2.164}$$

x_0 and y_0 are the distances of the reference source to the z-axis in x- and y-direction. The resulting interference pattern in the hologram plane is

$$I(x, y, z_0) = |E_P(x, y, z_0)|^2 + \frac{\mathrm{i}}{\lambda z_0} \mathcal{E}_P(u, v) \mathrm{e}^{\mathrm{i}2\pi(ux_0 + vy_0)} + \frac{\mathrm{i}}{\lambda z_0} \mathcal{E}_P^*(u, v) \mathrm{e}^{-\mathrm{i}2\pi(ux_0 + vy_0)} \tag{2.165}$$

This expression is similar to (2.160). The effect of the spherical phase factor associated with the near-field Fresnel diffraction pattern here is eliminated by the spherical reference wave with the same curvature.

If the object is focused into the plane of the hologram, Fig. 2.24, we speak of an *image hologram*. Here the real image of the object is recorded instead of the wave field reflected or scattered by the object. The advantage of image holograms is that they can be reconstructed by an incoherent light source of appreciable size and spectral bandwidth and they still produce acceptably sharp images. Also the image luminance is increased.

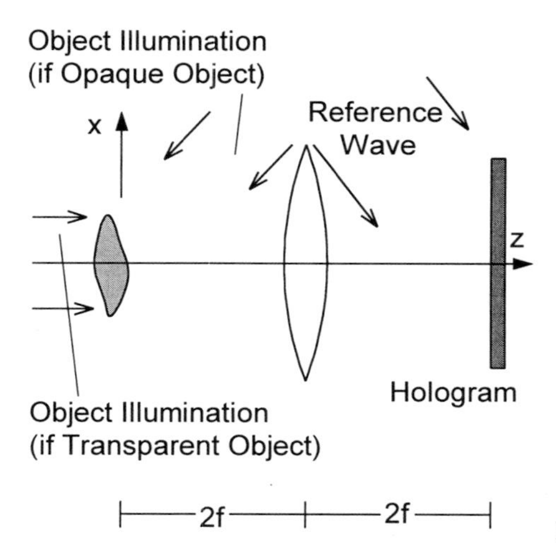

Figure 2.24: Recording of an image hologram

But we have to pay with an observation angle limited by the angular aperture of the used lens.

For the just mentioned image hologram the distance between object and hologram appears to be zero. The most general case is the holographic plate in the *near-field* or *Fresnel diffraction region*, Fig. 2.25, then we speak of a *Fresnel hologram.*

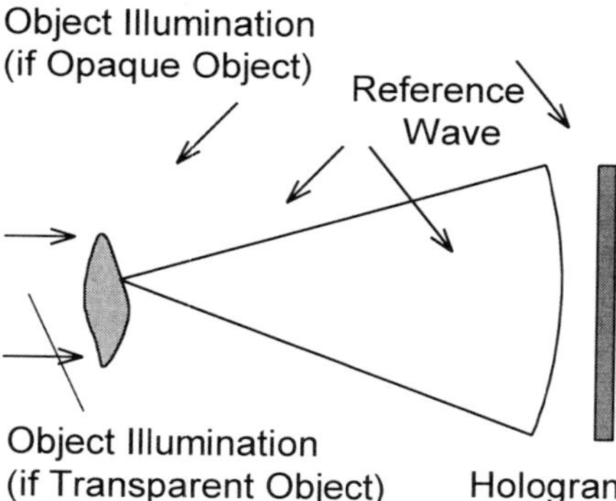

Figure 2.25: Recording of a Fresnel hologram

As the distance between object and hologram increases we reach the *far-field* or *Fraunhofer diffraction region* and get a so called *Fraunhofer hologram.* In this case either the object has to be small compared to the dimensions of the holographic arrangement

$$z_0 \gg \frac{x_0^2 + y_0^2}{\lambda} \tag{2.166}$$

where x_0 and y_0 are the maximal lateral dimensions of the object, or the object must be in the focus of a lens, Fig. 2.26.

Of course holographers have attempted early to get rid of the coherence requirements, at least in the reconstruction stage. If a usual hologram is illuminated with white light,

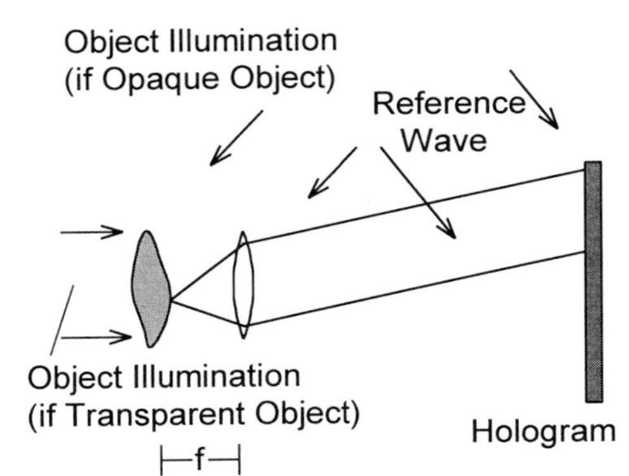

Figure 2.26: Recording of a Fraunhofer hologram

all the frequencies of the white light are diffracted in different directions. At each point different frequencies stemming from different points in the hologram are superimposed, so normally we will not recognize a reconstruction of an object wave field. The quality may be improved a little bit by a colour filter, but then the brightness decreases drastically.

The so called *white light holograms*, which can be reconstructed with white light, use the finite thickness of the photographic emulsion in the hologram [33]. Up to now we have only considered the lateral distribution of the phase or transmittance of the hologram. If the thickness of the sensitive layer is much greater than the distance between adjacent surfaces of the interference maxima, then the hologram should be considered as a *volume* or *three-dimensional hologram*. Generally we may speak of a volume hologram, if the thickness d_H of the layer is

$$d_H > \frac{1.6d^2}{\lambda} \tag{2.167}$$

where d is the distance between adjacent interference planes [34].

But if now the coherent object- and reference-waves impinge on the hologram from opposite sides we get interference layers nearly parallel to the hologram surface. The distance between subsequent interference layers is $\lambda/(2\sin\frac{\theta}{2})$ according to (2.30) and Fig. 2.3. For reconstruction this thick hologram is illuminated with white light which is reflected at the layers. Dependent on the wavelength the reflected waves interfere constructively in defined directions, an effect called *Bragg reflection*. Let the distance of the interference layers be d and illuminate by an angle α, then we find the n-th diffraction order for the wavelength λ in the observation direction of angle β

$$d = \frac{n\lambda}{\sin\alpha + \sin\beta} \tag{2.168}$$

The most intense wave is that in the first diffraction order $n = 1$. So if we look under the angle β onto the hologram illuminated under the angle α, we see a clear image with colour λ. The parallel layers modulated by the information about the image react like an interference filter for the specific wavelength λ.

An arrangement for recording such a white light hologram is given in Fig. 2.27. The expanded and collimated laser beam is directed through the hologram plate onto the

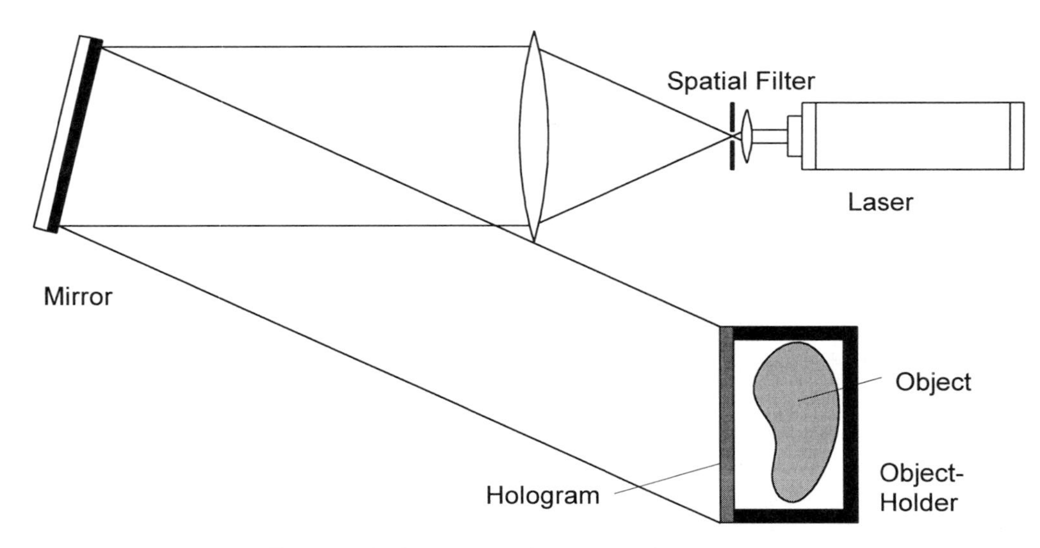

Figure 2.27: Recording of a white light hologram

object. Reference wave is the light coming directly from the laser, object wave is the wave passing through the plate and reflected by the object. For good results we need strongly reflecting objects and a hologram plate close to the object.

Another way to prevent the different diffracted colours from overlaying during the reconstruction with white light is the exchange of the variation of the vertical parallax against a variation of the wavelength [16]. For this task first a master hologram of the object is produced as usual, Fig. 2.28a. By reversing the direction of the reference beam

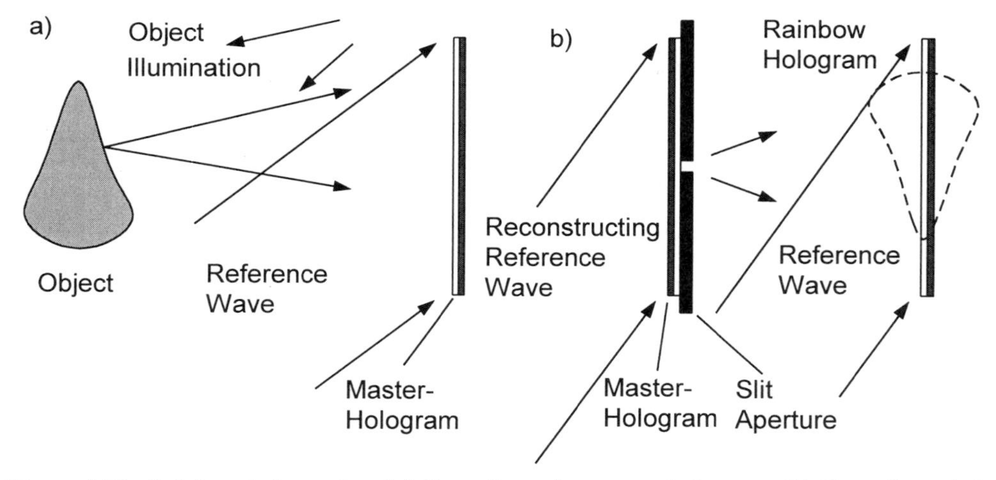

Figure 2.28: Rainbow holography: (a) Recording of a master hologram, (b) Recording of the rainbow hologram

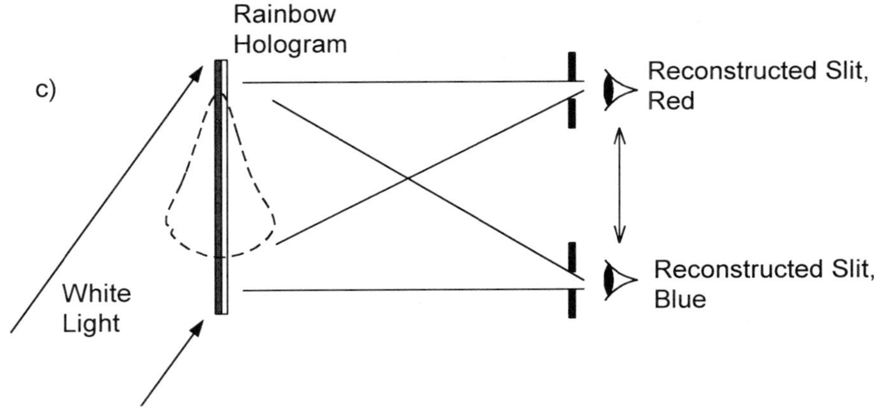

Figure 2.28: Rainbow holography: (c) Reconstruction with white light

or by turning the hologram by 180° a real pseudoscopic image is reconstructed. In front of the hologram now a horizontal slit aperture is placed and a hologram of the wave field passing through this slit is recorded, Fig. 2.28b. This has the first effect that the vertical parallax is lost, but this is not recognized immediately, as long as the eyes of the observer are horizontally arranged. The second effect is that the different colours reconstructed from the second hologram still overlay in space. But the colours converge to different reconstructed slits. Although neighboring colours are overlayed in the reconstructed slits, the range from blue to red can be stretched over a broad area so that each reconstructed slit produces a sharp image. The eyes of the observer are placed in one reconstructed slit and see the object in one colour. If the head is moved in vertical direction the object is seen in another spectral colour than before, Fig. 2.28c. Since in this way the object can be observed in the successive colours of the rainbow, the secondary hologram is called *rainbow hologram* and we speak of *rainbow holography*.

2.7 Elements of the Holographic Setup

2.7.1 Laser

Optical holography and holographic interferometry in the visible range of the spectrum became possible with the invention of a source radiating coherent light, the *laser*. The basic principle behind the laser is the *stimulated emission* of radiation. Contrary to the ubiquitous *spontaneous emission* here the emission of photons is triggered by an electromagnetic wave. All photons generated this way have the same frequency, polarization, phase, and direction as the stimulating wave.

Normally in a collection of atoms each one tends to hold the lowest energy configuration. Therefore in thermal equilibrium or even when excited the most atoms are in the *ground state*. Only if one succeeds in bringing a larger part of atoms into a higher excited state than remain in a lower state, which may be the ground state or an excited state, then an impinging wave can stimulate the emission of an avalanche of waves, all with equal phase and propagating in the same direction. This stimulated emission takes place as long as the *population inversion* between the lower and the higher energy states is maintained. To achieve the inversion, energy must be provided to the system by a process called *pumping*. So the laser can be regarded as an amplifier since an impinging wave generates a manifold of waves of the same direction, frequency, and phase.

To prevent this amplifier from amplifying only noise, a feedback is introduced by installing two mirrors on opposite sides of the active medium. If plane mirrors are adjusted exactly parallel, photons are reflected back and forth, what we get is an oscillator of high quality. Standing waves will be formed between the two mirrors. If one of the mirrors is semi-transparent, a part of the photons can leave the laser as the so called *laser beam* of coherent radiation. The resonant frequencies possible in a *cavity* of length L, the separation between the mirrors, are

$$\nu_n = \frac{nc}{2L} \tag{2.169}$$

with n an integer and c the speed of light.

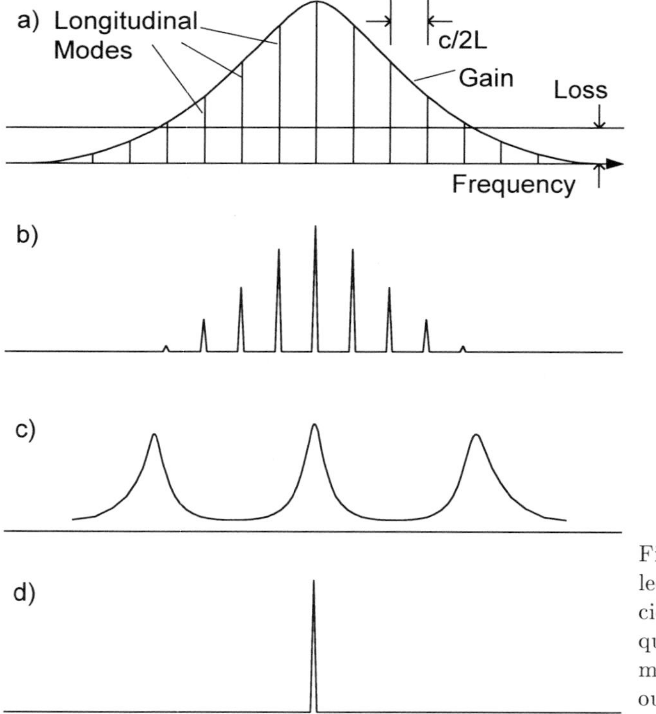

Figure 2.29: Single frequency selection: (a) Oscillation frequencies, gain, and loss, (b) Multifrequency output, (c) Etalon transmittance, (d) Single frequency output

Since the active medium possesses a gain curve around a central line, amplification takes place only at those frequencies, where the gain is higher than the cavity losses, Fig. 2.29a. If the medium allows for several central lines in the spectrum, these are selected by a *wavelength selector prism*, Fig. 2.30. The existence of several frequencies under the

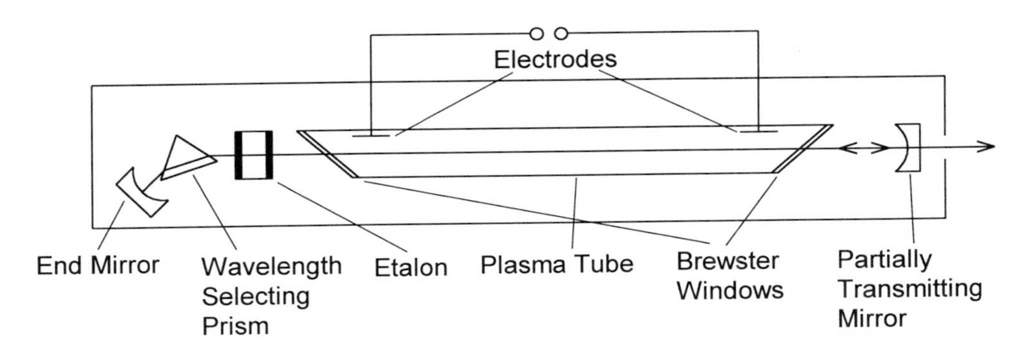

Figure 2.30: Typical gas laser

gain curve,- one speaks of the *longitudinal modes*, - results in a coherence length too short for most holographic applications. On the other hand short cavities would produce the desired single longitudinal mode under the gain curve, but with extremely low power. The solution is an intracavity *etalon*, Fig. 2.30, which is basically a *Fabry-Perot interferometer.* Only those modes are allowed to oscillate which match into the long laser cavity as well as the short cavity of the etalon. The result is a single frequency radiation of high coherence length, Figs. 2.29c and d.

Although principally a laser may emit in various *transverse modes*, which describe the intensity variation across the diameter of the laser beam, in holography only the TEM_{00}-mode, rendering the best spatial coherence, has to be used. This mode is achieved by inserting an aperture of small diameter into the resonator, the pinhole in Fig. 2.31. In the TEM_{00}-modethe intensity is Gaussian distributed, which constitutes the Gaussian background illumination of most holographic interferograms.

There is a vast amount of materials showing laser activity, pumped in various ways. But only some of them have gained importance for holographic applications, Table 2.1. The *ruby laser*, Fig. 2.31, is the most widely used *pulsed laser* in optical holography [35], mainly because of its output energy of up to about 10 J per pulse. Pumping is performed with xenon flashlamps. To achieve a single pulse of short duration an intra cavity *Q-switch* is employed. This is a fast-acting optical shutter, normally realized by a *Pockels cell.* The Pockels cell is a crystal showing birefringence when an electric field is applied. If its principal axes are oriented at 45° to the direction of polarization of the laser beam, it produces a mutual phase shift of 90° between the two polarization components when the voltage is applied. A wave transmitted by the activated Pockels cell after reflection at the end mirror is polarized 90° to its original direction and thus blocked by a polarizer. As soon as the flashlamp is fired, the Q-switch is closed. A large population inversion is built up. To the end of the flashlamp pulse, the voltage is switched off, Pockels cell and polarizer transmit freely, and the oscillation can take place. A short pulse, typically of 10 ns to 20 ns duration, is emitted. The wavelength of the ruby emission is 694 nm, which fits well to the sensitivity of photographic emulsions used in holography. One or more additional ruby rods normally follow the cavity to boost the output while preserving spatial and temporal coherence, Fig. 2.31.

Table 2.1: Lasers used in holographic interferometry

Laser	Wavelength	Pulsed/CW	Typical Power/Energy (single mode)	Excitation
Ruby	694 nm	Pulsed	10J/pulse	Flash lamp
He-Ne	633 nm	CW	2 mW - 50 mW	Electrical
Ar^+	458 nm	CW	200 mW	Electrical
Ar^+	488 nm	CW	1000 mW	Electrical
Ar^+	514 nm	CW	2000 mW	Electrical
Kr^+	647 nm	CW	500 mW	Electrical
He-Cd	325 nm	CW	25 mW	Electrical
He-Cd	442 nm	CW	25 mW	Electrical
Nd^{3+}:YAG	1060 nm	CW/Pulsed		
	530 nm (Frequency doubled)	CW/Pulsed	400 mW	Diode Laser
Dye	217 nm - 960 nm	CW/Pulsed	100 mW	Optical (Ar^+-Laser)
Laser Diode	670 nm	CW/Pulsed	25 mW	Electrical

The most common continuous wave lasers (*CW-lasers*) used for holography are the *gas-lasers*. For many applications the *helium-neon-laser* (He-Ne-laser) is the most economical choice. It emits red coherent light of 633 nm, has a long life and needs no water cooling. More power is supplied by the *argon-ion-lasers*, they are used if large areas have to be illuminated. They can be adjusted to operate in one of several lines, mainly in the blue and green. Sometimes they are replaced by the *neodym:YAG-lasers* which when frequency doubled also give green coherent light, but are compactly build and only need air-cooling, while the argon-laser is water-cooled [36, 37]. Frequency doubled Nd:YAG-lasers can be used in pulsed operation. The holograms are reconstructed with the same laser, ensuring identical wavelengths during recording and reconstruction [38]. The *krypton-laser* is similar to the argon-laser, but emits in the red. The *helium-cadmium-laser* has a short wavelength matching the sensitivity of photoresists. *Laser diodes* with good coherence and radiation in the visible are getting into the market now, their use in holography will increase in the future. An interesting feature of the laser diodes is their tunability, in which the wavelength of a single-mode laser can be changed continuously by variation of the injection current and/or the temperature of the active region [39, 40]. The *dye-lasers*, although relatively cumbersome to handle, offer the possibility of continuous variation of the wavelength over typically about 50 nm for a single dye. By changing the dyes more than the whole visible spectrum can be covered.

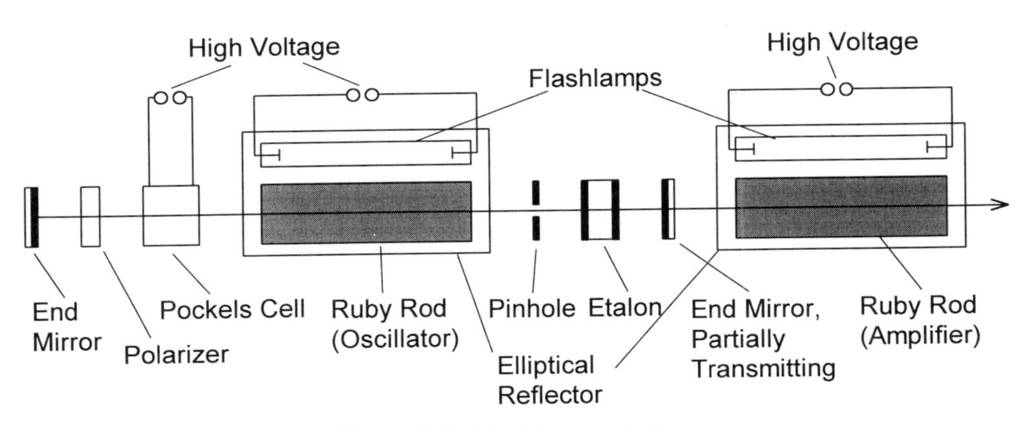

Figure 2.31: Q-switched ruby laser

2.7.2 Recording Media

Recording materials suitable for holographic interferometry should exhibit a number of properties, like a spectral sensitivity well matched to available laser wavelengths, a linear transfer characteristic, high resolution, low noise. They should be easy to handle, reusable or at least recycable, and inexpensive. No material has been found so far that meets all these requirements, so for a special application one has to find the optimal choice. A selection of recording materials is given in Table 2.2 [25].

The most widely used recording materials still are the *silver halide photographic emulsions*. These emulsions are commercially available on glass plates or film. They show a high sensitivity as well as a high spatial resolution. The main disadvantages are the wet chemical processing and the single use. Normally an *amplitude hologram* is recorded but by bleaching also a *phase hologram* can be obtained.

Dichromated gelatin for holographic recording is used as a gelatin layer containing a small amount of a dichromate. By a photochemical reaction this medium becomes harder on exposure to light. If the unhardened gelatin is washed out with warm water, a relief image is formed. If the gelatine film is processed properly a refractive index modulation is obtained [25].

Photoresists are light-sensitive organic films which yield a relief image after exposure and development. Although they are relatively slow and show nonlinear effects at diffraction efficiencies greater than about 0.05, they have found some use where the possibility of easy replication and the absence of any grain structure are important.

The *photopolymers* are organic materials which can be activated through a photosensitizer to exhibit thickness and refractive index changes due to photopolymerization or cross-linking. Thick layers can be produced to yield volume phase holograms with high diffraction efficiency and high angular selectivity, which can be viewed immediately after exposure [25].

Photochromics undergo reversible changes in colour when exposed to light. While being grain free and offering high resolution, photochromics have limited use due to their

Table 2.2: Recording materials used in holographic interferometry

Medium	Hologram Type	Processing	Reusable	Required Exposure $[J/m^2]$	Spectral Sensitivity [nm]	Resolution $[mm^{-1}]$
Silver halides	Ampl./ Phase	Wet chemical	No	5×10^{-3} -5×10^{-1}	400-700	1000-10000
Dichr. gelatin	Phase	Wet chemical	No	10^2	350-580	> 10000
Photoresists	Phase	Wet chemical	No	10^2	UV-500	3000
Photopolymers	Phase	Post exposure	No	$10 - 10^4$	UV-650	200-1500
Photochromics	Ampli.	None	Yes	$10^2 - 10^3$	300-700	> 5000
Photothermo-plastics	Phase	Charge	Yes	10^{-1}	400-650	500-1200 (bandpass)
$LiNbO_3$	Phase	None	Yes	10^4	350-500	> 1500
$Bi_{12}SiO_{20}$	Phase	None	Yes	10^1	350-550	> 10000
CCD	Ampli.	None	Yes	10^{-4} -10^{-3}	400-1000	75

low diffraction efficiency and low sensitivity. Dichromated gelatin, photoresists, photopolymers, and photochromics are not available ready to use, but must be prepared by the user before application.

Photothermoplastics are widely used in holographic interferometry [41], mainly because they are reusable and avoid the wet-chemical processing. The recording unit is a transparent stack of four layers: the glass substrate, a transparent conducting layer acting as a heating element, a photoconductor and the photothermoplastic film itself. A typical cycle of operation consists of five steps [25], Fig. 2.32:

- Charging: A uniform electrical charge is applied in darkness. Positive ions are sprayed onto the top surface, a uniform negative charge is induced in the photoconductor.

- Exposure: Charge carriers are produced in the photoconductor where light impinges. The carriers migrate to the oppositely charged surfaces and neutralize part of the charge from the first charging. This reduces the surface potential but does not change the surface charge density and the electric field.

- Second charging: By a second charging to a constant potential additional charges are deposited where the exposure caused a migration of charge. The electric field increases in these regions, producing a spatially varying field and hence a latent image.

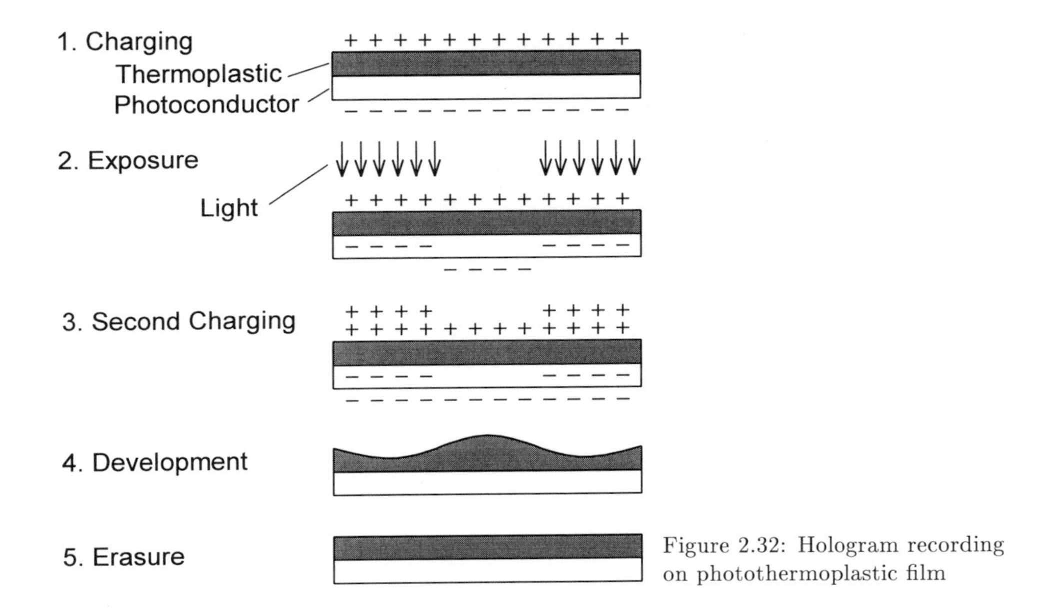

Figure 2.32: Hologram recording on photothermoplastic film

- Development: The thermoplast is heated nearly to its melting point. In regions of high electric field strength the layer gets thinner, these are the illuminated areas. The layer gets thicker in the unexposed areas. Cooling to room temperature freezes the thickness variation with the result of a phase hologram.
- Erasure: Heating to even higher temperatures than those used for development now smoothes out the thickness variations by surface tension. The hologram is erased, the thermoplastic is ready for a new recording cycle.

Photorefractive crystals have been investigated in laboratory tests. Phase volume holograms are recorded through the photorefractive effect. Most experiments used the electrooptic and photoconductive bismuth silicon oxide *(BSO) crystals* [42, 43, 44].

TV-camera-tubes and *CCD-targets* generally do not have the resolution required by off-axis holography. This on the one hand had led to the development of the *ESPI-methods*, Sec. 6.2, on the other hand to the *digital holography*, Sec. 4.7. Digital holography utilizes all the advantages the CCD-targets are offering: Fast acquisition of the primary holograms, rapid digital storage, numerical evaluation instead of optical reconstruction, thus not suffering from optical imperfections or limited diffraction efficiency.

2.7.3 Optical Components

Here we have not the room to treat the fundamentals of all optical components used in holographic interferometry, these can be found in every optics textbook [16, 45, 46]. Only some hints and practical considerations should be given. To produce good quality holograms, the mechanical and optical components always have to be fixed in a way to prevent spurious motions or vibrations. Sources of vibrations like transformers, ventilators, cooling with streaming water, etc. should be kept away from the holographic arrangement. A mechanical shutter should not be mounted on the same vibration isolated table as the other components. If motions cannot be avoided, a pulsed laser or one of the methods described in Sec. 5.1.4 have to be applied.

The quality of the expanded laser beam illuminating the object can be significantly improved by positioning a *pinhole* in the focal point of the magnifying lens. Dust and scratches on optical surfaces produce spatial noise. The pinhole acts as a *spatial filter* that allows to pass only the dc-term of the spectrum. Thus a clean illumination of nearly Gaussian characteristic will result. Magnifying the beam by a positive lens and spatial filtering is not possible when using a pulsed ruby-laser. Here the power concentration in the focus would be high enough to ionize the air, the resulting plasma is opaque to light. Therefore with pulsed ruby-lasers only negative lenses should be used. Employing adequate optics, objects of varying size and form can be investigated. Even the interior walls of pipes have been holographically analysed by utilizing panoramic annular lenses [47].

Polarization plays a minor role in holographic interferometry. Of course one has to guarantee that object and reference wave are polarized at identical angles to obtain good contrast holographic fringes. Holographic setups most often have optical axes only in one horizontal plane, vertical deflections are rather rare. It is recommended to use the lasers in such an orientation that the normals to the Brewster windows of the laser are in a vertical plane. This prevents any reflecting mirror from being accidentally positioned at the Brewster angle, thus cutting off the reflection [48].

Optical fibers are a means to conduct the laser light along paths differing from the straight propagation [49, 50]. Thus the laser, the splitting of the primary beam into object and reference beam, as well as parts of the paths of these beams can be decoupled from the rest of the holographic arrangement. Areas which are unaccessible by straight rays become accessible to holographic interferometry if optical fibers are used. Furthermore an optical fiber can be used as a sensor for pressure or temperature, since any expansion of the fiber due to the pressure- or temperature-variations influences the measurable optical path length.

Generally it can be stated that the influence of optical fibers to the temporal coherence can be neglected. But the spatial coherence is affected significantly if *multimode fibers* are used. Many modes can propagate in the fiber which may interfere and produce speckles. This may be tolerated for object illumination, since the diffusely scattering object also will degrade the spatial coherence. But a good spatial coherence in the reference wave during recording as well as for reconstruction is crucial. *Monomode fibers* with a core-diameter less than 50 μm should be employed [51]. Also results are reported on the use of coherent multimode fiber bundles to transmit the image of the test object from the test site to the holographic plate [52].

Optical fibers are already used in holographic interferometry in combination with CW-lasers. The use of Q-switched lasers with fibers is under investigation [53]. The problems here are that the high energies can destroy the faces of the fiber and may stimulate Brillouin scattering in the fiber volume.

2.7.4 Beam Modulating Components

A number of holographic interferometric methods require the modulation of the laser beam, like the *phase sampling* methods which employ a *phase shifting* device or the *heterodyne methods* which make use of a *frequency shift*. The fast shutter realized by a Pockels cell was already introduced in the context of Q-switched lasers.

A *phase shift* in a beam can be introduced by rotating a half-wave plate, moving a grating or employing an acoustooptical modulator, tilting a glass plate or shifting a mirror, Fig. 2.33, [54]. The effect of a $\lambda/2$-plate on circularly polarized light, Fig. 2.33a, is best

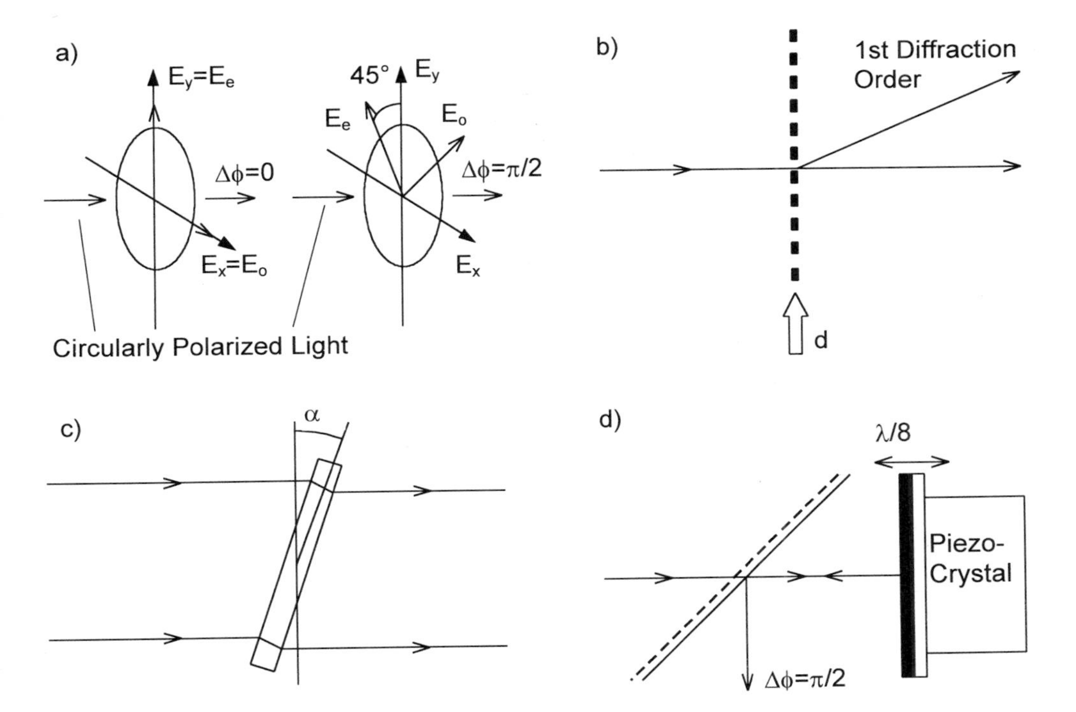

Figure 2.33: Phase shifting by (a) rotating half-wave plate, (b) shifted diffraction grating, (c) tilted glass plate, (d) translated mirror

described in the formalism of *Jones matrices*: Let the circularly polarized wave E be

$$E = \frac{1}{\sqrt{2}} \begin{pmatrix} e^{-i\frac{\pi}{4}} \\ e^{i\frac{\pi}{4}} \end{pmatrix} \tag{2.170}$$

The $\lambda/2$-plate oriented with the ordinary ray in the x-direction and the extraordinary ray in the y-direction is described by the matrix

$$M = \begin{pmatrix} e^{-i\frac{\pi}{2}} & 0 \\ 0 & e^{i\frac{\pi}{2}} \end{pmatrix} \tag{2.171}$$

The wave after passing the $\lambda/2$-plate is

$$M\,E = \begin{pmatrix} e^{-i\frac{\pi}{2}} & 0 \\ 0 & e^{i\frac{\pi}{2}} \end{pmatrix} \frac{1}{\sqrt{2}} \begin{pmatrix} e^{-i\frac{\pi}{4}} \\ e^{i\frac{\pi}{4}} \end{pmatrix} = \frac{1}{\sqrt{2}} \begin{pmatrix} e^{-i\frac{3\pi}{4}} \\ e^{i\frac{3\pi}{4}} \end{pmatrix} \tag{2.172}$$

The $\lambda/2$-plate after undergoing a rotation R of 45° has the matrix

$$R^{-1}MR = \frac{1}{\sqrt{2}} \begin{pmatrix} 1 & -1 \\ 1 & 1 \end{pmatrix} \begin{pmatrix} e^{-i\frac{\pi}{2}} & 0 \\ 0 & e^{i\frac{\pi}{2}} \end{pmatrix} \frac{1}{\sqrt{2}} \begin{pmatrix} 1 & 1 \\ -1 & 1 \end{pmatrix} = \begin{pmatrix} 0 & e^{-i\frac{\pi}{2}} \\ e^{-i\frac{\pi}{2}} & 0 \end{pmatrix} \tag{2.173}$$

The wave E after passing the rotated half-wave plate is

$$R^{-1}MR\,E = \begin{pmatrix} 0 & e^{-i\frac{\pi}{2}} \\ e^{-i\frac{\pi}{2}} & 0 \end{pmatrix} \frac{1}{\sqrt{2}} \begin{pmatrix} e^{-i\frac{\pi}{4}} \\ e^{i\frac{\pi}{4}} \end{pmatrix} = \frac{1}{\sqrt{2}} \begin{pmatrix} e^{-i\frac{\pi}{4}} \\ e^{-i\frac{3\pi}{4}} \end{pmatrix} = \frac{1}{\sqrt{2}} \begin{pmatrix} e^{i(-\frac{3\pi}{4}+\frac{\pi}{2})} \\ e^{i(\frac{3\pi}{4}+\frac{\pi}{2})} \end{pmatrix} \tag{2.174}$$

which is the wave after going through the unrotated plate but now shifted by $\pi/2$ [55].

A lateral displacement d of a *diffraction grating* [56, 57] shifts the phase of the n-th diffraction order by $\Delta\phi = 2\pi n d f$, where f is the spatial frequency of the grating, Fig. 2.33b. A moving diffraction grating also is realized in an *acoustooptical modulator* (AOM) that can be employed for phase shifting likewise.

If a plane parallel glass plate is tilted, the path through the plate changes and a phase shift is produced that depends on the thickness of the plate, its refractive index, the wavelength and the tilt angle, Fig. 2.33c. Because the exact phase shift is strongly influenced by the quality of the plate, this approach is not frequently used.

Most often the phase is shifted by a reflecting mirror mounted on a *piezoelectric transducer*, Fig. 2.33d. The piezo-crystal can be controlled electrically with high precision. A mirror shift of $\lambda/8$ corresponds to a pathlength change of $2\pi/8$ and due to the double pass results in a phase shift of $\pi/2$ in the reflected wave. If the light for illumination and/or the reference wave is transmitted through optical fibers, a common way to perform the phase shift is to wrap a portion of the fiber firmly around a piezoelectric cylinder that expands when a voltage is applied [58]. The stretching of the fiber results in a phase shift.

A *frequency shift* in principle can be produced the same way as the phase shift with the only difference that a continuous motion of the phase shifting component is required instead of a single step. Practically the frequency shift is realized by a diffraction grating moving with continuous velocity. This may be achieved by a rotating *radial grating*, or an *acoustooptical modulator* (AOM). An AOM or *Bragg cell* is a quartz through which an ultrasonic wave propagates. Because this is a longitudinal or compression wave, the index of refraction of the material varies sinusoidally with the same wavelength. Incident light that is diffracted or deflected into the Bragg angle gets a Doppler shift that is equal to the sound frequency. Wavelength shifts also can be induced by applying a ramp current to a laser diode [59].

3

Holographic Interferometry

In the preceding sections the interference effect was explained which occurs if two mutually coherent waves are superposed. So we have a means to compare two or more wave fields by checking the resulting interference pattern. Furthermore holography was introduced as a method for recording and reconstructing optical wave fields. These concepts now can be put together to the method of holographic interferometry. Holographic recording and reconstruction of a wave field is precise enough, that holographically reconstructed fields can be compared interferometrically either with a wave field scattered directly by the object, or with another holographically reconstructed wave field. Accordingly we define *holographic interferometry* as the interferometric comparison of two or more wave fields, at least one of which is holographically reconstructed [24].

As will become clear in the sequel only slight differences between the wave fields to be compared by holographic interferometry are allowed. This first concerns the objects, we even have to demand the same microstructure, second the geometry which must be the same for all wave fields to be compared, third the wavelength and coherence requirements for the optical laser radiation used, and fourth the change of the object which is to be measured. All these items: microstructure, geometry, wavelength, and the physical quantities of the object should be changed in such a small range, that only the phase of the scattered wave field is varied, but an alteration of the amplitude can be neglected. If furthermore the change of the wave field is spatially homogeneous enough to vary the phase smoothly from object point to object point, we recognize a macroscopic interference pattern which will be referred to as the *holographic interferogram* or *holographic interference pattern*. To prevent confusion between the stationary interference pattern stored as the hologram and the holographic interferogram, the first mentioned is sometimes called the *microscopic interference pattern* contrary to the observed and evaluated *macroscopic interference pattern* produced by holographic interferometry. Because holographic interferometry can bring to interference two or more wave fields simultaneously which existed to different times, and since it does not require specular reflecting surfaces, this method has found wide application as a measurement method. In this book we only deal with laser assisted holographic interferometry, that means, we stay in the near infrared, visible or near ultraviolet range of the spectrum. Nevertheless it should be mentioned that holographic interferometry can be performed using microwave or ultrasonic radiation [60, 61].

3.1 Generation of Holographic Interference Patterns

3.1.1 Recording and Reconstruction of a Double Exposure Holographic Interferogram

In the *double exposure method* of holographic interferometry two wavefronts scattered by the same object are recorded consecutively onto the same holographic plate [62]. The two wavefronts correspond to different states of the object, one in an initial condition, Fig. 3.1a, and one after the change of a physical parameter, Fig. 3.1b, e. g. by altering the object loading.

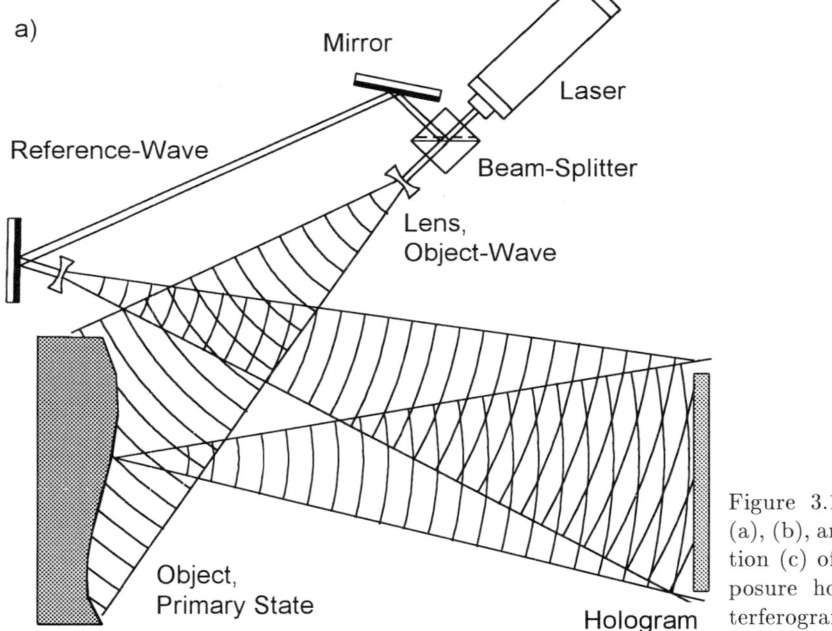

Figure 3.1: Recording (a), (b), and reconstruction (c) of a double exposure holographic interferogram

Let the complex amplitude of the first wavefront at an object point P be

$$E_1(P) = E_{01}(P)\, \mathrm{e}^{\mathrm{i}\phi(P)} \tag{3.1}$$

which is holographically recorded. E_{01} is the real amplitude and $\phi(P)$ is the phase distribution, (2.15). $\phi(P)$ varies spatially in a random manner due to the microstructure of the diffusely reflecting or refracting object. P identifies an object point. For the moment we do not have to discriminate between points of the object and the corresponding points in the image plane.

The variation of a physical parameter to be measured, e. g. the object shape due to deformation of an opaque object or a change in the refractive index distribution of

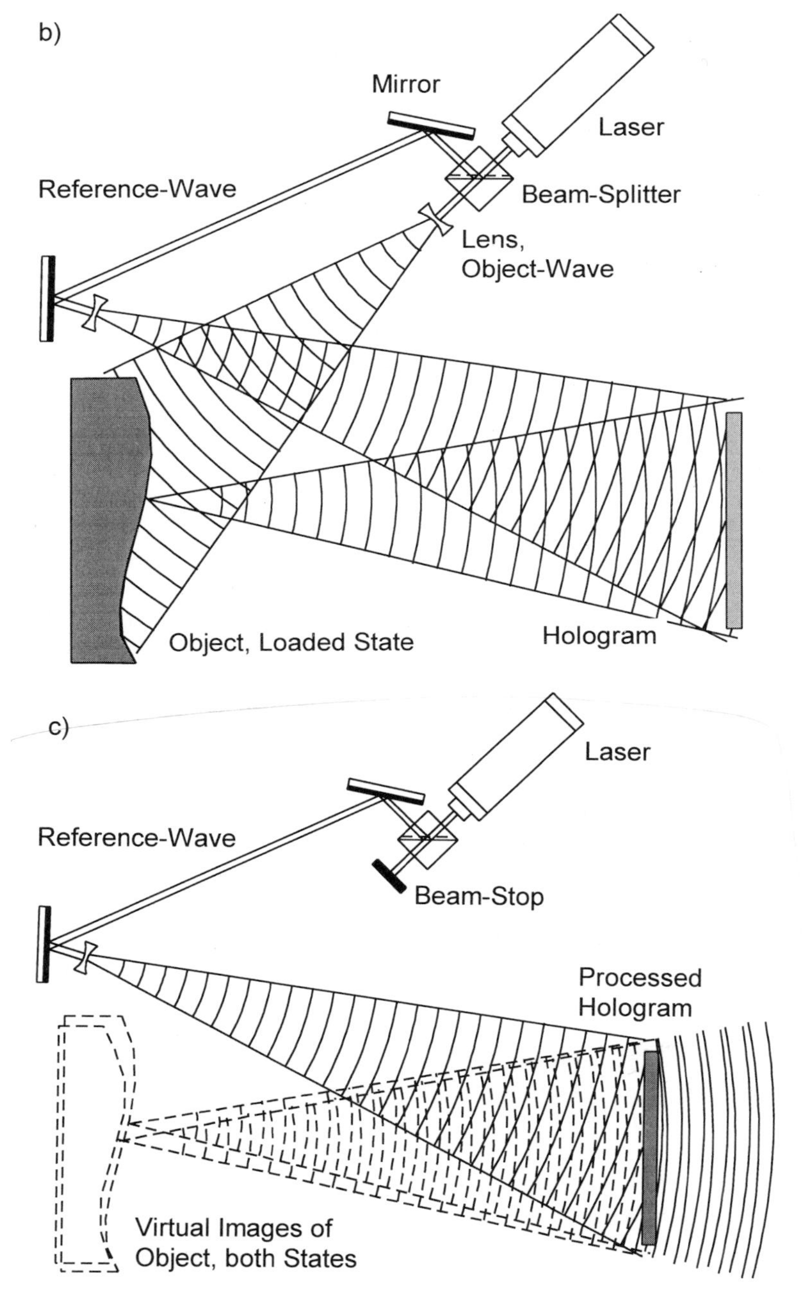

Figure 3.1: Cont.

a transparent object changes the phase distribution at P by $\Delta\phi(P)$. So the complex amplitude of the second wavefront to be recorded holographically onto the same plate is

$$E_2(P) = E_{02}(P)\, \mathrm{e}^{\mathrm{i}(\phi(P) + \Delta\phi(P))} \tag{3.2}$$

After development of the holographic plate both wavefronts are reconstructed simultaneously, Fig. 3.1c. They interfere and give rise to a stationary intensity distribution

$$\begin{aligned} I(P) &= |E_1(P) + E_2(P)|^2 \\ &= (E_{01}(P)\mathrm{e}^{\mathrm{i}\phi(P)} + E_{02}(P)\mathrm{e}^{\mathrm{i}(\phi(P)+\Delta\phi(P))})(E_{01}(P)\mathrm{e}^{-\mathrm{i}\phi(P)} + E_{02}(P)\mathrm{e}^{-\mathrm{i}(\phi(P)+\Delta\phi(P))}) \\ &= I_1(P) + I_2(P) + \sqrt{I_1(P)I_2(P)}\left(\mathrm{e}^{-\mathrm{i}\Delta\phi(P)} + \mathrm{e}^{\mathrm{i}\Delta\phi(P)}\right) \\ &= I_1(P) + I_2(P) + 2\sqrt{I_1(P)I_2(P)}\cos[\Delta\phi(P)] \end{aligned} \tag{3.3}$$

For identical amplitudes, $E_{01}(P) = E_{02}(P)$, we get

$$I(P) = 2I_1(P)\{1 + \cos[\Delta\phi(P)]\} \tag{3.4}$$

The change of the phase $\Delta\phi$ is called the *interference phase difference* or shortly *interference phase*. If the spatial variation of the interference phase over the observed reconstructed surface is low, the intensity distribution (3.3) represents the irradiance of the object, modulated by a cosine-shaped fringe pattern. Bright centers of fringes are the contours, where the interference phase is an even integer multiple of π, dark centers of fringes correspond to odd integer multiples of π. Clearly, if the interference phase changes too rapidly from one observable point to the next, say more than π, so that the sampling theorem is violated, we will recognize only a more or less random intensity distribution, which cannot be evaluated any more.

3.1.2 Recording and Reconstruction of a Real-Time Holographic Interferogram

In *real-time holographic interferometry* only one wavefront, belonging to a reference state of the tested object, is holographically recorded, Fig. 3.2a. After processing, the hologram is replaced exactly in its initial recording position. This replacement has to be performed within sub-wavelength precision. During illumination of the hologram with the original reference wave, the reconstructed virtual image wavefront coincides with the wavefront scattered directly by the object which still is in its original position. A change of the object now changes the scattered wavefront and the superposition with the holographically reconstructed original wavefront produces an interference pattern which can be observed in real time, Fig. 3.2b. Dynamic variations of the object lead to simultaneously observable variations of the interference pattern.

In both real-time and double-exposure holographic interferometry, the intensity variation in the fringe pattern has a cosine shape. However while in the double-exposure technique we have bright fringes where the interference phase is an even integer multiple of π, using amplitude holograms in the real-time method we get bright fringes where

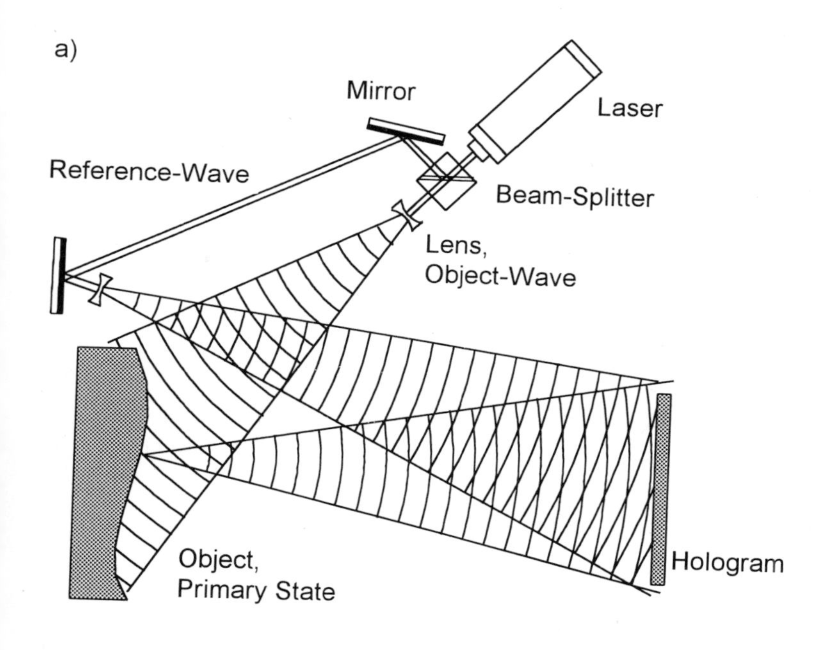

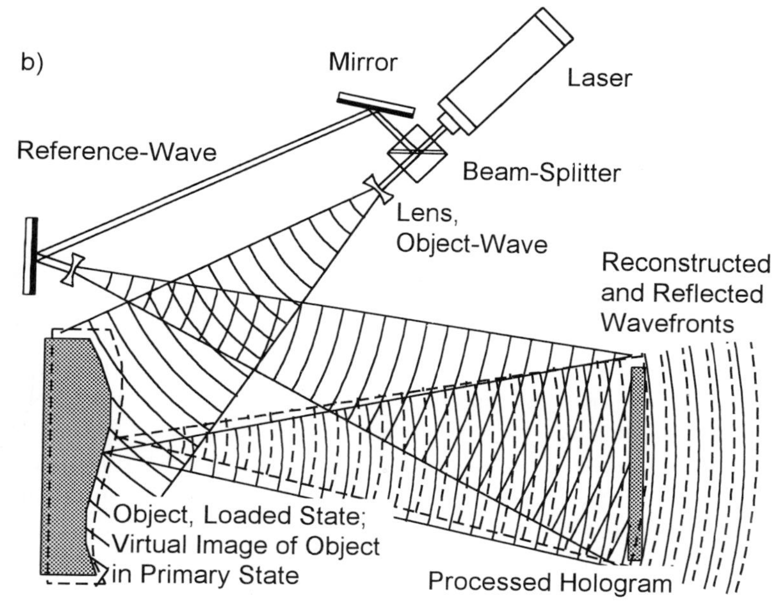

Figure 3.2: Recording (a) and reconstruction (b) of a real-time holographic interferogram

$\Delta\phi(P)$ is an odd integer multiple of π. This is due to the negative sign before the third term of (2.125) which describes the reconstructed virtual image. In the double exposure technique both reconstructed wavefronts carry the same sign. Using a phase hologram in the real-time technique, we have the same positive sign for the directly scattered and the reconstructed wavefronts, now leading to bright fringes at even integer multiples of π like with the double exposure technique. As a consequence loci where there is no change, $\Delta\phi(P) = 0$, are the centers of a bright fringe in the double exposure and in the real-time technique with phase holograms, these loci are the centers of a dark fringe for the real-time technique utilizing an amplitude hologram.

3.1.3 Time Average Holography

An in-depth discussion of *holographic vibration analysis* will be given in Sec. 5.5. Here we only show how other fringe characteristics than the above mentioned cosine shapes arise. One can use the real-time method with a holographically recorded and reconstructed wavefront representing the object in its rest state. Now consider a *harmonic vibration* which gives rise to an interference phase

$$\Delta\phi(P)\sin(\omega t) \tag{3.5}$$

ω is the angular frequency of the vibration, $\Delta\phi(P)$ is related to the maximal amplitude of the vibration at object point P. If we assume illumination and observation in normal direction, and the maximal amplitude as $Z(P)$, we get $\Delta\phi(P) = 4\pi Z/\lambda$, since the light has to travel to and from P along $Z(P)$. During vibration the real-time technique at each instant of time generates a cosine shaped pattern

$$I(P,t) = 2I_1(P)\{1 - \cos[\Delta\phi(P)\sin(\omega t)]\} \tag{3.6}$$

If the frequency ω is higher than the temporal resolution of the eye (with an assumed average response time 40 ms) or the applied sensor, a time averaged intensity is seen, which according to (C.2) is

$$\begin{aligned} I(P) &= 2I_1(P)\lim_{T\to\infty}\frac{1}{T}\int_0^T\{1 - \cos[\Delta\phi(P)\sin(\omega t)]\}dt \\ &= 2I_1(P)\{1 - \mathrm{J}_0[\Delta\phi(P)]\} \end{aligned} \tag{3.7}$$

where J_0 is the zero-order *Bessel function* of the first kind. These fringes have low contrast, Fig. 3.3. The method can be used for the identification of resonant frequencies by monitoring the fringe pattern while varying the excitation frequency.

The most frequently applied method for holographic vibration analysis is the *time average method*. Here the vibrating surface is recorded holographically with an exposure time which is long compared with the period of the vibration, $T \gg 2\pi/\omega$. Let us consider again the harmonic vibration as given in (3.5). Then we record and reconstruct holographically many waves of the form

$$E_{01}(P)\mathrm{e}^{\mathrm{i}\Delta\phi(P)\sin(\omega t)} \tag{3.8}$$

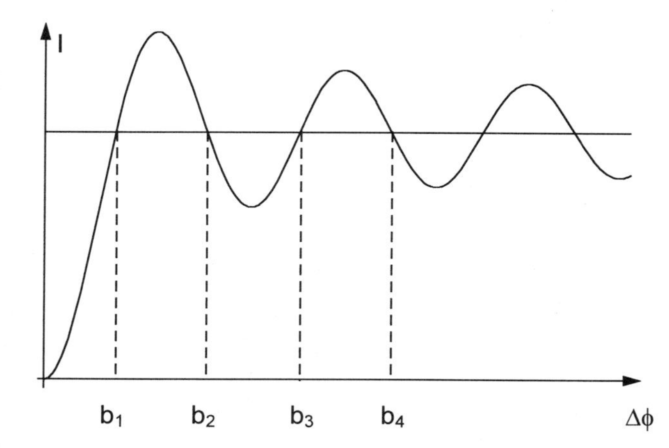

Figure 3.3: Time averaged real-time intensity

The continuum of wavefronts reconstructed simultaneously will interfere to

$$\begin{aligned} E_{av}(P) &= \lim_{T\to\infty} \frac{E_{01}(P)}{T} \int_0^T \mathrm{e}^{\mathrm{i}\Delta\phi(P)\sin(\omega t)} dt \\ &= E_{01}(P) \mathrm{J}_0 \left(\Delta\phi(P)\right) \end{aligned} \tag{3.9}$$

The observable intensity in the reconstructed image then is

$$I(P) = I_1(P) \mathrm{J}_0{}^2 \left(\Delta\phi(P)\right) \tag{3.10}$$

The fringes now are contours of equal vibration amplitudes of the spatial vibration modes. Maximal intensity belongs to $\Delta\phi(P) = 0$, which occurs at the nodes of the vibration mode. Dark centers of the fringes refer to zeros of the Bessel function J_0. The contrast of the fringes decreases with increasing order, Fig. 3.4, see also Fig. C.1.

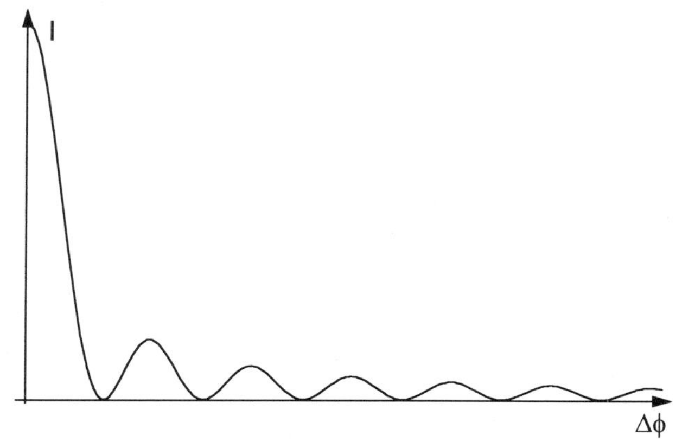

Figure 3.4: Time averaged intensity

3.1.4 Interference Phase Variation Due to Deformation

In holographic interferometric measurements of the *deformation* of diffusely reflecting opaque object surfaces, the displacement of each surface point P gives rise to an *optical*

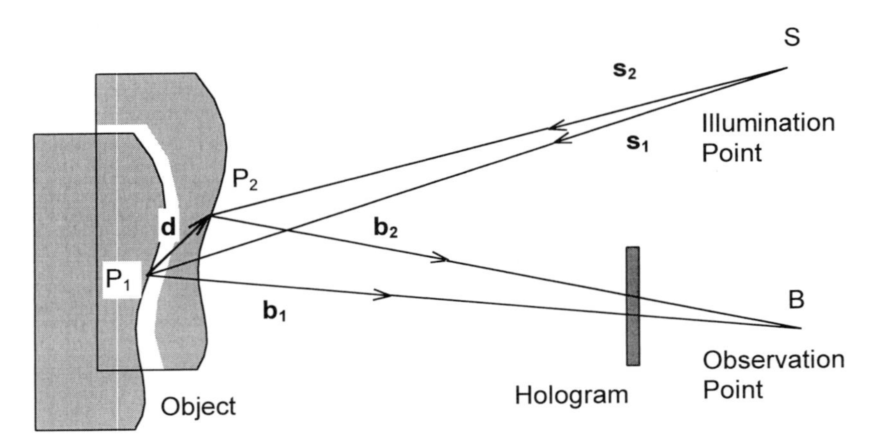

Figure 3.5: Holographic arrangement for measuring the deformation of an opaque surface

path difference $\delta(P)$. This is the difference between the paths from the source point S of the illuminating wavefront over the surface point P to the observation point B before and after changing the deformation state [63]. The interference phase $\Delta\phi(P)$ is related to this path difference by

$$\Delta\phi(P) = \frac{2\pi}{\lambda}\delta(P) \tag{3.11}$$

The observed intensity belonging to this interference phase is already given in (3.4).

In a holographic arrangement with diverging illumination and converging observation wave fronts, Fig. 3.5, let $S = (x_S, y_S, z_S)$ be the *illumination point* and $B = (x_B, y_B, z_B)$ be the *observation point* given in a Cartesian coordinate system. When the object is deformed, the surface point P moves from $P_1 = (x_{P1}, y_{P1}, z_{P1})$ to the new position $P_2 = (x_{P2}, y_{P2}, z_{P2})$, thus defining the *displacement vector*

$$\boldsymbol{d}(P_1) = (d_x(P_1), d_y(P_1), d_z(P_1)) = P_2 - P_1 \tag{3.12}$$

The optical path difference $\delta(P_1)$ now is expressed as

$$\begin{aligned} \delta(P_1) &= \overline{SP_1} + \overline{P_1B} - (\overline{SP_2} + \overline{P_2B}) \\ &= \boldsymbol{s_1} \cdot \boldsymbol{SP_1} + \boldsymbol{b_1} \cdot \boldsymbol{P_1B} - \boldsymbol{s_2} \cdot \boldsymbol{SP_2} - \boldsymbol{b_2} \cdot \boldsymbol{P_2B} \end{aligned} \tag{3.13}$$

where $\boldsymbol{s_1}$ and $\boldsymbol{s_2}$ are unit vectors in the illumination direction, $\boldsymbol{b_1}$ and $\boldsymbol{b_2}$ are unit vectors in the observation direction, and the $\boldsymbol{SP}_i$ and $\boldsymbol{P}_i\boldsymbol{B}$ are the vectors from S to P_i or P_i to B, resp. In analogy to Sec. 2.2.1 let $\boldsymbol{s}(P_1, P_2)$ be the bisector of the unit vectors $\boldsymbol{s_1}$ and $\boldsymbol{s_2}$ in illumination direction and $\boldsymbol{b}(P_1, P_2)$ the bisector of the unit vectors in observation direction. $\boldsymbol{\Delta s}(P_1, P_2)$ and $\boldsymbol{\Delta b}(P_1, P_2)$ are the half differences of the unit vectors

$$\begin{aligned} \boldsymbol{s}(P_1, P_2) &= \frac{1}{2}[\boldsymbol{s_1}(P_1) + \boldsymbol{s_2}(P_2)] \qquad & \boldsymbol{\Delta s}(P_1, P_2) &= \frac{1}{2}[\boldsymbol{s_1}(P_1) - \boldsymbol{s_2}(P_2)] \\ \boldsymbol{b}(P_1, P_2) &= \frac{1}{2}[\boldsymbol{b_1}(P_1) + \boldsymbol{b_2}(P_2)] \qquad & \boldsymbol{\Delta b}(P_1, P_2) &= \frac{1}{2}[\boldsymbol{b_1}(P_1) - \boldsymbol{b_2}(P_2)] \end{aligned} \tag{3.14}$$

By definition of the displacement vector $\boldsymbol{d}(P_1)$ we have

$$\begin{aligned} \boldsymbol{P_1B} - \boldsymbol{P_2B} &= \boldsymbol{d}(P_1) \\ \text{and} \quad \boldsymbol{SP_2} - \boldsymbol{SP_1} &= \boldsymbol{d}(P_1) \end{aligned} \tag{3.15}$$

Inserting this into (3.13) gives

$$\begin{aligned} \delta &= (\boldsymbol{s} + \Delta\boldsymbol{s}) \cdot \boldsymbol{SP_1} + (\boldsymbol{b} + \Delta\boldsymbol{b}) \cdot \boldsymbol{P_1B} - (\boldsymbol{s} - \Delta\boldsymbol{s}) \cdot \boldsymbol{SP_2} - (\boldsymbol{b} - \Delta\boldsymbol{b}) \cdot \boldsymbol{P_2B} \\ &= \boldsymbol{b} \cdot \boldsymbol{d} - \boldsymbol{s} \cdot \boldsymbol{d} + \Delta\boldsymbol{b} \cdot (\boldsymbol{P_1B} + \boldsymbol{P_2B}) + \Delta\boldsymbol{s} \cdot (\boldsymbol{SP_1} + \boldsymbol{SP_2}) \end{aligned} \tag{3.16}$$

where the arguments (P_1) have been omitted for clarity.

Now the displacements are far smaller than the dimensions of the arrangement geometry - $|\boldsymbol{d}(P_1)|$ is in the micrometer range, the $\overline{SP_i}$ and $\overline{P_iB}$ are in the meter range - so the same relation holds between the lengths of $\Delta\boldsymbol{s}$ and $\Delta\boldsymbol{b}$ compared to the lengths 1. of the unit vectors $\boldsymbol{s_i}$ and $\boldsymbol{b_i}$. Furthermore the vector $\Delta\boldsymbol{s}$ is nearly orthogonal to $\boldsymbol{SP_1} + \boldsymbol{SP_2}$ and vector $\Delta\boldsymbol{b}$ is nearly orthogonal to $\boldsymbol{P_1B} + \boldsymbol{P_2B}$, with the consequence that their scalar products are nearly zero. These scalar products can be neglected and we do not have to distinguish between P_1 and P_2 any more. Altogether the following relation holds for the macroscopic point P

$$\delta(P) = \boldsymbol{d}(P) \cdot [\boldsymbol{b}(P) - \boldsymbol{s}(P)] \tag{3.17}$$

For divergent illumination and observation the unit vectors $\boldsymbol{s}(P)$ and $\boldsymbol{b}(P)$ at surface point P are computed by

$$\boldsymbol{s}(P) = \begin{pmatrix} s_x(P) \\ s_y(P) \\ s_z(P) \end{pmatrix} = \frac{1}{\sqrt{(x_P - x_S)^2 + (y_P - y_S)^2 + (z_P - z_S)^2}} \begin{pmatrix} x_P - x_S \\ y_P - y_S \\ z_P - z_S \end{pmatrix} \tag{3.18}$$

and

$$\boldsymbol{b}(P) = \begin{pmatrix} b_x(P) \\ b_y(P) \\ b_z(P) \end{pmatrix} = \frac{1}{\sqrt{(x_B - x_P)^2 + (y_B - y_P)^2 + (z_B - z_P)^2}} \begin{pmatrix} x_B - x_P \\ y_B - y_P \\ z_B - z_P \end{pmatrix} \tag{3.19}$$

These unit vectors together with the factor $2\pi/\lambda$ (3.11) form the so called *sensitivity vector* $\boldsymbol{e}(P)$

$$\boldsymbol{e}(P) = \frac{2\pi}{\lambda}[\boldsymbol{b}(P) - \boldsymbol{s}(P)] \tag{3.20}$$

so that we get

$$\Delta\phi(P) = \boldsymbol{d}(P) \cdot \boldsymbol{e}(P) \tag{3.21}$$

This means that the interference phase at each point is given by the scalar product of the displacement vector and the sensitivity vector. The sensitivity vector is defined only by the geometry of the holographic arrangement. It gives the direction in which the setup has maximal sensitivity. At each point we measure the projection of the displacement vector onto the sensitivity vector. For displacements orthogonal to the sensitivity vector the resulting interference phase is always zero, independent of the magnitude of the displacement.

The above formula (3.21) is the basis of all quantitative measurements of the deformation of opaque bodies by holographic interferometry and will be used frequently in the following chapters. Throughout this book we assume an illumination point source. For collimated illumination this is infinitely far away. Holographic interferometry with diffuse illumination from a circular diffuser is investigated in [64].

3.1.5 Interference Phase Variation Due to Refractive Index Variation

Phase objects are *transparent objects* which do not significantly affect the amplitude of an optical wavefront passing through, but only the phase of the wavefront. Holographic interferometry has widely replaced Mach-Zehnder interferometry for analyzing phase objects, e. g. in applications like flow visualization, plasma diagnostics, or heat transfer analysis, to name just a few [65]. In the analysis of *transparent media* the optical path difference responsible for the interference fringes is generated by a change in the *refractive index distribution* along the optical path. This change may be due to absence and presence of a phase object or due to a variation of the phase object under test. As discussed before the double exposure or the real time method can be applied to compare the states before and after the variation of the refractive index field.

If the medium to be tested is hold in a transparent container, which is not altered between the recording of the two wavefronts, imperfections in the walls of the container affect both wavefronts in the same way. But since only the differences between the wavefronts generate the holographic interference pattern, the influence of the wall imperfections is canceled out.

A typical off-axis configuration with collimated beams for holographic interferometric measurements at a phase object is shown in Fig. 3.6. Employing the double exposure method, the first recording is done with the refractive index distribution $n_1(x, y, z)$, the second is performed with $n_2(x, y, z)$. Let the rays propagate along straight lines parallel to the z-axis. Then the interference phase distribution $\Delta\phi(x, y)$ in the plane perpendicular to the z-axis is

$$\Delta\phi(x, y) = \frac{2\pi}{\lambda} \int \Delta n(x, y, z) dz \qquad (3.22)$$

with $\Delta n(x, y, z) = n_2(x, y, z) - n_1(x, y, z)$. The resulting intensity distribution in the holographic interferogram is (3.4)

$$\begin{aligned} I(x, y) &= 2I_1(x, y)\{1 + \cos[\Delta\phi(x, y)]\} \\ &= 2I_1(x, y)\left\{1 + \cos\left[\frac{2\pi}{\lambda}\int \Delta n(x, y, z) dz\right]\right\} \end{aligned} \qquad (3.23)$$

Eqs. (3.22) and (3.23) will play the central role in all quantitative measurements of refractive index distributions at phase objects.

Although the configuration of Fig. 3.6 is conceptually easy, the plane waves have several drawbacks [24]: Any dust particle or scratch on optical elements will diffract light into a nearly spherical wave. These unwanted waves then will interfere with the object wavefront and give annoying concentric ring patterns. Direct observation will exhibit a bright spot

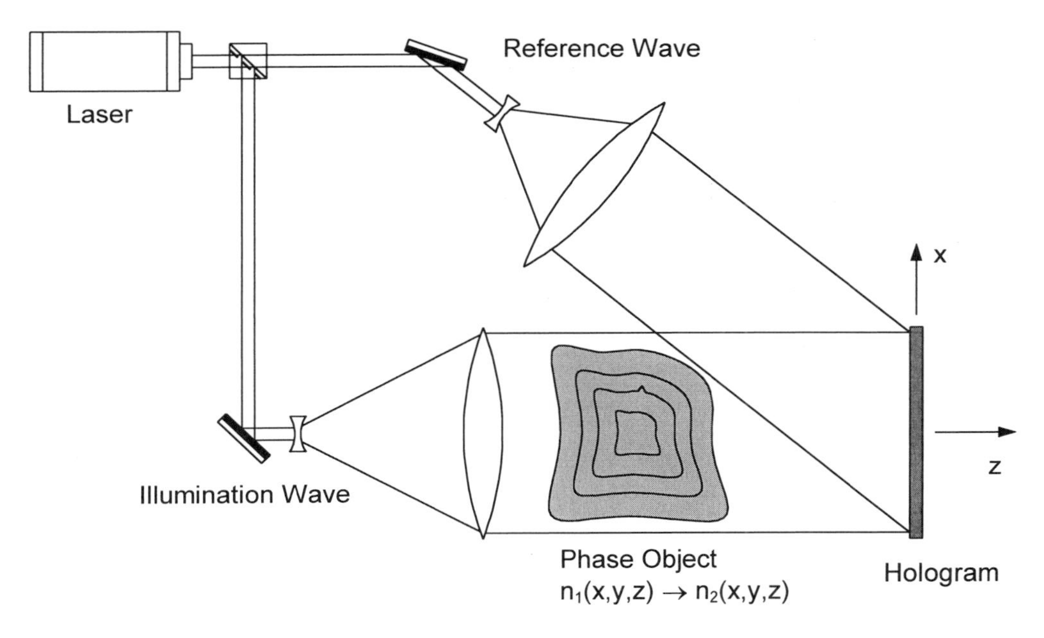

Figure 3.6: Configuration for holographic interferometry at a phase object

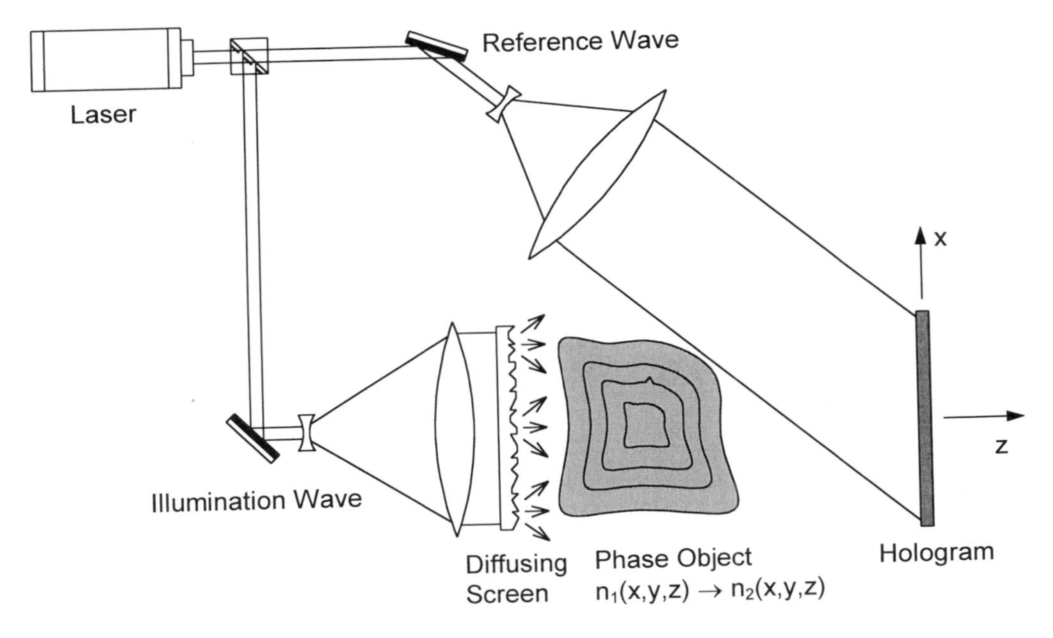

Figure 3.7: Configuration for diffuse illumination holographic interferometry

at the pinhole of the object beam and the field of view is limited to an area the size of the iris. Hence the fringes have to be projected onto a screen to be observed. All these disadvantages are avoided by using *diffuse illumination holographic interferometry* [66, 67]. A ground glass plate is used to diffusely scattering the object illumination wave, Fig. 3.7. Now the average irradiance is nearly uniform over the hologram, the influence of diffraction rings is minimized, the hologram can be observed with the unaided eye or recorded with a TV-camera, and the hologram can be viewed in multiple directions. This last mentioned property will become important for three-dimensional quantitative evaluations, see Sec. 5.8.

3.1.6 Computer Simulation of Holographic Interference Patterns

Based on the equations (3.3) to (3.23) on can calculate the expected holographic interference pattern for a given setup geometry, including laser wavelength and object shape, as well as the loading applied to the object or the deformation or refractive index distribution [68, 69]. The simulation program basically is a loop over all pixels of the interferogram to be calculated. If we want an interference pattern of a deformed opaque surface, first each pixel is identified with the Cartesian coordinates of the corresponding point of the surface. With (3.18), (3.19), and (3.20) the sensitivity vectors for these surface points and thus for each pixel are calculated. The scalar multiplication (3.21) with the corresponding displacement vector is performed. The displacement vector has been determined from the loading parameters and the mechanical laws or has been taken from computer memory. While rigid body translations are easiest to handle, - the displacement vector remains constant, - normally we have a combination of a rigid body motion, a deformation, and local deformation variations induced by defects or material inhomogeneities. The cosine (3.4) is applied to the calculated interference phase and gives the intensity distribution. Distortions like Gaussian background or speckle noise also can be taken into account. A Gaussian background may be simulated as an additive term, while the speckles are multiplicative [70]. Their stochastic nature is simulated by a random generator, which mimics the negative exponential probability distribution (2.99).

For the simulation of a real-time or a time average interferogram occuring in vibration analysis, the displacement vector is given by the maximal amplitude at each point, but then the Bessel function (3.7) or the squared Bessel function (3.10) is applied instead of the cosine (3.4). To get a better contrast in the simulated interferogram, the high intensity at the nodal lines, where the interference phase is near at zero, should be truncated, the remaining intensity is mapped onto the full gray-scale. This results in brighter higher-order intensity maxima.

Fig. 3.8 shows some simulated holographic interference patterns, where in Fig. 3.8a a rectangular plate was subjected to strain and torsion; the inhomogeneity in the fringe pattern results from an anticipated subsurface void in the plate. Fig. 3.8b displays a holographic interferogram of the same kind but with simulated background and speckle noise. Fig. 3.8c gives the real-time and Fig. 3.8d the time average interferogram of a vibrating rectangular plate clamped at all four edges.

The simulation of holographic interference patterns is helpful in a number of ways:

Figure 3.8: Simulated holographic interferograms: (a) double-exposure interferogram, (b) double-exposure interferogram with background and speckles, (c) real-time interferogram of a vibrating plate, (d) time average interferogram of a vibrating plate

- It helps in the planning phase of a holographic interferometric measurement to design optimized holographic arrangements. One can check whether an expected deformation produces fringes of sufficient but not too high density on the basis of cheap computer experiments instead of expensive practical experiments. The optimal loading amplitudes can be found this way.

- If it is impossible to perform a three-dimensional evaluation with multiple interferograms, one can compare the calculated interferogram of an expected three-dimensional deformation field with the experimentally produced single interference pattern. Although this does not substitute a complete three-dimensional evaluation, it may confirm the results, if also the interferograms of all other possible deformations have been simulated and they differ significantly from the measured one.

- In developing new evaluation methods, e. g. adapted methods for special measurement problems, these methods can be tested first at simulated patterns. Especially the behaviour of new methods, when evaluating more or less speckled patterns, can be tested systematically.

- In the application of advanced evaluation methods, e.g. fault detection with neural networks, test data or samples for network training are needed in large numbers and with high diversification [71]. These cannot be all generated experimentally. Based on some experimental examples the vast majority of the training samples are generated by computer simulation.

3.2 Variations of the Sensitivity Vectors

In Sec. 3.1.4 the foundations for holographic interferometric deformation measurements at diffusely reflecting opaque surfaces have been laid. The displacements of the object surface points together with the fixed sensitivity vectors give rise to the optical path differences which lead to the observable fringe patterns. Since the sensitivity vectors define the directions of the displacement components which are measured with the highest sensitivity, they should be examined in the stage of planning a holographic interferometric measurement experiment. The goal is to find an optimal arrangement that gives maximal accuracy and requires minimum effort for solving a given problem.

Contrary to deformation measurement holographic interferometry can also be applied by leaving the object surface unchanged but altering the sensitivity vectors between the construction of the two wavefronts to be compared interferometrically. This aspect is further addressed in Sec. 5.6, where it is used for holographic contouring.

3.2.1 Optimization of the Holographic Arrangement

The sensitivity vector is defined by (3.20) where the unit vectors $\boldsymbol{s}(P)$ in illumination direction and $\boldsymbol{b}(P)$ in observation direction are given by (3.18) and (3.19) for the case of divergent illumination and observation beams with well defined source point S and observation point B. For collimated illumination and observation, Fig. 3.9a, we assume the central points of the collimating lenses as source S and observation point B, and the unit vectors $\boldsymbol{s}$ and $\boldsymbol{b}$ constant over all surface points P. If at least one of the illumination or the observation beam is divergent, then the sensitivity vector varies over the surface, Fig. 3.9b. Only if both the illumination and the observation beam are collimated, the sensitivity vector is the same for all points of the investigated surface, Fig. 3.9a.

The consequences of these two cases can be investigated by looking at the variation of the interference phase (3.21) in e. g. the x-direction

$$\begin{aligned} \frac{\partial}{\partial x}\Delta\phi &= \frac{\partial}{\partial x}\left[d_x e_x + d_y e_y + d_z e_z\right] \\ &= \frac{\partial d_x}{\partial x}e_x + d_x\frac{\partial e_x}{\partial x} + \frac{\partial d_y}{\partial x}e_y + d_y\frac{\partial e_y}{\partial x} + \frac{\partial d_z}{\partial x}e_z + d_z\frac{\partial e_z}{\partial x} \end{aligned} \tag{3.24}$$

The arguments P have been omitted for clarity. We see that a change in the interference phase may stem from a variation of the displacement vector as well as from a variation of the sensitivity vector. For a constant sensitivity vector the interference phase depends solely on the displacement vector variation.

From a theoretical standpoint, only in the case of rigid body translations to be measured, it seems desirable to have varying sensitivity vectors to cover the object with fringes, in all other cases of deformations and rotations the evaluation will become easier with constant sensitivity vectors. But practically this is only possible for objects or illuminated areas on the objects less or equal the collimating lens dimensions. Nevertheless if the illumination and observation points are far away from the examined surface, compared with the surface dimensions, we can assume nearly constant sensitivity vectors. The remaining errors may be estimated with the help of (3.24) by taking the maximum anticipated values for all variables and summing the absolute values.

In this way the problem of measuring only the out-of-plane displacements is attacked as follows. By defining the Cartesian coordinate system so that for a central surface point P we have $P = (0,0,0)$ and furthermore $y_S = y_B = 0$, $x_S = -x_B$, and $z_S = z_B$, then the sensitivity vector at least at P reduces to $\boldsymbol{e}(P) = (0,0,e_z(P))^T$ with

$$e_z(P) = \frac{2\pi}{\lambda}[z_B(P) - z_S(P)] = \frac{4\pi\cos\Theta}{\lambda} \tag{3.25}$$

where Θ is the angle between the z-axis and the illumination direction which due to the assumptions agrees with the angle between the z-axis and the observation direction. The condition $x_S = -x_B$ and $z_S = z_B$ can be relaxed to only identical angles Θ for both directions. For constant or for nearly constant sensitivity vectors we now can determine the z-component of the displacement with good accuracy by

$$d_z(P) = \frac{\lambda\Delta\phi(P)}{4\pi\cos\Theta} \tag{3.26}$$

The basic ideas contained in this reduction of complexity can be worked out to a concept of optimizing the holographic setup [72]. In nearly all metrologic problems, we have a lot of previous information, e. g. about the directions of the deformations

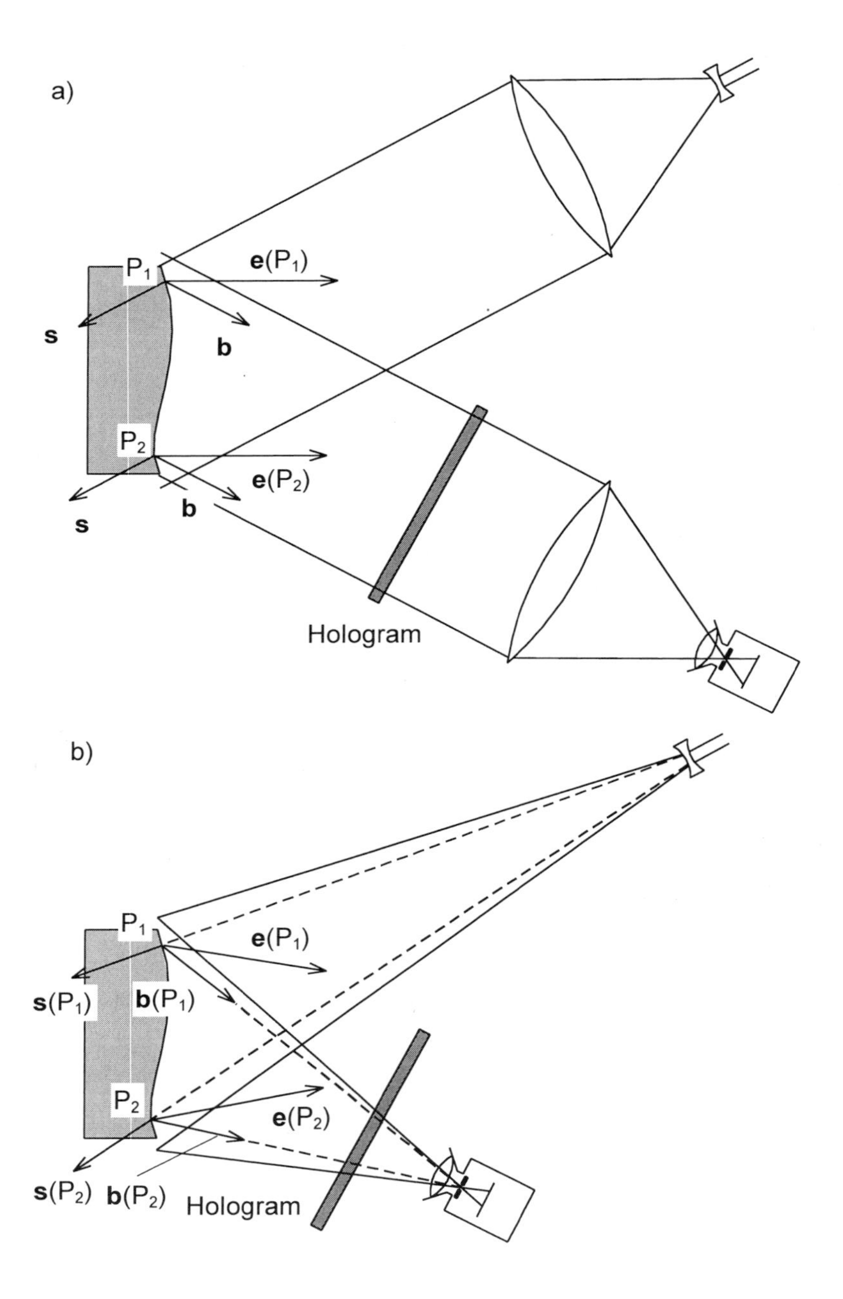

Figure 3.9: Constant (a) and not constant (b) sensitivity vectors

we expect. Then it is good practice to configure a holographic setup with maximum sensitivity in the expected direction and minimum sensitivity in directions of eventual distortions. As a help for this aim Jüptner et al. [73] have introduced the sensitivity functions and Abramson [74, 75, 76, 77, 78] designed the so called *holo-diagram*.

In [73] a Cartesian coordinate system based on the positions of the illumination source point and the observation point is defined. All lengths are normalized by the distance between these two points. The *sensitivity functions* indicate the sensitivity with respect to the different displacement components. Now if the direction of the object deformation can be anticipated, one has to search for regions where the sensitivity function for this direction varies linearly and the sensitivity functions for the other directions remain as close as possible to zero. It is worth mentioning that in this approach sensitivity is a property varying in space, only depending on the illumination and observation points, but there is no object already in the holographic arrangement.

The other aid for optimizing the holographic arrangement is the *holo-diagram* [74, 75, 76, 77, 78]. This consists of the two-dimensional projection of confocal ellipsoids with the focal points being the point source of illumination S and the point of observation B. By definition the path length from S to B is constant when travelled via any point along one ellipsoid. Adjacent ellipsoids are given path lengths from S to B via this ellipsoid which differ by λ. The distance between adjacent ellipses, the projections of the ellipsoids, varies with the directions to S and B, respectively. But it is constant along arcs of circles, or toroids in the three-dimensional case, known as k-circles. A displacement of an object point P from one ellipse to the next corresponds to one wavelength and thus to one fringe in the interference pattern. Displacements along the ellipses will cause no fringes. On the other hand the required displacement of P for inducing one fringe is minimum when its motion is along the normal to the ellipse. This normal corresponds to the sensitivity vector $\boldsymbol{e}$. As before the holo-diagram depicts the sensitivity of the holographic setup in space without the presence of an object.

For optimization of the holographic configuration the object has to be positioned in the holo-diagram in an orientation that the main direction of the anticipated displacement becomes orthogonal to the ellipses. Nevertheless the conditions have to be fulfilled that the surface under study should not be in shadow and that it should be freely visible from B.

An increase in sensitivity, that is more fringes for the same displacement, can be obtained with the three-step approach described in [79]. A double exposure hologram is recorded, with the illumination beam of the second exposure slightly tilted with respect to the first. The reconstructed image consists of modulated Young's fringes, which are recorded on a photographic plate, that is developed in a nonlinear way. This plate now is a modulated diffraction grating, which is illuminated by two plane waves. By adjusting the angle between the propagation directions of these waves, a $+n$-order diffracted wave can be made to coincide with a $-m$-order diffracted wave. These waves are filtered by a pinhole in the Fraunhofer plane. Their interference has an intensity of the form [79]

$$I(P) = I_0(P) + I_1(P)\cos[(n+m)\Delta\phi(P)] \tag{3.27}$$

where the interference phase can be seen to be amplified. The fringe sensitivity increases by $(n + m)$ in comparison with ordinary holographic interferometry (3.4).

When measuring three-dimensional displacement fields, a further optimization criterion is the condition of the system of equations to be solved. This topic is treated in more detail in Sec. 5.2.3.

3.2.2 Two Reference Beam Holographic Interferometry

In the preceding sections it was shown how the interference phase distribution giving rise to the observable holographic interferometric fringe pattern was affected by changing the optical pathlength. The change of the optical pathlength at each observed object point can be caused by a variation of a physical parameter of the object under test or by a modification of the sensitivity vector. In the next chapter evaluation methods will be introduced which employ an additional variation of the interference phase distribution independent from the interference phase due to the quantity to be measured. The phase of one of the interfering wave fields will be shifted relatively to the other. If this shifting is continuously in time, it is equivalent to a frequency shift, and is utilized in the *heterodyne method.* Contrary, if the shifting is in discrete steps, the many *phase shift* and *phase step methods* build on it.

To have independent access to the two reconstructed wave fields, in order to affect one independently from the other, a holographic setup employing two reference waves is required, Fig. 3.10 [80, 81]. The first exposure of the double exposure method is performed

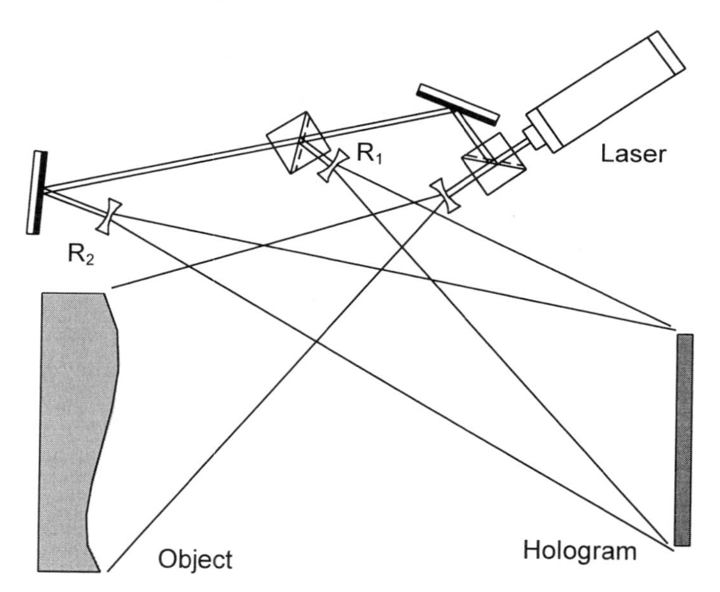

Figure 3.10: Two-reference-beam holographic arrangement with well separated reference sources

with object state E_1 and reference wave R_1 while R_2 is blocked. The second exposure, now of object state E_2 is done with the help of R_2 with R_1 being blocked. The reconstruction then is accomplished with R_1 and R_2 together to generate the interference between E_1

and E_2, but now this interference pattern may be modified by varying one of the reference waves while leaving the other one unaffected.

Two reference beam holography requires special attention to the multiplicity of the reconstructed images. Illuminating the hologram with both reference beams R_1 and R_2 together yields not only two, but four reconstructions in the first diffraction order, namely the two desired reconstructions $R_1R_1^*E_1$ and $R_2R_2^*E_2$, see (2.125) and (2.128), which give rise to the interference pattern and additionally the two undesired *cross-reconstructions* $R_2R_1^*E_1$ and $R_1R_2^*E_2$. As shown in Sec. 2.6.3 reconstruction with other reference beams than used during recording will result in shifted and distorted images. Eqs. (2.148) - (2.150) with $\mu = 1$ may be used to plan a two reference wave arrangement where the cross-reconstructions do not overlap.

As an easy example let an object surface extend along the x-direction from $P_1 = (x_1, 0, z)$ to $P_1 = (x_2, 0, z)$ $x_2 > x_1$, and the reference sources be at $R_1 = (x_R, 0, z)$ and $R_2 = (x'_R, 0, z)$ $x'_R > x_R$. Then (2.148) for P_1 recorded with R_1 and reconstructed with R_2 gives the position $P'_1 = (x_1 + x'_R - x_R, 0, z)$. Non-overlapping here occurs for $x_1 + x'_R - x_R > x_2$. Generally, the two reference sources chosen on the same side of the object must have mutual separation larger than the angular size of the object in the corresponding direction.

However, the consequence of a large separation of the reference sources is high sensitivity to hologram misalignment in the repositioning after development and to changes of the wavelength between hologram recording and reconstruction [27]. The sensitivity of two reference beam holographic interferometry to repositioning and wavelength changes occurs because the propagation of the two reconstructed wave fields are differently affected as can be checked quantitatively with the help of (2.148) - (2.150) using a coordinate system fixed to the hologram plate.

For reference beams assumed as plane waves in the directions $\boldsymbol{k_1}$ and $\boldsymbol{k_2}$, the additional phase difference $\Delta\phi$ at a point $\boldsymbol{r}_H$ in the hologram plane is [81]

$$\Delta\phi(\boldsymbol{r}_H) = [(\boldsymbol{k_1} - \boldsymbol{k_2}) \times \boldsymbol{\omega}] \cdot \boldsymbol{r}_H + \frac{\Delta\lambda}{\lambda}(\boldsymbol{k_1} - \boldsymbol{k_2}) \cdot \boldsymbol{r}_H \tag{3.28}$$

where $\boldsymbol{\omega} = (\Delta\xi, \Delta\eta, \Delta\chi)$ is the rotation vector for small hologram rotations around the x-, y-, and z-axes, respectively, and $\Delta\lambda$ is the change in wavelength. It is seen that both contributions, small rotations and wavelength shift, depend on the difference vector $\boldsymbol{k_1} - \boldsymbol{k_2}$. So the sensitivity to misalignment or wavelength change is drastically decreased if the reference sources are close together [82]. In such an arrangement, Fig. 3.11, all reconstructions overlap, but only $R_1R_1^*E_1$ and $R_2R_2^*E_2$ will give a stationary interference pattern as long as the cross-reconstructions are shifted laterally by more than the average pixel size. The distortions induced by the cross reconstructions are merely a tolerable decrease in the contrast of the interference pattern. Wavelength changes between recording and reconstruction cannot be avoided if recording was done with a pulsed laser but reconstruction has to be performed with a continuously emitting laser.

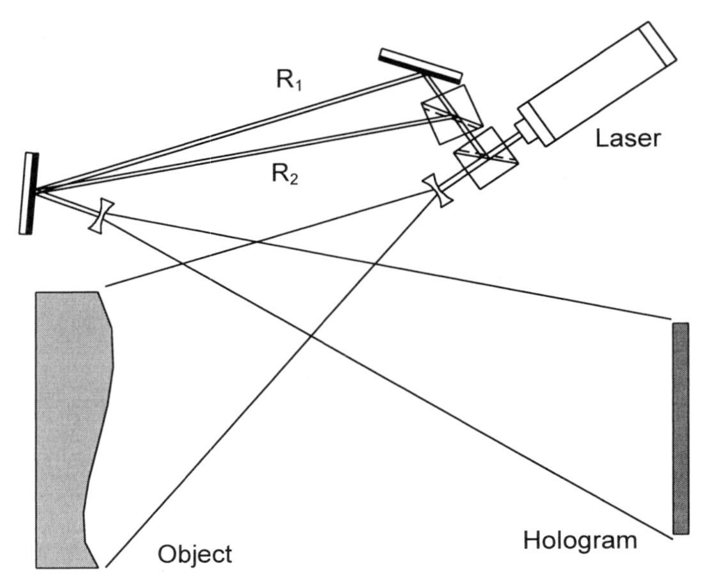

Figure 3.11: Two-reference-beam holographic arrangement with reference sources close together

3.3 Fringe Localization

3.3.1 Fringe Formation with Diffusely Scattering Surfaces

In Sec. 2.5 the speckle effect was introduced which always occurs if rough diffuse scattering surfaces reflect coherent light. In holographic interferometry wave fields interfere, which represent a diffusely reflecting surface in different deformation states. The light from each point of the undeformed surface will interfere with light from each point of the deformed surface. This produces a speckle pattern consisting of random irradiance variations of relatively high spatial frequency. As outlined in Sec. 2.5 the speckle pattern primarily is defined by the microscopic contour of the reflecting surface and by the size of the viewing aperture.

But additionally there is a systematic low-frequency variation which constitutes the holographic interferometric fringes conveying information about the displacement and deformation of the object between the exposures. Fig. 3.12 shows a rough surface before and after deformation, where P_1 and P_2 denote the same point of the surface before and after its displacement [24]. The same holds for Q_1 and Q_2. Light scattered by the surface in the neighborhood of point P_1 gives rise to a complex wave field, the light scattered at a neighborhood of P_2 forms another wave field. The wave fields are nearly identical but are mutually displaced and travel in slightly different directions. The resulting irradiance in an observation plane is shown, Fig. 3.12. The systematic low frequency variation is only induced by interference of light from corresponding points P_1 and P_2 or Q_1 and Q_2. Noncorresponding point pairs like (P_1, Q_1), (P_1, Q_2), (P_2, Q_1), or (P_2, Q_2) do not contribute to the formation of the holographic interference pattern. This is the concept of *homologous points* or *rays* [83]. These are the same points on the surface before and after

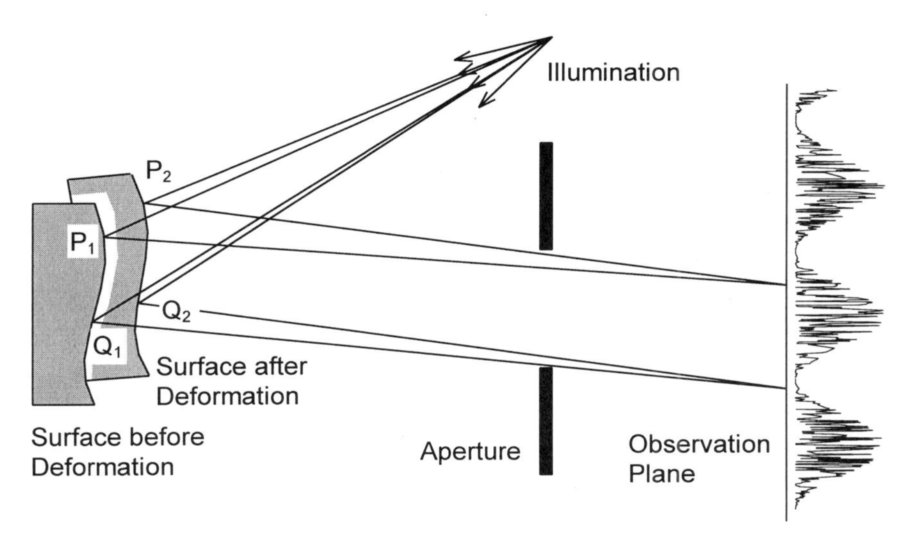

Figure 3.12: Scheme of fringe formation

deformation. Non-homologous points possess neighborhoods with distinct microstructures which are not correlated, thus not leading to a low-frequency interference fringe pattern but only contributing to the statistical speckle field. Altogether, for computation or for evaluation of fringe patterns, we need to consider only the change in optical pathlength δ of light scattered by homologous points such as P_1 and P_2. The pathlength δ is the one defined in (3.13).

The simultaneously present wavefronts in holographic interferometry freely propagate in space. They only interfere if they are imaged onto a detector surface, such as a sheet of film, a ground glass screen, a CCD-target or the retina of the eye. If a person looks directly through a double-exposure hologram to see the reconstructed virtual images, or looks through the single exposed hologram and additionally sees the illuminated object for the real-time method, the optical interference occurs on his or her retina, not at some other location in space. In the terminology of Sec. 2.5.2, Fig. 3.12 shows the objective fringe formation, while in practice we have the subjective fringe formation.

A simple system for viewing a double exposure hologram is given in Fig. 3.13. The hologram here only acts as a means for delivering the wavefronts simultaneously. Since the optics are focused onto the surface, the image of the composite field at P is formed at P' at the detector. The resulting interference pattern should depend on the optical pathlength δ, (3.17), but now we have a manifold of observation directions, two of them, $\boldsymbol{b}$ and $\boldsymbol{b}'$, are shown in Fig. 3.13. If the pathlength δ, resulting from all these directions, varies too much, no interference pattern will be observable. Clearly, shutting down the aperture and thus increasing the depth of focus may minimize the $\boldsymbol{b}$-variation and eventually will produce observable fringes.

Practically it has been found that instead increasing the focus a shift of the plane, where it is focused on, often exhibits observable fringes. Thus one recognizes that the

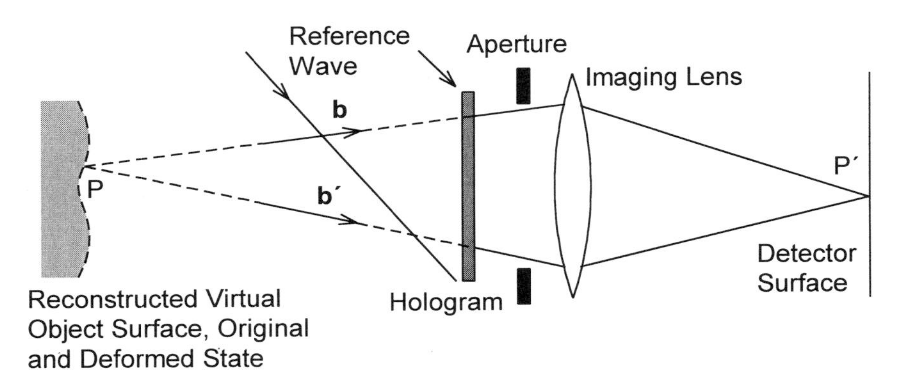

Figure 3.13: Fringe formation in double exposure holographic interferometry, focused on surface

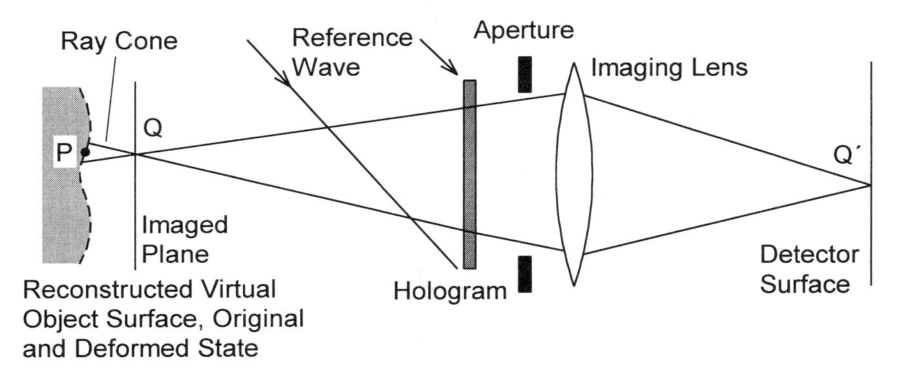

Figure 3.14: Fringe formation in double exposure holographic interferometry, focused on fringe localized outside the surface

fringes not generally are fixed to the object surface but may localize in space. This *localization* depends on the type of deformation, the observation direction, and as will be shown in the sequel, on the illumination direction. As we have seen already, a further important parameter for fringe formation is the aperture.

The focusing on a fringe localized outside the object surface is shown in Fig. 3.14. The optical pathlength of all rays from Q to Q' and thus the relative phases are equal, independent from the individual $\boldsymbol{b}$. So the interference pattern at Q' is identical to that which would be formed at Q, if a detector would be placed there. At Q the light is arriving from a ray cone centered on the axis $\overline{PQ}$. The angular extent of this cone is determined by the aperture of the viewing system. The interference pattern in the neighborhood of Q is the superposition of the interference patterns due to the motion of all surface points P subtended by this cone. If Q is where the fringes are localized, the value of δ (3.17) is nearly constant over the cone. But for arbitrary Q the pathlength δ may vary significantly, so that the many corresponding fringe patterns average out and only the random speckle pattern is observed.

For an analytic description of the fringe localization the point along a given viewing direction is determined, for which the variation of δ over a small cone of observation rays is minimized [84]. Therefore a Cartesian coordinate system fixed to the surface with the x-y-plane tangential and the z-axis normal at at least one surface point is defined. Thus the condition for localization of the fringes is that the expression for the optical pathlength δ (3.17) is minimized [24]

$$d\delta = \frac{\partial \delta}{\partial x}dx + \frac{\partial \delta}{\partial y}dy = 0 \tag{3.29}$$

where dx and dy are differential changes in the object point which is viewed from Q. The distance z from the object surface to Q is a parameter in the expression for δ. The values of z for which (3.29) is satisfied, define the curve or surface of fringe localization.

It should be mentioned that for this definition of localization the visibility of the holographic fringes defined by

$$V = \frac{\langle I \rangle_{max} - \langle I \rangle_{min}}{\langle I \rangle_{max} + \langle I \rangle_{min}} \tag{3.30}$$

becomes maximal [85, 86]. Here $\langle \ \rangle$ describes averaging over an ensemble of surfaces with different microstructures undergoing the same deformations.

3.3.2 Fringe Localization with Collimated Illumination

For small objects or at least for illumination sources sufficiently far from the object surface the curvature of the illumination wavefront at the object surface is quite small [24]. The $\boldsymbol{s}$ (3.18) can be assumed constant, but the $\boldsymbol{b}$ and $\boldsymbol{d}$ of (3.17) are functions of the surface point $P = (x, y, 0)$ and the observer position $B = (x_B, y_B, z_B)$. Then (3.29) becomes

$$\begin{aligned}
&\left(\frac{\partial b_x(P)}{\partial x}d_x(P) + \frac{\partial b_y(P)}{\partial x}d_y(P) + \frac{\partial b_z(P)}{\partial x}d_z(P)\right.\\
&\left. +[b_x(P) - s_x(P)]\frac{\partial d_x(P)}{\partial x} + [b_y(P) - s_y(P)]\frac{\partial d_y(P)}{\partial x} + [b_z(P) - s_z(P)]\frac{\partial d_z(P)}{\partial x}\right) dx\\
&+\left(\frac{\partial b_x(P)}{\partial y}d_x(P) + \frac{\partial b_y(P)}{\partial y}d_y(P) + \frac{\partial b_z(P)}{\partial y}d_z(P)\right.\\
&\left. +[b_x(P) - s_x(P)]\frac{\partial d_x(P)}{\partial y} + [b_y(P) - s_y(P)]\frac{\partial d_y(P)}{\partial y} + [b_z(P) - s_z(P)]\frac{\partial d_z(P)}{\partial y}\right) dy = 0
\end{aligned} \tag{3.31}$$

The vector $\boldsymbol{b}$ is defined in (3.19), its derivatives with respect to x and y, the corresponding components of P, are easily calculated to be

$$\begin{aligned}
\frac{\partial}{\partial x}b_x(P) &= -\frac{b_z(P)}{z_B}(b_y^2(P) + b_z^2(P)) & \frac{\partial}{\partial y}b_x(P) &= \frac{b_z(P)}{z_B}b_x(P)b_y(P))\\
\frac{\partial}{\partial x}b_y(P) &= \frac{b_z(P)}{z_B}b_x(P)b_y(P)) & \frac{\partial}{\partial y}b_y(P) &= -\frac{b_z(P)}{z_B}(b_x^2(P) + b_z^2(P))\\
\frac{\partial}{\partial x}b_z(P) &= \frac{b_z(P)}{z_B}b_x(P)b_z(P)) & \frac{\partial}{\partial y}b_z(P) &= \frac{b_z(P)}{z_B}b_y(P)b_z(P))
\end{aligned} \tag{3.32}$$

Combination with (3.31) yields

$$\begin{aligned} & \left(\frac{b_z}{z_B}\left[(b_y^2+b_z^2)d_x - b_x b_y d_y - b_x b_z d_z\right] - (b_x - s_x)\frac{\partial d_x}{\partial x} - (b_y - s_y)\frac{\partial d_y}{\partial x} - (b_z - s_z)\frac{\partial d_z}{\partial x}\right) dx \\ + & \left(\frac{b_z}{z_B}\left[(b_x^2+b_z^2)d_y - b_x b_y d_x - b_y b_z d_z\right] - (b_x - s_x)\frac{\partial d_x}{\partial y} - (b_y - s_y)\frac{\partial d_y}{\partial y} - (b_z - s_z)\frac{\partial d_z}{\partial y}\right) dy \\ = & \ 0 \end{aligned} \tag{3.33}$$

The arguments (P) have been omitted for clarity.

If the viewing aperture is of roughly the same dimension in all directions, as is the case for square or circular apertures, dx and dy can be varied independently, so the factors to dx and dy must each be identically zero, leading to the two conditions

$$z_B = \frac{b_z\left[(b_y^2+b_z^2)d_x - b_x b_y d_y - b_x b_z d_z\right]}{(b_x - s_x)\frac{\partial d_x}{\partial x} + (b_y - s_y)\frac{\partial d_y}{\partial x} + (b_z - s_z)\frac{\partial d_z}{\partial x}} \tag{3.34}$$

$$z_B = \frac{b_z\left[(b_x^2+b_z^2)d_y - b_x b_y d_x - b_y b_z d_z\right]}{(b_x - s_x)\frac{\partial d_x}{\partial x} + (b_y - s_y)\frac{\partial d_y}{\partial x} + (b_z - s_z)\frac{\partial d_z}{\partial x}} \tag{3.35}$$

Since the observation point B of (3.19) here is equivalent to the point Q in Fig. 3.14, each of these two equations describes a surface $z_B = z_Q = z_Q(\boldsymbol{b}, \boldsymbol{s}, \boldsymbol{d})$ in space. Both equations must be satisfied, therefore the fringes are generally localized along a curve which is the intersection of these two surfaces. Only in special cases (3.34) and (3.35) define identical surfaces, where then the fringes localize.

In the following the localization is determined for some special displacement fields of the object. If the object undergoes a *rigid body translation*, the displacement components d_x, d_y, and d_z are identical at all points of the investigated surface. Their derivatives are identically zero, so (3.34) and (3.35) give a localization at infinity $z_B = \infty$. Indeed the fringes can be observed with good visibility only at infinity, meaning the back focal plane of a lens, Fig. 3.15 [86].

For general deformations or *rigid body rotations* not all term in the denominators of (3.34) and (3.35) vanish, thus giving fringe localization at some finite distance from the surface.

As a first example let the surface rotate by a small angle θ about the y-axis, then

$$\boldsymbol{d}(x, y, 0) \approx (0, 0, \theta x) \tag{3.36}$$

and (3.34) yields

$$z = -\frac{b_x b_z^2}{b_z - s_z} x \tag{3.37}$$

This in the following is considered for some representative holographic arrangements, Fig. 3.16. If the observation is along the surface normal, the z-axis, Figs. 3.16a and b, then we have $b_x = 0$, yielding $z_B = 0$, meaning localization in the object surface. If we have an illumination direction inclined 45° to the z-axis and the observation is under

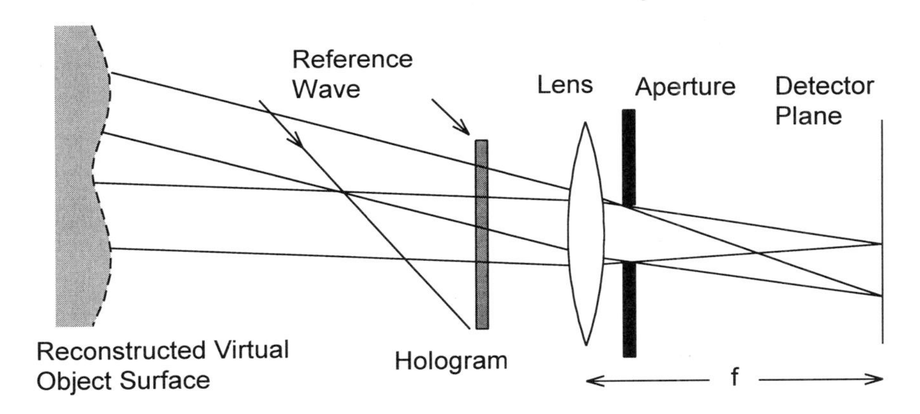

Figure 3.15: Observation of fringes localized at infinity

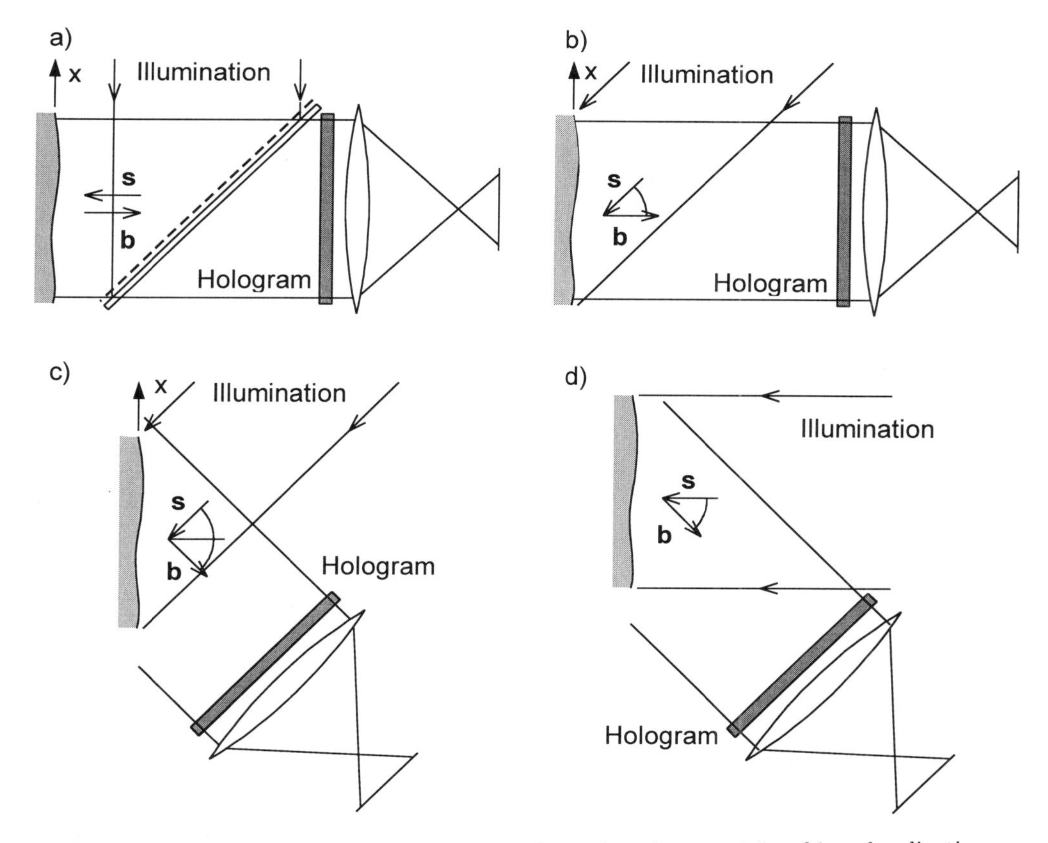

Figure 3.16: Representative holographic configurations for examining fringe localization

-45° with respect to the z-axis, Fig. 3.16c then $b_x = b_z = -s_x = \sin 45^o = \sqrt{2}/2$ and the localization plane is defined by

$$z = -\frac{1}{4}x \tag{3.38}$$

This defines a plane intersecting the observed surface along the y-axis. For normal illumination, $s_z = 1$, and observation under -45°, Fig. 3.16d, the localization is at

$$z = -\frac{1}{2(1+\sqrt{2})}x \tag{3.39}$$

If the object is rotated by an angle θ about an axis parallel to the surface, but not lying in it, the fringes localize off the surface. Let the rotation axis be parallel to the y-axis and lying a distance r behind the surface which is in the x-z-plane, then

$$\boldsymbol{d}(x, y, 0) \approx (r\theta, 0, -\theta x) \tag{3.40}$$

Let us furthermore for simplicity only consider illumination-, observation-, and object points in the x-z-plane, then $b_y = s_y = 0$ and from (3.34) we get

$$z = -\frac{b_z[(b_y^2 + b_z^2)r + b_x b_z x]}{(b_z - s_z)} \tag{3.41}$$

which again is a plane like in (3.37). If viewed normally, $b_z = 1$, $b_x = b_y = 0$, Figs. 3.16a and b, the fringes are in a plane parallel to the object surface but behind it, namely by $-r(1 - s_z)$.

A rotation about the z-axis by an angle θ gives

$$\boldsymbol{d}(x, y, 0) \approx (-\theta y, \theta x, 0) \tag{3.42}$$

and the localization conditions (3.34) and (3.35) are

$$\begin{aligned} z &= -\frac{b_z\left[(b_y^2 + b_z^2)y + b_x b_y x\right]}{b_y - s_y} \\ z &= -\frac{b_z\left[(b_x^2 + b_z^2)x + b_x b_y y\right]}{b_x - s_x} \end{aligned} \tag{3.43}$$

If now the object is illuminated and viewed in normal direction, Fig. 3.16a, the displacement vector is orthogonal to the sensitivity vector, no fringes are formed at all. For illumination at an angle of 45° and normal observation, Fig. 3.16b, we have $s_x = -s_z = \sqrt{2}/2$, $b_x = b_y = 0$, and $b_z = 1$, now the fringes are localized in the line

$$z = \sqrt{2}x \quad , \quad y = 0 \tag{3.44}$$

The configurations c and d of Fig. 3.16 give $z = x/2$ and $z = x$, respectively.

When deformations are involved, the displacement vector $\boldsymbol{d}$ may be a nonlinear function of the object surface coordinates and the fringes appear to localize along a curve in space, or in a curved surface. An elongation of e. g. a tensile test specimen is described by

$$\boldsymbol{d}(x, y, 0) \approx (\varepsilon x, 0, 0) \tag{3.45}$$

where ε is the linear strain. For the configurations a and c of Fig. 3.16 we will get no fringes, since there is orthogonality between displacement- and sensitivity vectors. For configuration b the localization is at $z = -\sqrt{2}x$ and for d in $z = x/2$.

The bending of a cantilever beam of length L fixed at one end is described by

$$\boldsymbol{d}(x, y, 0) \approx \left(0, 0, \frac{d}{2}\left[3(\frac{x}{L})^2 - (\frac{x}{L})^3\right]\right) \tag{3.46}$$

with d the deflection at the loose end. The fringes here are localized in $z = 0$ for arrangements a and b of Fig. 3.16, in

$$z = \frac{L}{6\sqrt{2}} \frac{3(\frac{x}{L}) - (\frac{x}{L})^2}{2 - (\frac{x}{L})} \tag{3.47}$$

for c and in

$$z = \frac{-L}{6(1 + 1/\sqrt{2})} \frac{3(\frac{x}{L}) - (\frac{x}{L})^2}{2 - (\frac{x}{L})} \tag{3.48}$$

for configuration d.

Generally it is possible to bring a fringe pattern that is localized far from the object surface to it by tilting the reference beam adequately [87].

3.3.3 Fringe Localization with Spherical Wave Illumination

If the curvature of the illuminating wavefront at the object surface cannot be neglected, differentiation of the sensitivity vectors in (3.29) leads to additional $\partial s_i(P)/\partial x$- and $\partial s_i(P)/\partial y$-terms in (3.31). The derivatives of $\boldsymbol{s}$ are calculated analogously to (3.32) and with the abbreviation $R = \sqrt{(x_P - x_S)^2 + (y_P - y_S)^2 + (z_P - z_S)^2}$ as

$$\begin{aligned}
\frac{\partial}{\partial x} s_x(P) &= -\frac{s_y^2(P) + s_z^2(P)}{R} & \frac{\partial}{\partial y} s_x(P) &= \frac{s_x(P)s_y(P)}{R} \\
\frac{\partial}{\partial x} s_y(P) &= \frac{s_x(P)s_y(P)}{R} & \frac{\partial}{\partial y} s_y(P) &= -\frac{s_x^2(P) + s_z^2(P)}{R} \\
\frac{\partial}{\partial x} s_z(P) &= \frac{s_x(P)s_z(P)}{R} & \frac{\partial}{\partial y} s_z(P) &= \frac{s_y(P)s_z(P)}{R}
\end{aligned} \tag{3.49}$$

and instead of (3.33) now we get

$$\left(\left[\frac{b_z}{z_B}(b_y^2 + b_z^2) - \frac{1}{R}(s_y^2 + s_z^2)\right] d_x - \left[\frac{b_z}{z_B} b_x b_y - \frac{1}{R} s_x s_y\right] d_y - \left[\frac{b_z}{z_B} b_x b_z - \frac{1}{R} s_x s_z\right] d_z \right.$$
$$\left. - \quad (b_x - s_x)\frac{\partial d_x}{\partial x} - (b_y - s_y)\frac{\partial d_y}{\partial x} - (b_z - s_z)\frac{\partial d_z}{\partial x}\right) dx$$

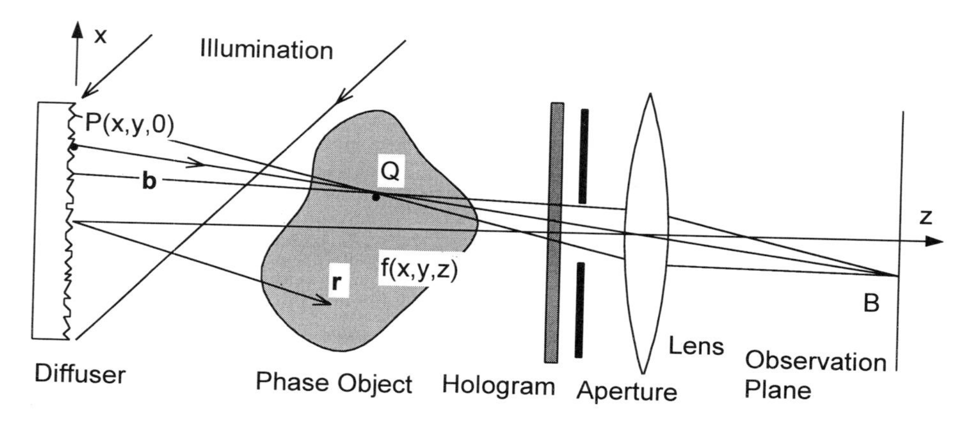

Figure 3.17: Fringe localization of phase objects

$$
\begin{aligned}
+ & \left(\left[\frac{b_z}{z_B}(b_x^2 + b_z^2) - \frac{1}{R}(s_x^2 + s_z^2) \right] d_y - \left[\frac{b_z}{z_B} b_x b_y - \frac{1}{R} s_x s_y \right] d_x - \left[\frac{b_z}{z_B} b_y b_z - \frac{1}{R} s_y s_z \right] d_z \right. \\
- & \left. (b_x - s_x)\frac{\partial d_x}{\partial y} - (b_y - s_y)\frac{\partial d_y}{\partial y} - (b_z - s_z)\frac{\partial d_z}{\partial y} \right) dy \\
= & \ 0
\end{aligned}
\tag{3.50}
$$

This is the most general condition for fringe localization and can be applied to the different rigid body motions and deformations as well as to the varied holographic configurations in the same manner as (3.33) did. Then one recognizes that the variation in the sensitivity vector introduces curvature into the surface or curve of localization and can lead to localization at finite distances even for rigid body translations. A detailed analysis of this case shows that the fringes now localize in a curved surface near $z = R$.

3.3.4 Fringe Localization with Phase Objects

Using the same approach as in the sections before now we investigate the localization of the holographic interference fringes arising from refractive index changes in phase objects [24]. Here only diffuse illumination holographic interferometry is considered, Fig. 3.7, for it is the practically relevant case. The analysis in the sequel will employ straight rays, ray bending due to refractive index gradients will be neglected. Fig. 3.17 shows the geometry the following analysis is based on. In front of the illuminated diffuser is a phase object introducing the refractive index variation

$$
f(\boldsymbol{r}) = f(x, y, z) = \frac{2\pi}{\lambda}\Delta n(x, y, z) \tag{3.51}
$$

where $\Delta n(x, y, z)$ was already defined in (3.22). Without loss of generality a Cartesian coordinate system is fixed to the diffuser with the x-y-plane tangential and the z-axis normal to the diffuser surface. The viewing system should be focused on a plane containing the point Q. As already seen, the holographic interference fringes are localized at points

Q where the phase difference $\Delta\phi$ is nearly constant over the cone of ray pairs passing through Q and imaged onto the observation plane.

In analogy to (3.22) the optical path difference is

$$\Delta\phi = \int_S f(x, y, z) ds \tag{3.52}$$

where s is along the ray of direction $\boldsymbol{b}$ through the phase object. It is assumed that $f(x, y, z) = 0$ outside a finite region S, - the intersection of the observation ray with the phase object, - and that f exhibits no discontinuities.

The condition for fringe localization is given analogously to (3.29) by

$$d(\Delta\phi) = \frac{\partial \Delta\phi}{\partial x} dx + \frac{\partial \Delta\phi}{\partial y} dy = 0 \tag{3.53}$$

Since the integration limits are fixed, we can interchange the order of integration and differentiation, and with the chain rule we get

$$\frac{\partial \Delta\phi}{\partial x} = \int_S \nabla f \cdot \frac{\partial \boldsymbol{r}}{\partial x} ds \tag{3.54}$$

with $\nabla f = (\partial f/\partial x, \partial f/\partial y, \partial f/\partial z)^T$. Since

$$\boldsymbol{r} = \begin{pmatrix} x \\ y \\ 0 \end{pmatrix} + \begin{pmatrix} b_x \\ b_y \\ b_z \end{pmatrix} s \tag{3.55}$$

gives the derivative

$$\frac{\partial \boldsymbol{r}}{\partial x} = \begin{pmatrix} 1 + \frac{\partial b_x}{\partial x} s \\ \frac{\partial b_y}{\partial x} s \\ \frac{\partial b_z}{\partial x} s \end{pmatrix} \tag{3.56}$$

(3.54) writes as

$$\frac{\partial \Delta\phi}{\partial x} = \int_S \left[\left(1 + \frac{\partial b_x}{\partial x} s \right) \frac{\partial f}{\partial x} + \frac{\partial b_y}{\partial x} \frac{\partial f}{\partial y} s + \frac{\partial b_z}{\partial x} \frac{\partial f}{\partial z} s \right] ds \tag{3.57}$$

In the same way the derivative with respect to y is

$$\frac{\partial \Delta\phi}{\partial y} = \int_S \left[\left(1 + \frac{\partial b_y}{\partial y} s \right) \frac{\partial f}{\partial y} + \frac{\partial b_x}{\partial y} \frac{\partial f}{\partial x} s + \frac{\partial b_z}{\partial y} \frac{\partial f}{\partial z} s \right] ds \tag{3.58}$$

Again Δx and Δy can be varied independently, so each of (3.57) and (3.58) must vanish. With the derivatives of the components of $\boldsymbol{b}$ given in (3.32) we get from (3.57)

$$\int_S \left[\left(1 - \frac{b_z}{z} (b_y^2 + b_z^2) s \right) \frac{\partial f}{\partial x} + \frac{b_z}{z} b_x b_y \frac{\partial f}{\partial y} s + \frac{b_z}{z} b_x b_z \frac{\partial f}{\partial z} s \right] ds = 0 \tag{3.59}$$

and from (3.58)

$$\int_S \left[\left(1 - \frac{b_z}{z}(b_x^2 + b_z^2)s\right) \frac{\partial f}{\partial y} + \frac{b_z}{z} b_x b_y \frac{\partial f}{\partial x} s + \frac{b_z}{z} b_y b_z \frac{\partial f}{\partial z} s \right] ds = 0 \tag{3.60}$$

For the distance s measured from the diffuser we can write $s = z/b_z$. Now splitting the integrals and solving for the z_l where the fringes localize gives the two conditions

$$z_l = \frac{\int_S \left[(b_y^2 + b_z^2)\frac{\partial f}{\partial x} - b_x b_y \frac{\partial f}{\partial y} - b_x b_z \frac{\partial f}{\partial z}\right] z\, dz}{\int_S \frac{\partial f}{\partial x} dz} \tag{3.61}$$

$$z_l = \frac{\int_S \left[(b_x^2 + b_z^2)\frac{\partial f}{\partial y} - b_x b_y \frac{\partial f}{\partial x} - b_y b_z \frac{\partial f}{\partial z}\right] z\, dz}{\int_S \frac{\partial f}{\partial y} dz} \tag{3.62}$$

(3.61) and (3.62) generally define two surfaces, the interference fringes localize along the intersection of these two surfaces.

For the special case of viewing along the z-axis we have $b_x = b_y = 0$ and $b_z = 1$, then the conditions (3.61) and (3.62) reduce to

$$z_l = \frac{\int_S \frac{\partial f}{\partial x} z\, dz}{\int_S \frac{\partial f}{\partial x} dz} \tag{3.63}$$

$$z_l = \frac{\int_S \frac{\partial f}{\partial y} z\, dz}{\int_S \frac{\partial f}{\partial y} dz} \tag{3.64}$$

Numerical and experimental examples depicting the localization of fringes due to refractive index variations are given in [24]. For single radially symmetric fields the analysis of (3.63) and (3.64) shows that the fringes localize in the center plane of the object. The plane is normal to the line of sight and contains the axis of symmetry. Looking through two identical radially symmetric objects, one placed behind the other, gives a localization just in the plane midway between the two objects. For more general phase objects the curves of fringe localization become quite convoluted.

3.3.5 Observer Projection Theorem

A concept for considering the effect of variations of the observation vector $\boldsymbol{b}(P)$ is given by the so called *observer projection theorem.* This theorem is implicit in the geometric optics approach to fringe localization which is followed in this chapter, but can also be derived by wavefront analysis in wave optics [88]. This theorem states that, if the holographic interferometric fringes are localized off the object surface, they can be projected onto the object surface radially from the center of the aperture of the viewing system.

The theorem is useful when fringes to be recorded by a CCD-camera are localized well off the object surface, especially if their spatial frequency is high. In order to record the fringes and the object surface simultaneously, the aperture of the camera must be small. But a small aperture will cause unacceptable noise in the form of speckles, as pointed out in Sec. 2.5. Following the observer projection theorem, however, one may record the fringes while focused with a large aperture in their localization surface. Then the object surface is recorded separately from the same camera position. After appropriate relative magnification these two stored images can be superimposed digitally. Nevertheless this last step can be omitted if the correspondence between the pixel coordinates and the object surface is known and considered in the further processing.

The treatment of localization of the holographic interferometric fringes based on geometric optics given here followed closely that of Vest [24]. Its aim was not to give a detailed description of all localization aspects, but to introduce quickly into the problems and possibilities of fringe localization, to present the basic mathematical relations, and to give some easy examples. A more thorough description can be found in [24] in the context of a global description of the fringe patterns. An introduction to localization may further be found in [89, 90]. An exhaustive study of localization was performed by Stetson, presented in a series of papers [88, 91, 92, 93, 94, 95, 96, 97]. His treatment is based on wavefront analysis. More facts about localization can be found in [83, 98, 99, 100, 101, 102].

3.4 Holographic Interferometric Measurements

3.4.1 Qualitative Evaluation of Holographic Interferograms

Holographic interferometry produces unique two-dimensional patterns of fringes, whose density and form depend on the loading of a tested structure, the structure itself and the geometry of the holographic arrangement. Since even minute loading amplitudes may cause deformation amplitudes that generate interference fringes, holographic interferometry is an extraordinary well suited tool for *nondestructive testing* (*NDT*): faults and flaws in a technical component like cracks, voids, thin areas, debonds, etc. under proper load induce a characteristic local deformation, which can be detected in the holographic interferogram [103, 104, 105]. A defect in a structure may be critical with regard to one loading

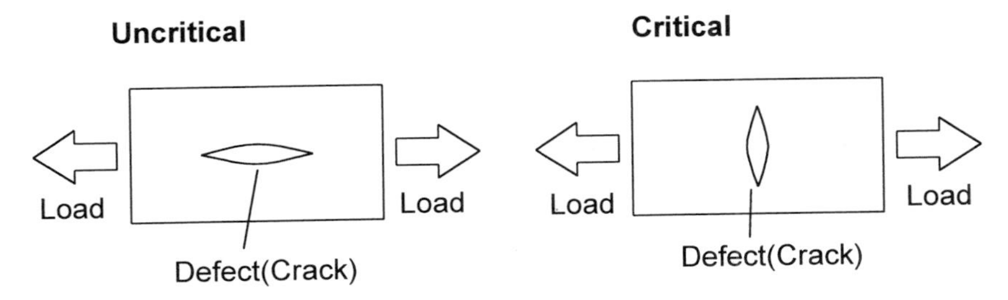

Figure 3.18: Critical and uncritical defects with regard to applied tensile stress

type or loading direction but not to another, Fig. 3.18. Therefore it is recommended to test a component with the type and direction of the intended operational load, but with low loading amplitude just sufficient to produce enough interference fringes to enable a reliable evaluation. This way we get a validation of the tested structure relative to the applied load. The defects are not visualized in their own but their response to the load is registered. The defects may reach to the surface, e. g. surface cracks, but also subsurface defects like voids or material inhomogeneities are detectable. All these potentialities and advantages lead to the field of *holographic nondestructive testing* (*HNDT*).

Development and utilization of an HNDT scheme or device for a particular application involves three interrelated tasks [106]: Selecting a loading, - type, direction, and amplitude - which causes a detectable anomaly in the fringe pattern, if a defect is present; designing the optical system, - sensitivity vectors in the proper direction, - and interpreting the results. Still for the most applications in HNDT the interferograms are observed and judged by skilled personnel. But there are attempts to perform an automatic computer controlled evaluation in HNDT, see Sec. 5.9. Since the interpretation of the interferograms consists only of the detection of typical patterns indicating the existence of a defect, but no quantitative interference phase determination is carried out, one speaks about *qualitative evaluation*.

Application areas for HNDT are manifold: components like tires or other automotive parts, turbine blades, honeycomb panels, pressure transducers, pressure vessels, aircraft parts, satellite-tanks, musical instruments, artwork, composites, and laminates are tested holographically, to name only a few [107, 108, 109, 110, 111, 112, 113, 114, 115, 116, 117, 118]. A 100 percent holographic inspection is performed e. g. at tires for special purposes, like retreaded aircraft tires, or at the nitrogen-tanks of European satellites.

3.4.2 Holographically Measurable Physical Quantities

Holographic interferometry measures optical pathlength differences, which affect the interference phase, e. g. according to (3.11) or (3.22), which in turn gives rise to the observable interference pattern, (3.4) or (3.10). There are a number of physical processes, where the optical path length is modified by various physical quantities. If the induced modification of the path length can be controlled to change only the phase of the light field and the lateral displacement to stay in the range less than the average speckle diameter, then these physical quantities are measurable by holographic interferometry.

The expression 'measurable' here can be understood in a very broad sense. It ranges from the quantitative derivation of the precise values of the physical quantity to a qualitative judgement by trained persons, where by looking at the interferogram areas of inhomogeneous variations of the underlying physical quantity can easily be detected, giving insight to the integrity of the object under test.

In this section only a short glimpse on the many holographically measurable physical quantities should be given, the detailed physical laws reigning the path length variations and the strategies how to evaluate them are presented in Chaps. 3 and 4.

One class of holographic interferometric measurements concerns the displacements and deformations of diffusely reflecting opaque surfaces, as introduced in Sec. 3.1.4. Some measurable quantities based on the displacement of surface elements are:

- *One-dimensional displacements* of the surface points in the direction of the sensitivity vector can be measured. If it is known that the displacement of all surface points is in the same direction, as is often the case, an optimized holographic setup as described in Sec. 3.2.1 should be used. The evaluation then is directed by (3.26).

- The *three-dimensional displacement* vector field of the surface points can be measured. This vector field may be defined by *rigid body translations*, *rigid body rotations*, general *deformations*, or a combination of these. For determining the three-dimensional vectors, a system of at least three linearly independent equations of the form (3.21) has to be solved for each surface point.

- Since holographic interferometry measures with high spatial resolution, numerical differentiation for the derivation of *strains* and *stresses* becomes possible. The *in-plane strains* in the case of *plane stress* can be determined [119]. Stresses are then calculated by the related proportionalities. *Bending moments*, *thermal expansion coefficients*, or the *Poisson ratio* have been successfully evaluated. *Surface tensions* of fluids have been measured by holographic interferometry [120].

- Contrary to these *static displacements* also *dynamic displacements* are measured holographically. The motions can be frozen by defined triggering of pulsed illumination for recording, or the integrated states interfere in the time average method. Motions with constant velocity or with constant acceleration as well as *transient events* as e. g. the propagation of *bending waves* after an impact are amenable to holographic interferometry.

- Mechanical *vibrations* are further dynamic motions to be investigated holographically. The *vibration amplitude* distributions of *harmonic vibrations*, *damped vibrations*, and other *nonlinear vibrations* can be determined. With refined procedures it is even possible to measure *phase relations* of the *vibration modes*.

The second class of holographic interferometric measurements is related to the *refractive index variation* of *transparent* or *phase objects*, Sec. 3.1.5. The refractive index distributions of gases, liquids, solids, or plasmas can be affected by a number of physical quantities which are measured via this effect. Some examples are:

- In *flow diagnostics* holographic interferometry today has widely replaced *Mach-Zehnder interferometry* [121, 122, 123, 124, 125, 126]. The *density of a gas* or a mixture of gases is related to the index of refraction by the *Gladstone-Dale equation* [127]. Therefore it is possible to measure holographically the density either in a steady or in a transient state.

- For *flow visualization* in *fluid dynamics* holographic interferometry is a means to depict the *streamlines* in the fluid under test. *Shock waves* [128, 129, 130, 131, 132, 133] as well as crack propagation in glass [134] are measured. Carrier fringes can be produced by exchanging a liquid to one with another refractive index.

- Combining the Gladstone-Dale relation, which relates the index of refraction to the density of a gas, with the *ideal gas equation*, one gets the dependence of the refractive index from the *pressure* and the *temperature* [135], taking additionally into account the molecular weight and the universal gas constant. Thus we have a means to measure the temperature distribution in a gas in *heat transfer* experiments [136].
- For liquids the Gladstone-Dale relation has to be replaced by the *Lorentz-Lorenz equation* which takes into account the specific refractivity of the substance under test.
- In *mass transfer* experiments holographic interferometry can be employed to measure the spatial or temporal distributions of *mass concentrations*. Due to the wavelength dependence of the Gladstone-Dale constant multiple wavelength interferometry is recommended to produce redundant data, especially if multiple parameters like temperature and concentrations are to be measured.
- Since the refractive index of an *electron gas* strongly depends on the *electron density*, this latter quantity can be measured holographically in *plasma diagnostics*. A two-wavelength method for the determination of the electron density is based on the strong *dispersion*, say, the strong dependence of the refractive index of the electron gas on the wavelength.
- The *stresses* in transparent solids can be measured holographically since due to the *stress-optical effect* the refractive index is a function of the model's state of stress. This function is expressed in the *Maxwell-Neumann stress-optical law* valid for *birefringent* as well as for *optically isotropic materials*.

In a third class of holographic interferometric measurements not the object to be measured is affected, but the holographic arrangement is modified to induce the optical pathlength variation. It will be explained in Sec. 5.6 how the resulting observable interference fringes are related to the *three-dimensional contours* of the object surface shape. Of course by measuring the contour twice, once before and once after a deformation of the object, this is a way to measure deformations also, at least in principle. This procedure may be of interest for large deformations, which would lead to lateral shifts of surface points larger than the average speckle size.

3.4.3 Loading of the Objects

Holographic interferometry can visualize the difference between different deformation states of an opaque surface. To change the deformation state between the generation of the two optical wave-fields to be compared, the object must be loaded, or an applied load must be varied. Many measurement problems to be solved holographically are of the form: Investigate and measure how a given component deforms in reaction to a specific load. The problem of HNDT normally is: Test a component with regard to possible defects. In this latter case one has to decide on the proper loading, which comprises the type, the direction, and the amplitude of the load [137]. Five basic types of loading play a dominant role: direct mechanical load, pressure load, vibrational load, impulse load, and thermal load [106, 138].

Direct *mechanical stressing* is applied by *bending moments*, *tensile stress*, *torsional stress*, *point load*, or *gravity*. If tensile stress should be utilized, one has to examine carefully the test equipment with regard to additional unwanted torsional stress. Since most holographic arrangements are much more sensitive to out-of-plane displacements than to in-plane-displacements, the fringes caused by torsion may dominate the fringe patterns of the in-plane-motions. The direction of direct mechanical stressing must be adapted to the expected flaw orientation in HNDT or to the structural orientation if we have to investigate heterogeneous material like fiber-reinforced plastics. Centrifugal forces in rotating components also may be regarded as causing mechanical stresses. Gravitational load can be performed by a 180°-tilt of the whole holographic arrangement contained in a suitable framework including the object.

Pressure loading is produced by internal high or low pressure in hollow components or by placing the test object into a *pressure-* or *vacuum-chamber*. Cylindrical or spherical pressure vessels as well as whole tube systems are tested by internal pressure. Containers may be filled with water to induce the pressure. Tires are inspected in a differential vacuum to detect ply separations, broken belts, debonds, or voids.

Vibrational load can be excited by the acoustic fields of loudspeakers or by electrodynamic shakers in point contact [139]. This load is used to find out areas of extraordinary sound emission in large structures - e. g. for noise reduction of cars - or to detect debonds, delaminations, or material inhomogeneities. A variation of amplitude, frequency, and sometimes phase of the excitation must be possible, e. g. for the detection of resonant modes. Rayleigh-waves may be excited by ultrasonic transducers [140] or by laser pulses [141]. The vibration mode structure visualized by their holographic interference patterns may disclose manufacturing errors like improper castings or incorrect or uneven wall thickness.

Impulse load by a local impact generates a travelling bending wave that can be recorded by employing a pulsed laser [142]. The impact may be generated by a pendulum or by a modified air-gun. Cracks or debonds obstruct the regular propagation of the bending waves and thus are detected.

There are a number of ways to apply *thermal load*: radiation sources like IR-lamps may heat the tested component, - often from the rear side of the inspected surface -, hot air jets may be generated by conventional hair-dryers, volatile fluids sprayed on a surface cool it during evaporation, high power DC generates heat in conductive material. Also heat may be brought to a component by induction heating or by microwaves, especially to materials containing water. Due to thermal expansion most materials are deformed when the temperature is changed [143, 144]. Furthermore the thermal conductivity locally varies at voids or debonds. Faulty electronic components have a thermal behaviour during operation different from intact components, so they can be identified holographically.

The loads may be categorized into *static loads* and *dynamic loads*. The direct mechanical loads and the pressure loads are static, while thermal load only in the equilibrium state can be viewed as static. The dynamic loads may further be divided into *periodic* and *transient loads.* Vibration loads are periodic while an impact causes a transient effect, thermal loading often must be regarded as transient. While static and periodic loads can be employed in conjunction with CW-lasers, the fast transient phenomena require a pulsed laser to be registered holographically.

Contrary to the temporal discrimination of the loads, they can be divided into *point loads* and *distributed loads.*

4

Quantitative Evaluation of the Interference Phase

Quantitative evaluation of holographic interference patterns for measurement purposes consists of the pointwise determination of the numerical value of the physical quantity which produced the optical path length change at each point and thus gave rise to the intensity distribution. The way how the physical quantity to be measured is contained in the observed intensity distribution was shown generally in Chap. 2. There the predominant role of the interference phase was discussed.

A computer aided quantitative evaluation thus is composed of two principal steps: First the interference phase distribution is determined from the recorded holographic interferogram, and second the interference phase is combined with the sensitivity vectors to achieve the spatial distribution of the physical quantity to be measured [108, 145, 146, 147, 148, 149, 150, 151, 152, 153, 154, 155, 156, 157]. In this chapter the first of these two steps is addressed and the different methods to perform this task are discussed. The determination of the physical quantities out of the interference phase distribution is postponed to Chap. 5.

4.1 Role of Interference Phase

In (3.4) it was shown how in the double-exposure, or in the real-time techniques the intensity distribution $I(x, y)$ of the holographic interference pattern depends on the interference phase distribution $\Delta\phi$ by the cosine-function. There identical amplitudes for all surface points have been assumed. If we want to recognize the more general case of differing amplitudes and modifications of the intensity by environmental effects, noise, or other distortions, we can write

$$I(x, y) = a(x, y) + b(x, y) \cos [\Delta\phi(x, y)] \tag{4.1}$$

This is the intensity distribution $I(x, y)$ that is recorded by e. g. a CCD-camera, digitized into an array of $N \times M$ pixels, quantized into L discrete gray-values and in this form stored digitally in the computer memory. The (x, y) denote the pixel coordinates. The mapping of object points to the individual pixel coordinates also is postponed to Chap. 5. The $a(x, y)$ and $b(x, y)$ contain the intensities of the interfering wave fields and the various disturbances. Generally one can say that $a(x, y)$ contains all additive contributions and $b(x, y)$ comprises all multiplicative influences.

The objective is to extract the interference phase distribution $\Delta\phi(x, y)$ from the more or less disturbed intensity distribution $I(x, y)$ of (4.1). Constant values of $\Delta\phi$ define fringe loci on the object's surface, therefore $\Delta\phi(x, y)$ is called *fringe-locus function* by some authors [88, 158].

4.1.1 Sign Ambiguity

When one tries to extract the interference phase $\Delta\phi(x, y)$ out of the intensity distribution $I(x, y)$ by a kind of inversion of (4.1), the problem arises that the cosine is not a one-to-one function, but is even and periodic.

$$\cos \Delta\phi = \cos(s\Delta\phi + 2\pi n) \qquad s \in \{-1, 1\}, \quad n \in \mathbf{Z} \tag{4.2}$$

An interference phase distribution determined from a single intensity distribution remains indefinite to an additive integer multiple of 2π and to the sign s.

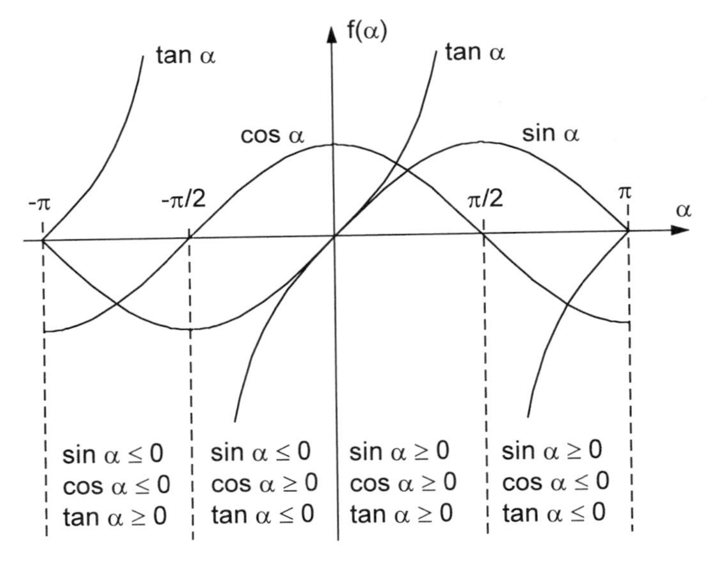

Figure 4.1: Signs of the trigonometric functions in $[-\pi, +\pi]$

Each inversion of expressions like (4.1) contains an inverse trigonometric function. All inverse trigonometric functions are expressed by the *arctan-function* like $\arccos(x) = \arctan(\sqrt{1 - x^2}/x)$. The arctan-function of a single variable has its *principal value* in the interval $[-\pi/2, +\pi/2]$. But in most algorithms for quantitative evaluation of holographic interferograms the argument of the arctan-function is accomplished by a quotient, where the numerator characterizes the sine of the argument α and the denominator corresponds to the cosine of α. Then it is good practice to consider the signs of the numerator and the denominator separately, as is done by the FORTRAN- or C-function ATAN2(X,Y), for in this case the principal value is determined consistently in the interval $[-\pi, +\pi]$. The four situations of the sine- and cosine-signs are shown in Fig. 4.1.

But there still remains a modulo 2π-uncertainty as well as the sign ambiguity. Fig. 4.2 displays a part of a graph which extends to infinity upward and downward. Each path

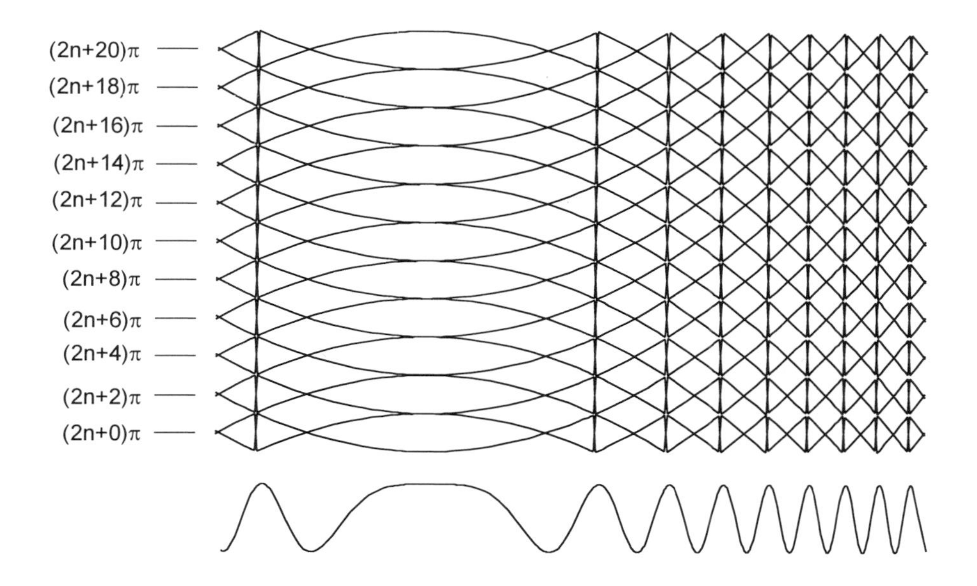

Figure 4.2: Ambiguity of the interference phase

from the left to the right through this graph represents a one-dimensional interference phase distribution belonging to the intensity distribution given at the bottom of the figure [159, 160].

A practical way to get rid of the *sign ambiguity* is to use side-informations about the experimental conditions leading to the measured optical path length changes and so to the interference phase distribution. In many applications one can assume that the measured interference phase distribution is not only continuous but as well differentiable, meaning a smoothly varying function. Depending upon the known direction in which the load acts on the object, e. g. compressive or tensile, a proper sign distribution assignment norm can be fixed based on the increase or decrease of the interference phase.

Reliable ways to eliminate the sign ambiguity without using pre-knowledge about the underlying experiment consist of either recording multiple phase stepped interferograms or introducing experimentally a linear phase carrier with a positive slope higher than the steepest descent of the interference phase, thus producing only increasing phase maps [161, 162]. The concepts also are known as *infinite fringes* and *finite fringes*. If for $\Delta\phi(x, y) = 0$ a field of uniform irradiance, that is, an infinitely wide fringe, results, one speaks of an *infinite fringe interferogram*. Since $-\Delta\phi(x, y) = 0$ and $+\Delta\phi(x, y) = 0$ yield the same fringe pattern, there remains a *sign ambiguity*. If reference fringes are introduced by an additional carrier phase gradient of known sign, the resulting interferograms are called *finite fringe interferograms* [24].

4.1.2 Absolute Phase Problem

The 2π-*ambiguity* manifests in the evaluated interference phase distributions by wrapping the phase *modulo* 2π. Since only principal values of the arctan-function in the interval $]-\pi, +\pi]$, - or equivalently in $]0, 2\pi]$, - are determined, as soon as an extreme value of the interval is reached, the phase jumps to the other extreme value, although the correct phase proceeds smoothly increasing or decreasing.

The modulo 2π effects are corrected by the processing step called *demodulation*, *continuation*, or *phase unwrapping*, see Sec. 4.9. By addition or subtraction of integer multiples of 2π the phase jumps are eliminated. The correct additive term in some applications can be determined if there is a point P in the pattern where the exact value of the displacement or equivalently the interference phase is known. Preferably this value is $\Delta\phi(P) = 0$. If a continuous variation of the interference phase can be assumed, the 2π-multiples at each point can be determined by counting the 2π-jumps from P to this point along an uninterrupted path. Sometimes an elastic ribbon is tied from one point P on the tested surface to a point in the holographic arrangement which has undergone no displacement but is lying in the observable interference pattern. Then the fringes along this ribbon are counted starting from zero [163].

Defining the interference phase at an arbitrary point of the investigated surface as zero, if one is only interested in the deformation relative to this point but not in the additional rigid body translation, is only admissible for constant sensitivity vectors. This approach can be envisaged as evaluating the variation of the displacement from the variation of the interference phase. Now (3.24) shows that the variation of the interference phase in the case of varying sensitivity vectors not only depends from the variation of the displacement vectors but also from the direct values of the displacements. Consequently, if the sensitivity vector varies over the surface, the constant additive term has to be taken into account, meaning that the *absolute phase* including the correct multiples of 2π must be evaluated. Sometimes it may be sufficient to estimate the maximum errors by assuming minimal and maximal multiples of 2π and put them into (3.21) and (3.24) using the extreme sensitivity vectors of the actual holographic setup.

The error of not recognizing the absolute phase is demonstrated in Fig. 4.3 for the example of a cantilever beam clamped at the left end and bended by a point load at the right end. Curve 1 in Fig. 4.3a gives the z-displacement $d_z(x)$ along the x-axis of the cantilever beam of length 100 mm. The z-component of the sensitivity vector is shown in Fig. 4.3b, when the coordinates of the illumination point are $S = (-200, 0, 250)$ mm, and those of the observation point are $B = (50, 0, 250)$ mm. The surface point P varies as $P = (x, 0, 0)$, with x running from $x = 0$ mm to $x = 100$ mm. The wavelength is 514.5 nm. The one-dimensional interference phase distribution corresponding to this deflection and arrangement is shown in curve 1 of Fig. 4.3c. Reflecting the fact that we do not know the additive constant, this interference phase is changed arbitrarily to curve 2

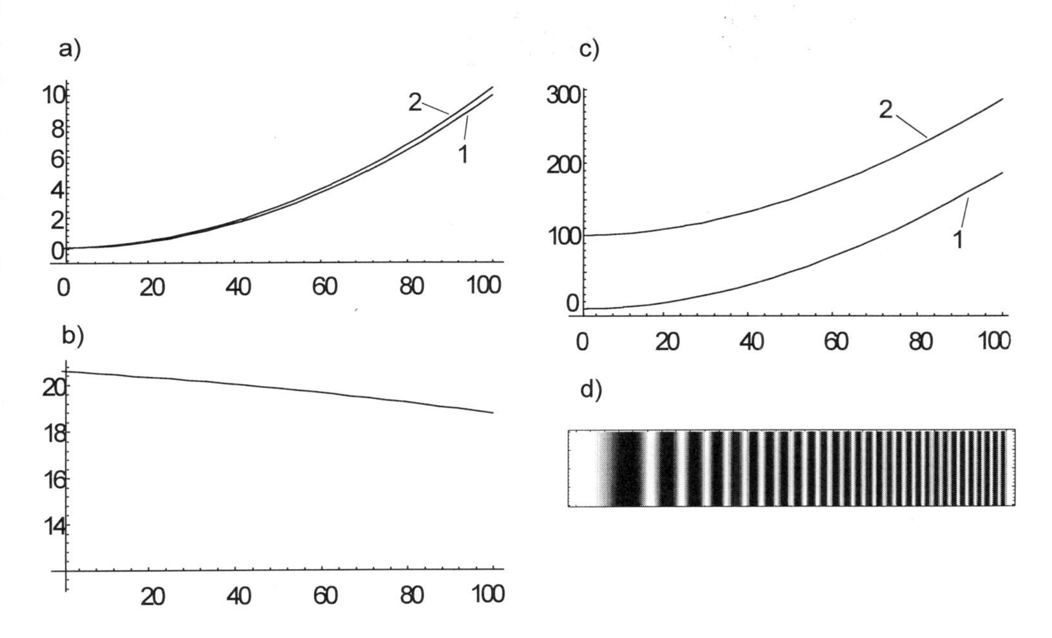

Figure 4.3: Effect of unknown 2π-multiples in the interference phase distribution

by adding a constant of 32π. The evaluation of this phase distribution gives a $d_z(x)$ which is displayed as curve 2 in Fig. 4.3a, after the evaluated z-displacement at the clamping point is brought to coincidence with the original one by subtraction of the displacement offset. Still there remains a difference at the other end of the cantilever which is of the order of one fringe period. Fig. 4.3d shows the expected holographic interference pattern.

4.2 Disturbances of Holographic Interferograms

As already taken into account in (4.1) holographic interferograms besides the ambiguities discussed in Sec. 4.1 suffer from a number of *distortions* degrading the interference pattern. The methods for computer aided evaluation have to deal with them in order to produce reliable results. A typical intensity distribution at one instant of time is of the form [164]

$$I(x,y) = I_0(x,y)[1 + V(x,y)\cos\Delta\phi(x,y)]R_S(x,y) + R_E(x,y) + R_D(x,y) \tag{4.3}$$

Here $I_0(x,y)$ denotes the low frequency background intensity caused by a varying illumination, e.g. a Gaussian profile of the enlarged laser beam, or a changing object reflectivity. $V(x,y)$ is the fringe visibility influenced mainly by speckle decorrelation and the ratio between the reference and object wave amplitudes. $R_S(x,y)$ describes the contrast variation caused by the speckles, which act as signal dependent coherent noise. $R_E(x,y)$ contains the time-dependent electronic noise due to the electronic components of the image processing equipment and $R_D(x, y)$ describes diffraction patterns of dust particles in the optical paths.

Of these degradations which are summarized in the additive noise $a(x, y)$ and the multiplicative noise $b(x, y)$ of (4.1), the speckles have been treated in Sec. 2.5, the others are described in the sequel.

4.2.1 Varying Background Illumination

Reasons for an uneven background illumination in the holographic interference patterns may be

- the beam profile of the illuminating laser beam as modified by the projecting optics, in most cases a Gaussian intensity profile,
- an uneven reflectivity of the object surface caused by spatially varying surface characteristics,
- an uneven sensitivity of the sensor (CCD-target) used for recording the interference pattern,
- additional diffraction patterns and parasitic interferences caused by dust particles in the optical paths.

The varying background illumination influences the maximal intensity which can be reached at a point, this effect is modeled by the multiplicative contrast V, as well as it contributes to the additive background I_0.

The background variations caused by expanded and projected laser beams have a low frequency but those caused by an uneven reflectivity of the object surface or by varying sensor sensitivity may occur with high spatial frequencies.

Spurious diffraction patterns should be avoided already in the experimental process, since they lie in the same frequency bands as the desired interference pattern. So they cannot be filtered out easily by a bandpass filter with proper cutoff frequencies.

4.2.2 Electronic Noise

The *electronic noise* in photodetectors is recognized as a random fluctuation of the measured voltage or current and is caused by the quantum nature of matter. A first category of noise sources is due to the photodetector as an electronic component: noise is generated even without impinging light. Thermal noise, shot noise, generation-recombination noise and the $1/f$-noise fall into this category. The second category contains the additional noise generated when photons impinge on the detector, especially the photon noise [165].

Electronic noise is a sum of numerous random processes obeying different statistical laws, but the central limit theorem of probability theory states that the overall process will be directed by a Gaussian distribution. This theoretical statement has been confirmed experimentally [166].

Electronic noise is a temporally variant process, so its influence may be diminished by averaging over a sequence of interference patterns recorded at different time instants. The *signal-to-noise ratio* SNR will increase by $\sqrt{n}$, if n patterns are averaged.

In practical applications of holographic interferometry electronic noise plays a minor role as compared to the speckle noise. So normally no special care is deserved to electronic noise. In the model describing the content of the interference fringes (4.3) it is recognized in the additive term R_E.

4.2.3 Speckle Decorrelation

Since in holographic interferometry a wavefront change is to be measured, the reconstructing pupil samples slightly different portions of the two wavefronts. Only identical parts interfere, the rest, presenting themselves in the form of noise, lead to a fall in fringe visibility [85]. The random amplitudes of the wavefronts constitute the speckle pattern, see Sec. 2.5. For generation of a holographic interference pattern only identical speckle fields interfere [86]. So the diminishing of the fringe visibility first is caused by a *speckle decorrelation* due to a variation of the interference phase $\Delta\phi$ across an individual speckle and second by speckle decorrelation due to a displacement of the speckles.

Normally in holographic interferometry the fringe spacing is much greater than the speckle size, meaning approximately constant fringe order or equivalently constant interference phase over one speckle. But it can be proved that the correlation and thus the visibility goes to zero when the fringe spacing approaches the speckle size. So it was shown [86] that the visibility of the fringe pattern is proportional to the correlation coefficient, which in turn is proportional to a quantity Z. For this Z it can be derived

$$|Z| = \mathrm{sinc}(\pi/N) \tag{4.4}$$

where N is the number of speckles per fringe. Obviously for $N = 1$ we get $|Z| = 0$.

The speckle decorrelation due to an in-plane translation Δx, - this motion is the most interesting one in this circumstance, - can be determined by considering that the speckle pattern in the image plane moves in the opposite direction in proportion to the magnification of the viewing system. A movement whose amount is small compared with the resolution element size, changes the speckle pattern at a given point in the image plane also by a small amount. It can be shown that the speckle pattern at a point is totally decorrelated when the object is translated by an amount Δx equal to the resolution element diameter, namely

$$\Delta x = \frac{1}{m}\frac{f}{a}\lambda \tag{4.5}$$

where m is the magnification, a is the diameter of the viewing lens aperture, and f is its focal length. This shows the larger the magnification of the viewing system, the smaller the in-plane translation which already decorrelates the speckle pattern [86].

The influence of speckle decorrelation is contained in the parameter V standing for the visibility in (4.3).

4.2.4 Digitization and Quantization

Generally for a computer aided quantitative evaluation the reconstructed holographic interference pattern is recorded by a CCD-camera and stored in the computer memory in digital format. That means the recorded intensity is digitized into an array of $M \times N$ image points, the so called *pixels* and quantized into L discrete *gray-values*. The implications of this *digitization* and *quantization* are discussed in the following.

The numbers M and N of digitization set an upper bound to the density of the interference fringes to be recorded. While of course the optical resolution of the imaging system has an influence which can be investigated with the help of the concepts of the *point spread function* and the *modulation transfer function* [167] here only the role of the *sampling theorem* should be discussed. The sampling theorem, App. A.5, demands more than two detection points per fringe. But this assumes ideal point detectors while here we are dealing with finite sized detector elements with a unity fill ratio [168]. Furthermore the effective size of a pixel is not necessarily equal to its spacing, and is usually larger than its physical size. This is most likely due to charge leakage to neighboring elements. So despite of the claims of the sampling theorem one has to supply at least 3 to 5 pixels per fringe period to yield a reliable evaluation. This requirement must be fulfilled particularly where the highest slope of the interference phase in the pattern occurs.

If very dense holographic interference patterns are present, the resolution should be adjusted by zooming into partial patterns and evaluating several of them successively.

In the discussion of resolution and digitization one should always keep in mind that the speckle size must be far less, at least a factor of 10 smaller, than the smallest fringe period. Otherwise the fringes cannot be sampled reliably.

When intensity frames are acquired, the analog video signal is usually converted to a digital signal of discrete levels by an *analog-to-digital converter* (ADC) [169]. In practice a quantization into 8 bits corresponding to 256 gray-values or into 10 bits giving 1024 values are most common. Experience has shown that 8 bits for a reliable evaluation of holographic interference patterns are sufficient. For phase shifting evaluation the resulting error due to quantization into 8 bits was calculated to $3.59 * 10^{-4}$ wavelengths [169].

The quantization error is affected by the modulation depth of the signal. The modulation of the signal times the number of quantization levels yields the effective number of quantization levels. Therefore, the lower the modulation, the fewer the number of quantization levels which span the signal from peak to valley, and the greater the resulting error in the evaluated phase. In order to minimize this effect, the fringe intensity should cover as much of the detector's dynamic range as possible.

4.2.5 Environmental Distortions

Environmental distortions like vibrations, acoustic noise, and air turbulence may cause degradations of the holographic interferograms. Especially if a two reference beam configuration or the real-time method are employed, as necessary for the phase sampling evaluation described in Sec. 4.5, errors may be generated. So care has to be taken to isolate the holographic arrangement from vibration. This is normally done by vibration isolated tables. Equipment which undergoes any mechanical motion should be removed from the table. If the object is too big to be placed on an optical table, the vibrations may be eliminated by optical compensation with a reference mirror fixed to the object, Sec. 5.1.4. Acoustic noise of the ambient may be coupled directly or by air into the holographic arrangement, which should be avoided by performing the experiment in a quiet room. Air turbulence may change the refractive index distribution of air and thus the optical path lengths randomly. It should be reduced by covering the optical paths and keeping away air ducts or fans from the setup.

Vibrations, acoustic noise, or air turbulence modify the interference phase distribution to be measured. For long exposure times the fluctuating interference phase is integrated, which leads to a contrast degradation up to a demolition of the interference pattern. This can be avoided by taking short exposures, e. g. with a pulsed laser. Then the instantaneous state of the object is recorded including the influence of the distortions. If these affect different areas of the object surface differently, they are rarely to detect.

4.3 Fringe Skeletonizing

The *fringe skeletonizing* methods are computerized forms of the evaluation by manual fringe counting in photographs of the interferograms done at the desktop in the early days of holographic interferometry [170]. These methods assume that the local extrema of the intensity distribution correspond to the maxima and minima of the cosine-function (4.1). In this case the interference phase at pixels, where an intensity maximum or minimum is detected, is an even or odd integer multiple of π.

The methods for fringe skeletonizing can be divided into those based on *fringe tracking*, those related to *segmentation*, and those falling not naturally in one of these two categories. A general processing scheme for an evaluation by skeletonizing of a digitally recorded and stored holographic interference pattern consists of the following steps [164, 171, 172]:

1. Improvement of the signal-to-noise-ratio in the interference pattern by spatial and temporal filtering.
2. Specification of the boundary of the fringe pattern to be analyzed in the whole stored frame.
3. Extraction of the raw skeleton by fringe tracking, pattern segmentation, or another method.
4. Enhancement of the skeleton by linking together interrupted lines, by adding missing points, and by removal of artifacts, line crossings or interconnections. This step may be performed interactively.
5. Numbering of the fringes by attaching interference order numbers to them.
6. Interpolation of the interference phase distribution between the skeleton lines.
7. Calculation of the physical values to be measured from the interference phase distribution.

4.3.1 Pattern Preprocessing

Before attempting to extract the skeleton out of the stored holographic interference pattern, it is advisable to minimize the disturbances, some of which described in Sec. 4.2 [173]. Of special importance are the *shading correction* to compensate for an uneven background intensity and the *smoothing* for eliminating the speckle influence.

Smoothing of the speckle noise in the recorded interferogram is mainly done by *linear low-pass filtering.* This process is described by the *convolution* of the intensity pattern $I(x,y)$ with the *impulse response* or *convolution kernel* $h(x,y)$ of the filter.

$$\begin{aligned} I'(x,y) &= I(x,y) \star h(x,y) \\ &= \sum_{x'}\sum_{y'} I(x,y)h(x-x',y-y') \end{aligned} \tag{4.6}$$

Common kernels are 3×3-kernels, which modify the intensity of each pixel in dependece from its own intensity value and the intensity values of the neighboring pixels. Examples are the averaging filters having the kernels

$$h_1(x,y) = \begin{cases} 1/9 & \text{if } \; -1 \le x \le 1 \text{ and } -1 \le y \le 1 \\ 0 & \text{else} \end{cases} \tag{4.7}$$

or

$$h_2(x,y) = \begin{cases} 1/16 & \text{if } \; (x,y) \in \{(-1,-1),(-1,1),(1,-1),(1,1)\} \\ 1/8 & \text{if } \; (x,y) \in \{(-1,0),(1,0),(0,-1),(0,1)\} \\ 1/4 & \text{if } \; (x,y) = (0,0) \\ 0 & \text{else} \end{cases} \tag{4.8}$$

These kernels are written shortly

$$h_1 = \frac{1}{9}\begin{pmatrix} 1 & 1 & 1 \\ 1 & 1 & 1 \\ 1 & 1 & 1 \end{pmatrix} \quad \text{and} \quad h_2 = \frac{1}{16}\begin{pmatrix} 1 & 2 & 1 \\ 2 & 4 & 2 \\ 1 & 2 & 1 \end{pmatrix} \tag{4.9}$$

Filters with larger windows than 3×3 are possible. Some of them can also be achieved by repeated application of a 3×3-filter to the pattern.

The presented linear filters are examples for *nonrecursive* or *finite impulse response filters*, meaning a vanishing impulse response outside a finite interval. No information on the application of *recursive filters* on speckled holographic interferograms has been found in the literature.

The *median filter* is an example for a *nonlinear filter* for speckle suppression. The grey value of each pixel here is replaced by the median of the gray-values of the neighboring pixels. Thus isolated points are removed, but steps or ramps are not blurred as with the spatial averaging filters [174].

Further filter strategies like homomorphic filtering [175, 176, 177], geometric filtering [178, 179], or Wiener filtering [180] have been suggested. Nevertheless temporal averaging of multiple frames of the same interference pattern but with varied speckles always is superior to spatial filtering of a single frame [177].

Shading correction is the removal of the low-frequency illumination variation mainly due to the Gaussian intensity distribution of the expanded laser beam. It can be performed by a linear high-pass filter which is accomplished by subtraction of a low-pass filtered version of the pattern from its unfiltered version. It can be done by homomorphic filtering meaning to apply the Fourier transform, then to remove the low frequency components of the spatial frequency spectrum, and finally to transform back into the spatial domain by the inverse Fourier transform. If it is possible to record the background separately, that is, the illuminated object wavefront before interfering, the pattern can be divided pointwisely by the background intensity [174, 181]. The background intensity distribution also may be obtained by summing two anti-phase reconstructions of the interference pattern [174] or by fitting a Gaussian to the pattern by the method of least squares, with the exponential in the Gaussian being linearized by taking the logarithm.

A histogram modification [173] may compensate detector nonlinearities or may improve the visual appearance of the interferogram, which can be important for interaction by the user, but seems not to be a prerequisite for computerized skeletonizing algorithms.

For the automatic determination of the *boundary of objects* not filling the full frame, common boundary following algorithms will be confused by the dark interference fringes [171]. So one possible approach is to apply heavy low-pass filtering to the pattern, which will blur out the gaps between the bright interference fringes, and then to detect the edge of the blurred object by differentiating the image or simply by applying an appropriate threshold [171]. It would also be possible to use an algorithm for finding the convex hull of a set of points. Both of these approaches will have difficulties with irregular boundaries and with artifacts such as nonuniform illumination. The easiest method is to have the user interactively specifying some sort of simplified boundary [171].

4.3.2 Fringe Skeletonizing by Segmentation

The *segmentation* techniques divide the pixels of the interference pattern into different regions representing bright and dark fringes or representing ridges, valleys, and slopes in the gray-value landscape [164, 182, 183, 184, 185, 186]. The extraction of the skeleton based on segmentation mostly consists of the steps:

- Segmentation of intensity into maxima, minima and slopes.
- Enhancement of the regions.
- Production of the fringe skeleton.

A direct approach is the production of a binary pattern by taking the average intensity as threshold [183]. This requires a good background correction or one has to use an adaptive threshold algorithm. After binarization, the margins of the regions have to be smoothed, normally done by shrinking and expanding operations, also called erosion and dilatation operations. These reduce isolated points, which are remnants of speckles, fill in broken lines or truncate line artifacts. The skeleton then is formed either by finding the centers of the regions corresponding to the dark and bright fringes, which may be performed by thinning the regions to lines of one pixel width [185], or by defining the points of transitions from bright to dark regions and vice versa as skeleton points [183].

Another approach moves a digital filter over the digital pattern, which decides whether a pixel belongs to the center of a bright fringe [186], or classifies the pixels into those belonging to ridges, valleys, slopes, or being undecidable [164, 187]. Again a proper filtering has to improve the skeletons. Anisotropic filters are suitable for this task, they avoid the melting of regions belonging to different fringes [188]. On the other hand interrupted skeleton lines have to be repaired. An algorithm connecting pairs of points with this intent should recognize the indicators least distance, approximately the same direction, and the absence of line crossings. Some of the steps of an evaluation of a holographic interferogram by skeletonizing based on segmentation are demonstrated in Fig. 4.4. The object is a valve loaded by internal pressure. Fig. 4.4a shows the holographic interferogram together with an intensity profile along the dotted horizontal line. The enhanced intensity distribution, after averaging and shading correction, is given in a gray-scale display in Fig. 4.4b. The enhanced intensity is segmented, Fig. 4.4c, where the ridges are white, the valleys are gray, and unidentified pixels are black. This pattern is enhanced by region-growing and binary filtering; the result is given in Fig. 4.4d. These regions are then thinned to a skeleton as shown in Fig. 4.4e. Numbering and interpolation lead to a continuous phase distribution as the one given in Fig. 4.4f.

4.3.3 Skeletonizing by Fringe Tracking

In methods based on fringe tracking, the algorithm looks for neighboring pixels which correspond to local maxima or minima in the gray-value distribution [171, 189, 190, 191, 192, 193]. The first step in recognizing the fringes is to locate a starting point on each fringe. Only for a restricted number of interference patterns it is sufficient to traverse a line which is known to cut all fringes, and then to take the maxima on this line as the starting points [171]. Due to the manifold of possible holographic interference patterns usually the starting points are manually defined by the user. If there are locally parallel fringes in the pattern, straight lines of finite length perpendicular to these fringes can be constructed. Along these lines the starting points and the direction of search can be defined [184].

The tracking then either follows the curved ridges characterized by local intensity maxima or traces the boundaries between adjacent fringes by using derivatives of gray levels. Since the starting point is not necessarily at either end of the fringe, the program has to trace in one direction from the starting point and afterwards has to go back to the starting point and trace in the other direction. A search for the next point in the range from -90° to $+90^o$ relative to the forward direction or even a smaller range prevents from getting caught in small loops around insufficiently filtered speckles [171]. This results in an automatic check of five of the eight nearest neighbors of the current pixel for the maximum intensity.

The tracking procedure stops if a pixel is hit that is already marked as belonging to a skeleton line. Reaching the starting point either means to have correctly traced a closed fringe, or the process is erroneously caught in a small loop. Another reason for hitting

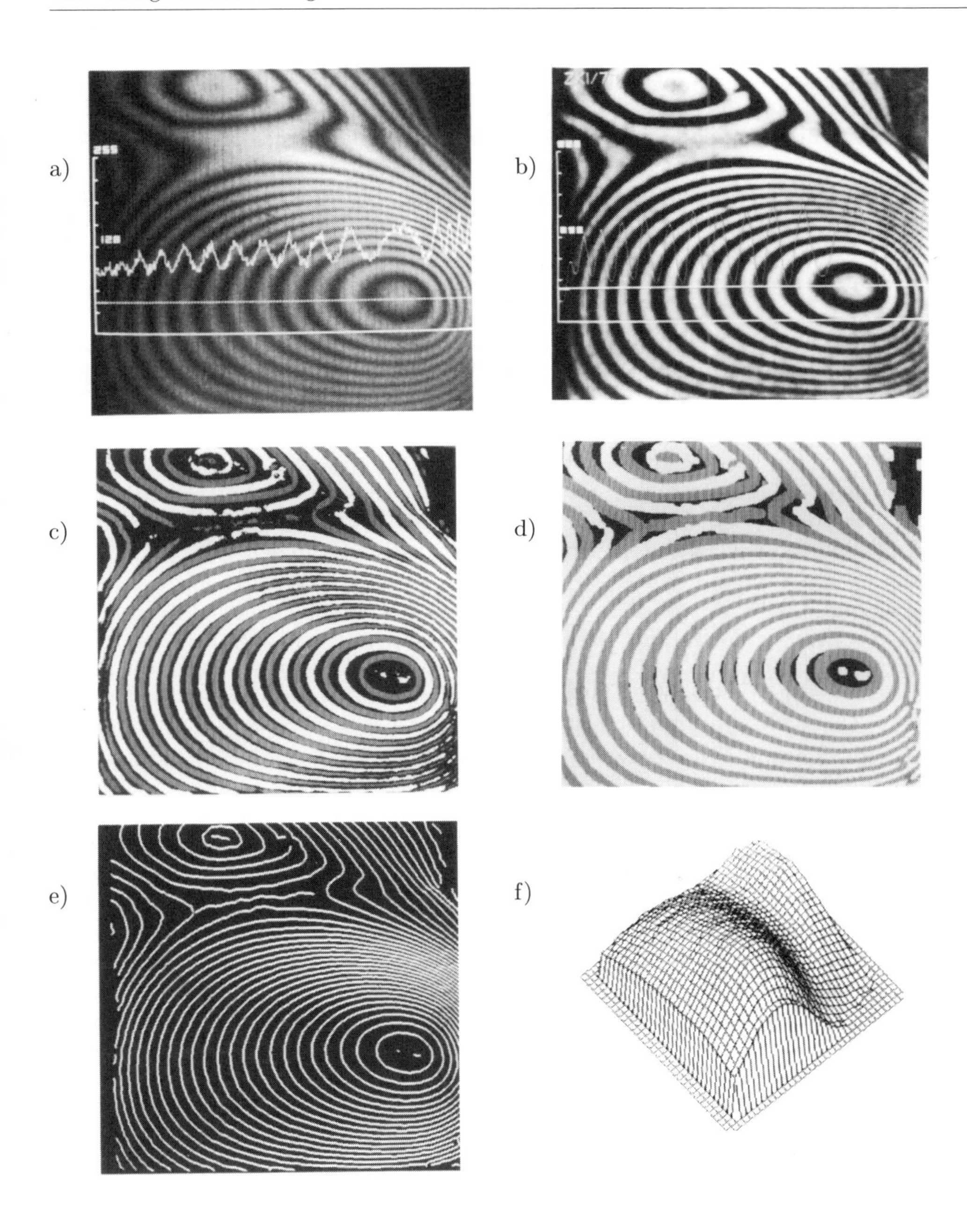

Figure 4.4: Interferogram evaluation by skeletonizing (Courtesy of W. Osten, BIAS)

an already marked skeleton point is the attempt to cross another fringe which normally is forbidden. In this case it has to be checked whether a bifurcation of the fringes or hyperbolically formed fringes due to a saddle point in the interference phase distribution have occurred. Most programs at this point require interaction by the user. Also the operator may link data points belonging together, continue if obstacles are met, correct wrong decisions, and finally checks whether no fringes have been overseen.

4.3.4 Other Fringe Skeletonizing Methods

The *phase lock method* uses a sinusoidal phase modulation obtained by, e.g., a piezoelectrically excited axially oscillating mirror [194]. The resulting intensity is written as

$$I(x, y, t) = a(x, y) + b(x, y) \cos[\Delta\phi(x, y) + L \sin \omega t]. \tag{4.10}$$

$L < \lambda/2$ is the amplitude and $\nu = \omega/2\pi$ the frequency of this oscillation. A bandpass filter, centered on $\sin \omega t$, determines the amplitude $U_\omega = 2b(x, y)J_1(L) \sin \Delta\phi(x, y)$, which is zero at the points (x, y), where $\Delta\phi(x, y) = N\pi$. These points can thus be detected and give a skeleton whose lines correspond to the interference phase differences of $\pi/2$.

A special fringe contour detection scheme is proposed in [195]. It only works with interferograms fulfilling the prerequisites: (1) the presence of a dominant *spatial frequency* associated with the fringe pattern, (2) the near invariance of this frequency with position. Then, along lines normal to the fringes, the one-dimensional *Fourier spectrum* is calculated by the *FFT algorithm.* The phase of the dominant spatial frequency computed for each image line is a quantitative measure of fringe displacement at each line.

Skeletonizing methods nowadays are used if there is no way to produce multiple phase shifted interferograms. An alternative in this case is the Fourier transform evaluation, Sec. 4.6, which in its general form also requires only a single interferogram. From the interference phase distribution determined by the Fourier transform algorithm, (4.52) to (4.54), a skeleton may be produced by taking only those pixels where the 2π-jumps occur. Even further intermediate skeleton lines are possible if the pixels nearest to certain phase values are selected. The evaluation of a single interferogram generally is lacking in the sign, but taking the absolute value circumvents this problem by producing two skeleton lines. As an example let the interference phase distribution $\Delta\phi(x, y)$ modulo 2π be in the interval $[-\pi, +\pi]$. Selecting all points (x, y) with $|\Delta\phi(x, y)| \approx \pi/2$ yields the two skeleton lines belonging to $-\pi/2$ and to $+\pi/2$, but no information, whether a part of a skeleton line belongs to $-\pi/2$ or to $+\pi/2$.

A rather new approach to skeletonizing uses concepts of *artificial neural networks* [196]. Points of the fringe skeleton are found by Kohonen's *self organizing feature map.* At the beginning a number of points, the *neurons*, are spread randomly over the interference pattern. The processing step consists of a random choice of an interferogram point, searching for the nearest neuron to this point, and moving this neuron towards the selected interferogram point. The amount of this motion is proportional to the distance between the point and the neuron, to the intensity of the interferogram point, and to an actual learning rate. This forces a higher probability for motion towards high intensity fringe centers than towards dark fringe areas. If this step is repeated sufficiently often, all neurons will concentrate at the bright fringe centers. If the number of neurons is high enough, the neurons in each fringe can be automatically connected by a nearest neighbor criterion without ambiguity, thus yielding the skeleton.

4.3.5 Fringe Numbering and Integration

After the skeleton lines are found we have to perform the *fringe numbering*, meaning to define a *fringe order* to each line. The integer fringe orders $n(x, y)$ correspond to the interference phase values $\Delta\phi(x, y)$ via $\Delta\phi(x, y) = 2\pi n(x, y)$. Even if we are not obliged to map the absolute fringe orders to the skeleton lines, see Sec. 4.1.2, the relations between the fringe orders must be fulfilled. So in particular local fringe order maxima and minima have to be uniquely detected. Generally, if a continuous interference phase distribution can be assumed, neighboring skeleton lines can differ in order only by -1, 0, or $+1$, lines of different order must not intersect or merge, lines do not end inside the field of view, and the skeleton line number differences, integrated along any closed line through the interferogram, always yield zero [183]. Automatic fringe numbering algorithms based on these constraints still may require manual interaction by the user [171, 183, 185, 186, 197]. A graph-theoretic approach to automatic fringe numbering is given in [198].

The problem is eased if a substantial degree of tilt can be added to the object deformation, leading to essentially parallel fringes with the measurement parameter being encoded as the deviation from straightness of the fringes [173]. The most important fact is that now the interference order behaves monotonically if moving roughly perpendicular to the skeleton lines, enabling an easy fringe numbering.

After assignment of the interference order to each skeleton line, the interference phase values are known along these lines, which represents a rather irregular distribution of points. To determine the values at all points of a regular grid, the interference phase values have to be interpolated for the grid points based on the phase values at the skeleton lines [199]. Three *interpolation* methods are presented in [200]:

Interpolation based on *one-dimensional splines* fits cubic polynomials to the skeleton points in horizontal and vertical directions through the grid point and gets the phase value for each grid point from the two spline values at this point.

In another approach at each grid point the four next neighboring skeleton points right and left as well as up and down are taken and from their phase values the interference phase at the grid point is calculated by *bilinear interpolation*.

Closely related to this scheme is the *interpolation by triangulation*, where the whole skeleton is covered with small triangles. Two vertices of each triangle lie on one skeleton line, the third vertex is on an adjacent skeleton line. The phase at each grid point is found by linear interpolation from the phases at the vertices of the triangle it is contained in.

A comparison of the resulting accuracy has shown that of these methods the interpolation by triangulation yields the highest accuracy [200].

4.4 Temporal Heterodyning

4.4.1 Principle of Temporal Heterodyning

The backbone of *temporal heterodyning* is the interference of two optical waves of different frequencies. As shown in 2.2.2, two mutually coherent harmonic waves, differing in frequency by $2\Delta f$, produce an intensity oscillating with the *beat frequency*, which equals the *frequency difference* $2\Delta f$.

To translate this principle into *heterodyne holographic interferometry*, the two interfering wavefields are both holographically reconstructed with different optical frequencies f_1 and f_2, or the reconstructed wavefield has another frequency than the reflected or refracted one [80, 82, 201, 202, 203, 204]. Due to the context with the interference phase, more often the angular frequencies $\omega_i = 2\pi f_i$, $i = 1, 2$ are employed.

The frequency of the holographically reconstructed wavefield is defined by the frequency of the reconstructing reference wave. To achieve a frequency shift between the two reconstructed wave fields in the *double exposure method*, they have to be recorded and reconstructed with a *two reference beam holography* arrangement as described in Sec. 3.2.2. Now the optical frequency of one of the two reference beams is shifted. This in most cases is done by a pair of *acoustooptical modulators* as is indicated in Fig. 4.5, which shows the typical holographic arrangement with two reference waves for performing the temporal heterodyne method.

The two reference beam method in conjunction with the double exposure holography is the standard for temporal heterodyning. *Real-time holographic interferometry* with a wave field reconstructed by a reference wave having a frequency shift relative to the wave illuminating and being reflected by the object is not feasible, due to the extreme stability requirements and the high sensitivity to distortions from outside.

The holographic interference pattern resulting from reconstruction with two reference waves having a mutual frequency shift of $\Delta\omega = \omega_2 - \omega_1$ is

$$I(x, y, t) = a(x, y) + b(x, y) \cos[\Delta\phi(x, y) + t\Delta\omega] \tag{4.11}$$

where $a(x, y)$ and $b(x, y)$ are the additive and multiplicative distortions and $\Delta\phi(x, y)$ is the interference phase distribution to be determined. If the frequency offset $\Delta\omega/2\pi$ was adjusted low enough to be resolved by opto-electronic sensors, say < 100 MHz, the interference phase can be measured with high accuracy by an electronic phasemeter. One would wish a two-dimensional sensor which should simultaneously detect oscillations of some MHz at a high number of pixels. Unfortunately such a sensor does not exist, therefore point-sensors fulfilling the demanded temporal resolution have to be scanned mechanically over the reconstructed real image.

The phase $\Delta\phi(x, y) + t\Delta\omega$ of a single oscillation bears no information, but from the phase difference of the oscillating intensities of two points

$$\Delta\Delta\phi(x_1, y_1; x_2, y_2) = [\Delta\phi(x_1, y_1) + t\Delta\omega] - [\Delta\phi(x_2, y_2) + t\Delta\omega] \tag{4.12}$$

the relative change $\Delta\Delta\phi$ in the interference phase $\Delta\phi$ can be determined. Therefore two detectors have to record the oscillating intensity at different points in the real image

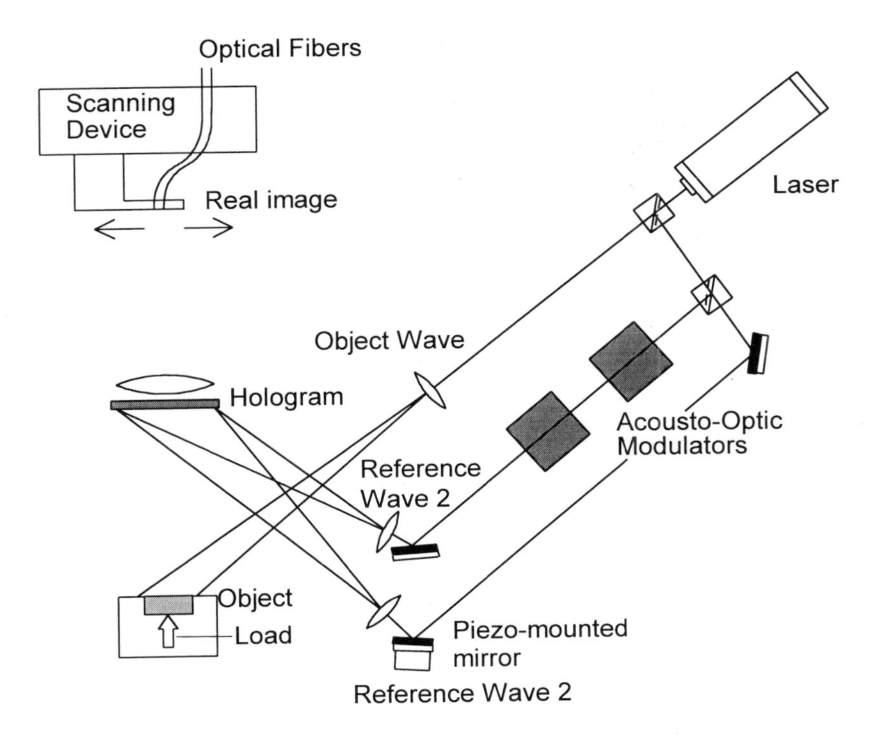

Figure 4.5: Two reference beam arrangement for double exposure temporal heterodyning

and an electronic phasemeter measures the interference phase difference $\Delta\Delta\phi$ modulo 2π between these two points. The interference phase differences can be summed up from point to point to yield the interference phase distribution along a line or across a plane.

4.4.2 Technical Realization of Temporal Heterodyning

Principally there are two ways to perform the temporal heterodyne evaluation. One is to evaluate the interference phase difference by keeping one detector fixed to a reference point whilst the other is scanned. This way one measures the phase difference modulo 2π with respect to the phase subsisting at the reference point. The other method is to scan a pair, triple, quadruple, or quintuple of photodetectors over the real image. In this way phase differences $\Delta\Delta\phi_x$ and $\Delta\Delta\phi_y$ between adjacent points with known separations are measured. Of course it is assumed that the x- and y-directions coincide with the scanning directions and the geometric arrangement of the photodetectors. In practice not the detectors are scanned over the real image, but the ends of optical fibers, which transmit the light to photodiodes. The signals of the photodiodes are amplified and after passing a narrowband filter are fed to the phasemeter. A technical realization of a detector with five optical fiber ends about 1 mm apart is shown in Fig. 4.6.

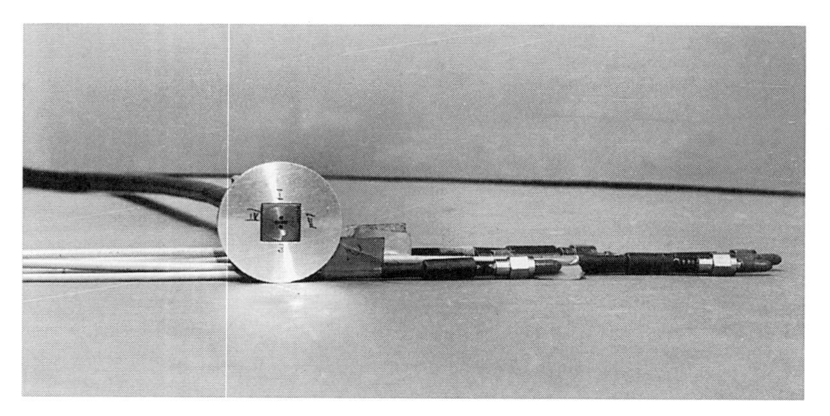

Figure 4.6: Five-fiber detector for temporal heterodyning (Courtesy of R. J. Pryputniewicz, WPI)

The determination of the mutual phase shift between two oscillating signals is done by the simultaneous measurement of the delay Δt between the time instants of the signals passing a *trigger level* in increasing direction and the period T of the oscillation. The phase shift difference $\Delta\Delta\phi$ then is calculated by

$$\Delta\Delta\phi = \frac{\Delta t}{T} 2\pi \tag{4.13}$$

A better accuracy and resolution is achieved by averaging over n periods, Fig. 4.7.

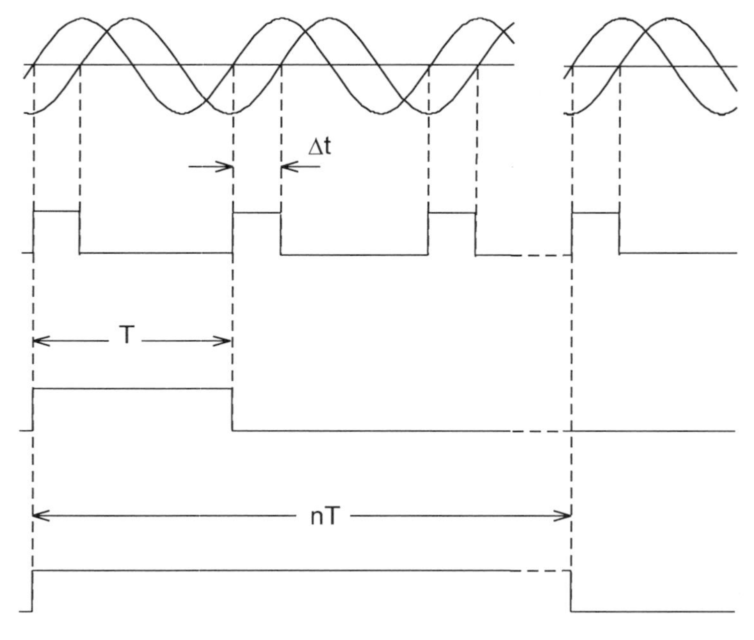

Figure 4.7: Measurement of time delay in temporal heterodyning

$$\Delta\Delta\phi = \frac{2\pi}{nT}\sum_{i=1}^{n}\Delta t_i \tag{4.14}$$

Most phasemeters allow a control of the measuring time t_m, then the number n of averaged periods depends on the beat frequency Δf by $n = \Delta f \; t_m$.

The interference phase distribution determined by temporal heterodyning is correct in the sign as long as the sampling theorem is fulfilled. The solution of the *sign ambiguity* is due to the phasemeter's ability to discriminate positive and negative phase shifts in the interval $]-\pi,\pi]$. The sampling theorem requires that the scanning steps, and the distance of the detectors if detector pairs are scanned, are smaller than half a fringe period. Conversely the fringe density of measurable interference patterns is bounded by the detector separation and the scanning steps.

4.4.3 Errors of Temporal Heterodyning

Temporal heterodyning measures interference phase differences by the phase shift between two signals oscillating with the beat frequency independently of the local mean intensity or local contrast. Electronic phase measurement has a resolution of typically 1/1000 of the full cycle 2π. Thus one may claim a resolution of 1/1000 of a fringe at least theoretically. But a number of systematic and statistical errors make these figures difficult to reach.

Good results presume a proper control of the trigger levels and nearly the same amplitudes of the signals to be compared. The trigger levels of both signals should be equal and near zero. Unequal trigger levels lead to false measured intervals $\Delta t'$ instead of the correct Δt, see Fig. 4.8.

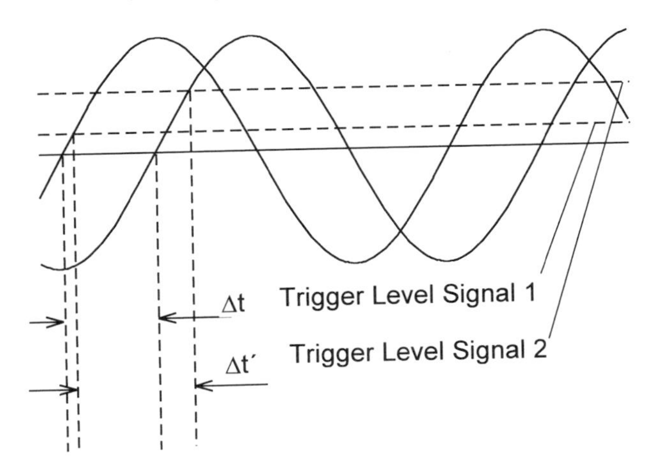

Figure 4.8: Error by unequal trigger levels

Different amplitudes of the signals lead to systematic errors even for identical trigger levels, Fig. 4.9. Due to the hysteresis band a trigger level adjusted to 0 mV can obtain ± 5 mV in reality. This leads to a maximal systematic phase error for sine waves of

$$\arcsin\left(\frac{5\text{mV}}{U_{max1}}\right) - \arcsin\left(\frac{5\text{mV}}{U_{max2}}\right) \tag{4.15}$$

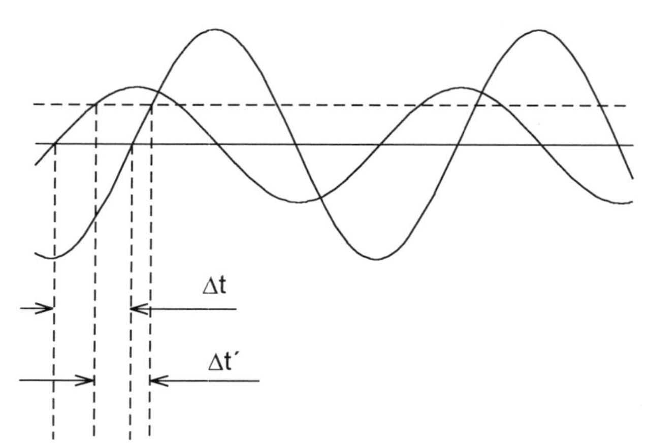

Figure 4.9: Error by unequal amplitudes

which may induce phase errors up to 1.5°. Unequal intensities between reference point and actually scanned point are always present in holographic interferometry, at least due to speckles or spatially varying reflection.

The influence of a frequency drift during the measuring time caused by improper driving of the frequency shifting components has been tested by numerical simulations [205]. These have shown that the error caused by frequency drift is small compared to the aforementioned errors.

A systematic error can arise with the scanning of detector pairs, if the scanning steps do not agree exactly with the detector spacing. The individual errors depend on the local slope of the interference phase distribution, they cumulate in the subsequent integration process. If the detector spacing and the scanning steps can be measured exactly, even if they do not agree, the resulting error can be corrected by numerical interpolation [205].

The most critical aspect of temporal heterodyning is the required high mechanical stability during the scanning of the reconstructed interference pattern. On the one hand the mechanical advance of the detectors can be achieved only with limited resolution, on the other hand scanning the whole pattern, especially along many parallel lines, lasts a long time during which environmental influences like noise, vibrations or temperature changes may falsify the measuring results.

These errors are nearly the same independent whether the double exposure method with two reference beam holography or the real-time method is employed. In the first case mutual variations of the optical path lengths cause the errors, while in the second case mutual variations between reference wave and reflected object wave are responsible. So after each mechanical scanning step the evaluation process has to wait until the whole arrangement has come to rest. Experiments have shown that the statistical errors occurring during the scanning are the most severe, especially if they go into the integrating process to calculate the overall interference phase distribution from the interference phase differences.

A long-term drift of the interference phase during the scanning process does not affect the result as long as this drift influences all evaluated points in the same way. This is because only interference phase differences are measured. But a temporal change of the interference phase which modifies the evaluation points differently may cause errors. In this regard the method with a fixed reference point is more critical than the method of simultaneously scanning several sensors if the phase difference at points far apart get different amounts of interference phase change. On the other hand measuring phase differences at neighboring points with scanned sensor pairs is much more sensitive to short-time errors, since these cumulate in the subsequent integration process.

4.4.4 Experimental Application of Temporal Heterodyning

Experiments employing temporal heterodyning have been carried out with the two reference beam holographic arrangement, shown schematically in Fig. 4.5 and in a photograph in Fig. 4.10. The arrangement consists of a 35mW He-Ne-laser, whose beam is splitted in

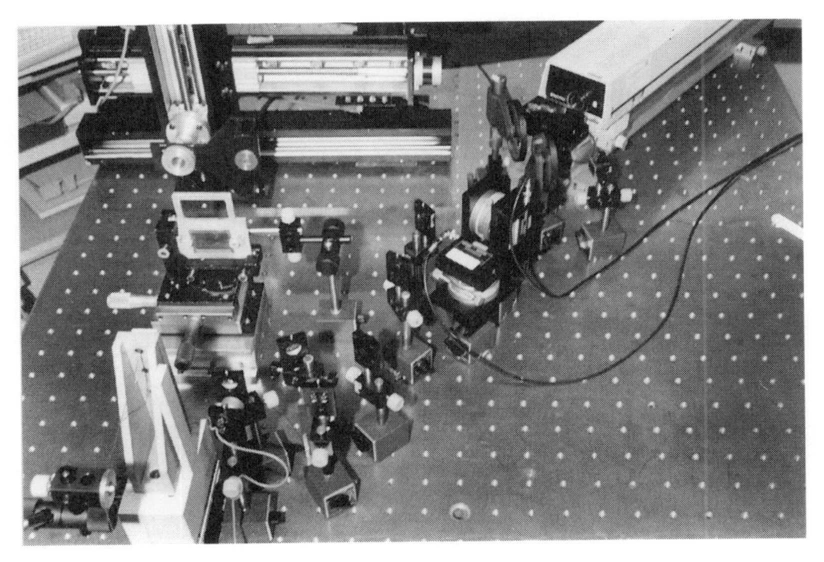

Figure 4.10: Two reference beam arrangement for double exposure temporal heterodyning

one object illumination wave and two reference waves. The two acoustooptical modulators are driven one with 40 MHz, the other with -40.08 MHz, producing a beat frequency of 80 kHz. The negative sign indicates the -1. diffraction order of the modulators, while the positive sign denotes the +1. order. The object is a bended cantilever, Fig. 4.11. The intensities are registered by the ends of two of the five optical fibers, Fig. 4.6, and transmitted to photodiodes. Their signals go via an amplifier and filter to the phasemeter, the evaluated phase difference is transmitted via IEEE-interface to the computer, which not only collects and evaluates the phase data, but also controls the stepper motors of the scanning device.

In this way 36 interference phase differences of point pairs have been recorded. This number is determined by the length of the object in the real image and the fixed distance

Figure 4.11: Test object

of the fiber ends. Integration yields the continuous interference phase distribution which is proportional to the deflection curve of the cantilever beam shown in Fig. 4.12. The

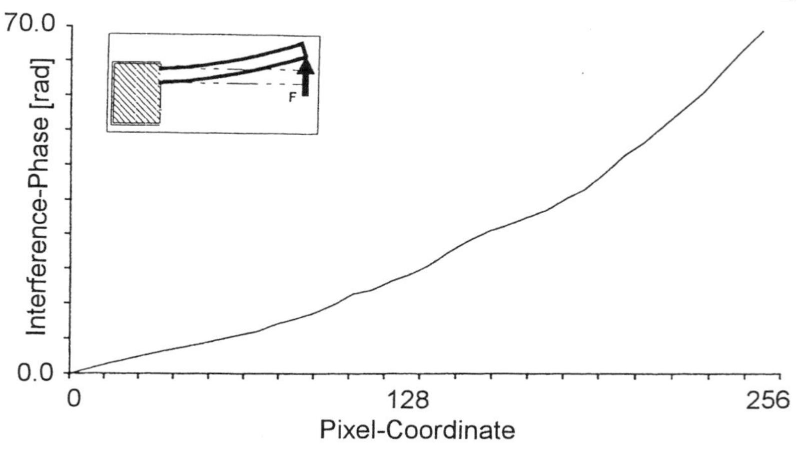

Figure 4.12: Evaluated interference phase by temporal heterodyning

small deviations from the theoretical deflection line are caused by air turbulences during the measuring time. After each scanning step some seconds have been waited until all mechanics came to rest. The errors cumulate to about 8% of the whole deflection [206].

4.5 Phase Sampling Evaluation

Although the heterodyne method is the widely spread state of the art in interferometric length measurement, in holographic interferometry it has only found limited application. This is mainly due to the necessity of sensors with high bandwidth, which is only fulfilled by point detectors. On the other hand the two-dimensional holographic interferograms appeal for image detectors like TV tubes or CCD arrays which do not reach the required temporal bandwidth. A way out of this dilemma is offered by the *phase step* or the *phase shift methods* [54].

The frequency shift in one of the interfering light waves of the heterodyne method can be envisaged as a continuous shift of the mutual phase between the light waves. Now this phase may be varied very slowly, in the extreme case stepwise, so that the intensity can be sampled corresponding to different values of this reference phase. The intensity distributions $I_n(x, y)$ recorded in this way are expressed by the so called *phase sampling equation*

$$I_n(x, y) = a(x, y) + b(x, y) \cos[\Delta\phi(x, y) + \phi_{Rn}] \qquad n = 1, \ldots, m \qquad m \geq 3 \tag{4.16}$$

where $a(x, y)$ and $b(x, y)$ are the additive and multiplicative distortions, $\Delta\phi(x, y)$ is the interference phase distribution to be determined, and ϕ_{Rn} is the shifted reference phase belonging to the n-th intensity distribution $I_n(x, y)$.

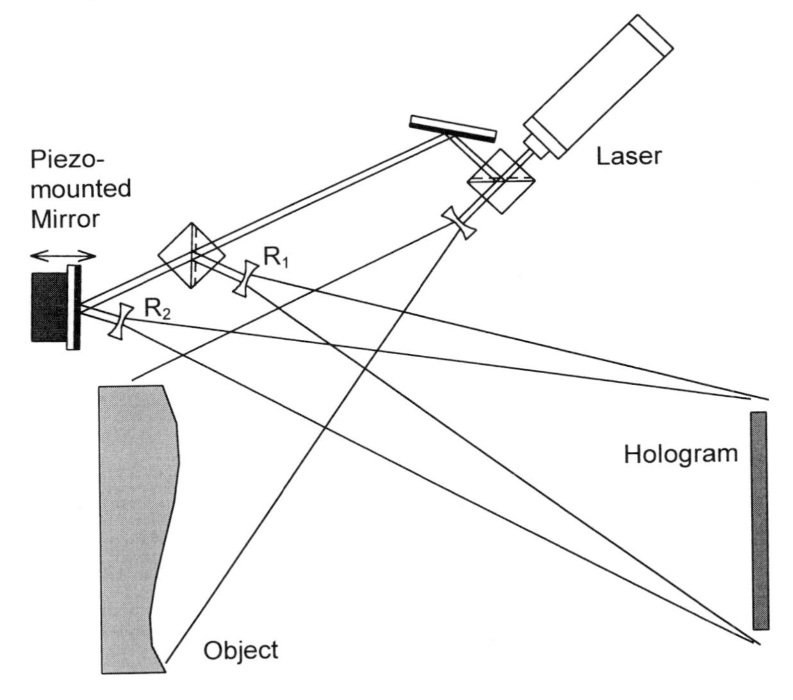

Figure 4.13: Two reference beam arrangement for double exposure phase sampling evaluation

The phase step and phase shift methods which record and evaluate a set of intensity distributions (4.16) represent the widely spread state of the art in the automatic evaluation of interference patterns especially of the holographic interference patterns. Due to the redundant information contained in the I_n's the interference phase is calculated with high accuracy at all pixels of the interference pattern and without sign ambiguity. Due to the above mentioned conceptual relation to heterodyning these methods sometimes are called *quasi heterodyne methods* [82, 207, 208, 209, 210, 211].

The different components one can use to perform the phase shifts are described together with the other components of the holographic setup in Sec. 2.7. A frequently used option is a mirror, which reflects the reference wave, mounted on a piezo crystal to shift the mirror by fractions of the used wavelength. This is depicted in Figs. 4.13 and 4.14 to indicate any phase shifting component.

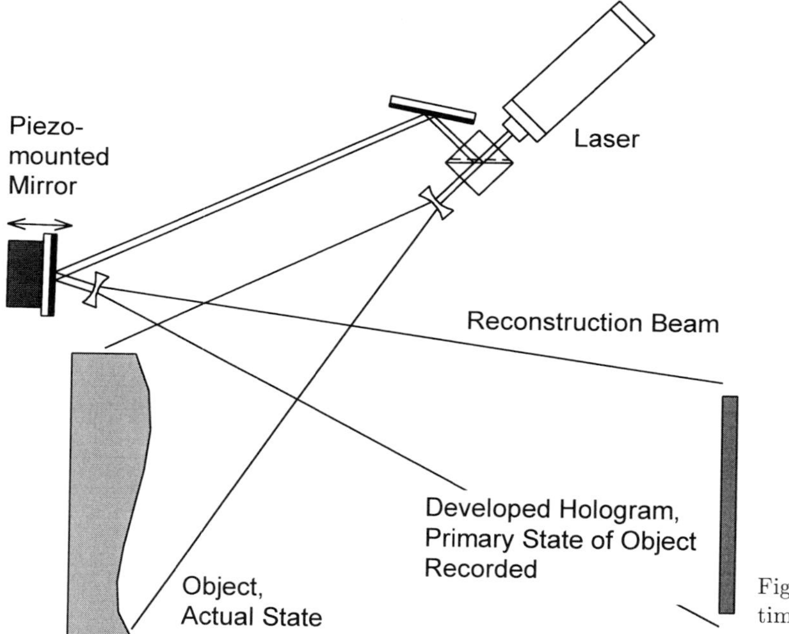

Figure 4.14: Real-time method for phase sampling evaluation

Since the holographically reconstructed wave field has an optical phase distribution defined by the phase of the reconstructing reference wave, phase shifting the reference wave during reconstruction of a double exposure hologram would shift both reconstructed images in phase and so no effect on the interference fringes will be seen. Therefore in the *double exposure method* of holographic interferometry the two states have to be recorded with different reference waves and reconstructed by both of them simultaneously, but this approach brings all the problems of *two reference beam holography*, see Sec. 3.2.2. The phase shift here is introduced by shifting only one of the two reference waves, Fig. 4.13.

The other way is to employ the real-time method, Fig. 4.14. The wave field coming directly from the object is not affected by the phase shift in the reference wave which only modifies the optical phase of the reconstructed wave field. Thus the reference phase ϕ_R

in (4.16) can be varied.

In most applications the phase shifted interferograms are recorded subsequently. The simultaneous recording of phase shifted interferograms, e. g. when measuring *transient events*, is enabled by the introduction of a *diffraction grating* in the object beam [57, 212]. The fringe patterns are obtained from direct interference of wavefronts propagating in the n-th diffraction order directions behind the grating and the hologram. For phase shifting the grating is shifted transversely between the two exposures of double exposure holography or between recording and reconstruction when employing real-time holographic interferometry, Sec. 2.7.4.

4.5.1 Phase Shifting and Phase Stepping

Although it is possible to perform arbitrary phase shifts and recognize these in the evaluation, it is recommended to use constant phase steps $\Delta\phi_R = \phi_{Rn+1} - \phi_{Rn}$ to keep the analysis easy. The phase can be shifted linearly in time, Figs. 4.15a and b, or in discrete

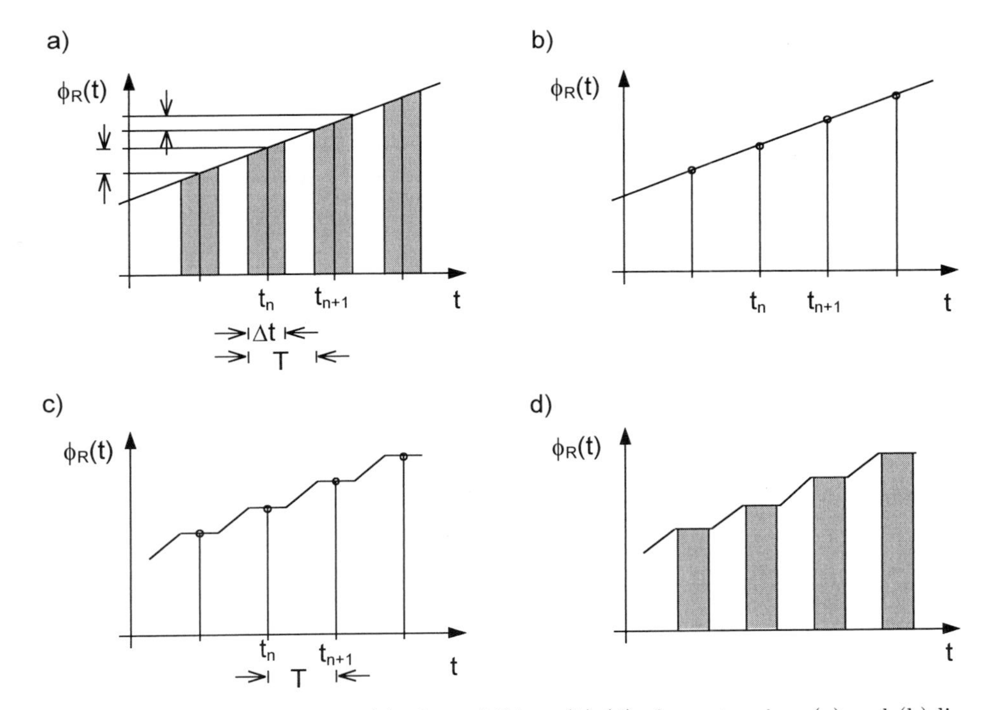

Figure 4.15: Phase shift vs. time. (a) phase shifting, (b)-(d) phase stepping, (a) and (b) linear phase shift, (c) and (d) stepwise phase shift

steps, Figs. 4.15c and d. In practical applications, the time delay T between the recording of the phase shifted intensities should be as short as possible, e. g. only limited by the video frequency. For short T spurious vibrations induced by the environment have a minimal influence on the interference phase, in a rather worse case they can be modeled as an additional constant phase shift.

Theoretically one may discriminate between *phase shifting*, Fig. 4.15a, where each intensity is integrated over the time interval Δt during which the phase varies linearly, and *phase stepping*, Figs. 4.15b to d, where we have fixed phase values as assumed in (4.16). The result of integration in phase shifting can be calculated

$$I_n(x,y) = \frac{1}{\Delta t} \int_{t_n-\Delta t/2}^{t_n+\Delta t/2} a(x,y) + b(x,y)\ \cos[\Delta\phi(x,y) + \phi_R(t)]\ dt \tag{4.17}$$

Assuming a linear variation of $\phi_R(t)$ with t, this integral can be evaluated applying a substitution and the formula $\sin(\alpha+\beta) - \sin(\alpha-\beta) = 2\cos\alpha\sin\beta$ to yield

$$I_n(x,y) = a(x,y) + \mathrm{sinc}\left(\frac{\Delta\phi_R}{2}\right) b(x,y)\ \cos[\Delta\phi(x,y) + \phi_{Rn}] \tag{4.18}$$

This expression corresponds to (4.16), only the contrast term $b(x,y)$ is modified by the constant factor $\mathrm{sinc}(\Delta\phi_R/2)$. Here $\Delta\phi_R$ denotes the phase shift during the time interval Δt. In this sense phase shifting is equivalent to phase stepping, both names in the following are used synonymously. One can see that for $\Delta\phi_R = 0$, the phase stepping of Figs. 4.15b to d, the sinc-function has a value of one. At the other extreme, for sampling over a whole period of $\Delta\phi_R = 2\pi$ the sinc is zero, we get no intensity modulation at all.

4.5.2 Solution of the Phase Sampling Equations with Known Phase Shifts

If several intensity distributions $I_n(x,y)$ with mutual phase shifts $\Delta\phi_R = \phi_{Rn+1} - \phi_{Rn}$ are recorded, a nonlinear system of equations of the form (4.16) has to be solved for each point (x,y). As long as the phase shifts are known, there remain three unknowns: $a(x,y)$, $b(x,y)$ and the desired interference phase $\Delta\phi(x,y)$. So at least $m = 3$ equations are necessary, meaning that at least three phase shifted intensities have to be recorded.

For solution of the nonlinear system of equations (4.16) the Gaussian least squares approach is used [213] by introducing

$$u(x,y) = b(x,y)\cos[\Delta\phi(x,y)] \qquad \text{and} \qquad v(x,y) = -b(x,y)\sin[\Delta\phi(x,y)] \tag{4.19}$$

Herewith (4.16) is rewritten, with the (x,y) omitted for clarity

$$\begin{aligned} I_n &= a + b\ \cos[\Delta\phi + \phi_{Rn}] \\ &= a + u\ \cos\phi_{Rn} + v\ \sin\phi_{Rn} \end{aligned} \tag{4.20}$$

The sum of the quadratic errors

$$\sum_{n=1}^{m} (a + u\ \cos\phi_{Rn} + v\ \sin\phi_{Rn} - I_n)^2 \tag{4.21}$$

has to be minimized. Partial differentiation of this functional with respect to a, u, and v and equating the derivatives to zero gives the linear system of three equations

$$\begin{pmatrix} m & \sum \cos\phi_{Rn} & \sum \sin\phi_{Rn} \\ \sum \cos\phi_{Rn} & \sum \cos^2\phi_{Rn} & \sum \sin\phi_{Rn}\cos\phi_{Rn} \\ \sum \sin\phi_{Rn} & \sum \sin\phi_{Rn}\cos\phi_{Rn} & \sum \sin^2\phi_{Rn} \end{pmatrix} \begin{pmatrix} a \\ u \\ v \end{pmatrix} = \begin{pmatrix} \sum I_n \\ \sum I_n \cos\phi_{Rn} \\ \sum I_n \sin\phi_{Rn} \end{pmatrix} \tag{4.22}$$

This system has to be solved pointwise for u and v, while the solution for a may be omitted. Having found $u(x,y)$ and $v(x,y)$, the interference phase is determined modulo 2π by

$$\Delta\phi(x,y) = \arctan\frac{-v(x,y)}{u(x,y)} \tag{4.23}$$

The pointwise solution of system (4.22) requires only a single inversion of the matrix to the left, provided the phase steps $\Delta\phi_R$ are the same for all pixels [55]. Simple solutions of (4.22) are given for phase steps which are constant and have special values like 30°, 45°, 60°, or 90° because the sines and cosines of these values are easy to handle. The general scheme for deriving the evaluation equation is given in the following at the example for $m = 4$ intensities with constant phase shift $\Delta\phi_R = 90^o$ and a starting value of $\phi_{R1} = 0^o$. The elements of the matrix in (4.22) are calculated by

	n	ϕ_{Rn}	$\sin\phi_{Rn}$	$\cos\phi_{Rn}$	$\sin^2\phi_{Rn}$	$\cos^2\phi_{Rn}$	$\sin\phi_{Rn}\cos\phi_{Rn}$
	1	0°	0.	1.	0.	1.	0.
	2	90°	1.	0.	1.	0.	0.
	3	180°	0.	-1.	0.	1.	0.
	4	270°	-1.	0.	1.	0.	0.
Σ			0.	0.	2.	2.	0.

The system of equations (4.22) now is

$$\begin{pmatrix} 4 & 0 & 0 \\ 0 & 2 & 0 \\ 0 & 0 & 2 \end{pmatrix} \begin{pmatrix} a \\ u \\ v \end{pmatrix} = \begin{pmatrix} I_1 + I_2 + I_3 + I_4 \\ I_1 - I_3 \\ I_2 - I_4 \end{pmatrix} \tag{4.24}$$

Its solution $u = (I_1 - I_3)/2$, $v = (I_2 - I_4)/2$ gives the interference phase modulo 2π at each point (x,y) by

$$\Delta\phi(x,y) = \arctan\frac{I_4(x,y) - I_2(x,y)}{I_1(x,y) - I_3(x,y)} \tag{4.25}$$

More such formulas, all derived by the same scheme, are given in Table 4.1.

Table 4.1: Phase shift methods for known phase shifts

Phase shift	Start	m	Evaluation formula
30°	0°	3	$\Delta\phi = \arctan \frac{(3\sqrt{3}-5)I_1 + (\sqrt{3}-2)I_2 + (7-4\sqrt{3})I_3}{(5-3\sqrt{3})I_1 + (2\sqrt{3}-3)I_2 + (\sqrt{3}-2)I_3}$
30°	-30°	3	$\Delta\phi = \arctan \frac{I_1 - I_3}{(2+\sqrt{3})(-I_1 + 2I_2 - I_3)}$
45°	0°	3	$\Delta\phi = \arctan \frac{(2+\sqrt{2})I_1 - (2+2\sqrt{2})I_2 + \sqrt{2}I_3}{-\sqrt{2}I_1 + (2+2\sqrt{2})I_2 - (2+\sqrt{2})I_3}$
45°	-45°	3	$\Delta\phi = \arctan \frac{\sqrt{2}(I_3 - I_1)}{(2+\sqrt{2})(I_1 - 2I_2 + I_3)}$
60°	0°	3	$\Delta\phi = \arctan \frac{2I_1 - 3I_2 + I_3}{\sqrt{3}(I_2 - I_3)}$
60°	0°	4	$\Delta\phi = \arctan \frac{5(I_1 - I_2 - I_3 + I_4)}{\sqrt{3}(2I_1 + I_2 - I_3 - 2I_4)}$
60°	0°	5	$\Delta\phi = \arctan \frac{\sqrt{3}(2I_1 - 3I_2 - 4I_3 + 5I_5)}{8I_1 + 3I_2 - 4I_3 - 6I_4 - I_5}$
90°	0°	3	$\Delta\phi = \arctan \frac{I_1 - 2I_2 + I_3}{I_1 - I_3}$
90°	45°	3	$\Delta\phi = \arctan \frac{I_3 - I_2}{I_1 - I_2}$
90°	0°	4	$\Delta\phi = \arctan \frac{I_4 - I_2}{I_1 - I_3}$
90°	0°	5	$\Delta\phi = \arctan \frac{7(I_4 - I_2)}{4I_1 - I_2 - 6I_3 - I_4 + 4I_5}$
90°	-180°	5	$\Delta\phi = \arctan \frac{2(I_2 - I_4)}{-I_1 + 2I_3 - I_5}$
120°	0°	3	$\Delta\phi = \arctan \frac{\sqrt{3}(I_3 - I_2)}{2I_1 - I_2 - I_3}$
α	$-\alpha$	3	$\Delta\phi = \arctan \frac{(1-\cos\alpha)(I_3 - I_1)}{\sin\alpha(I_1 - 2I_2 + I_3)}$
$\frac{2\pi k}{m}$	0°	m	$\Delta\phi = \arctan \frac{-\sum_{i=1}^{m} I_i \sin[\frac{2\pi k}{m}(i-1)]}{\sum_{i=1}^{m} I_i \cos[\frac{2\pi k}{m}(i-1)]}$

The so called *2+1-technique* [214] is contained implicitly in Table 4.1. In this technique the first two intensities $I_1(x,y)$ and $I_2(x,y)$ are taken very quickly with a $\Delta\phi_R = 90^o$ phase shift between them. If these two images are taken on either side of the interline transfer in a standard CCD video camera, 1 ms exposures can be taken as quickly as 1 μs apart. This will freeze the most vibrations or air turbulence affecting the measurement. The third intensity is the background $I_3'(x,y) = a(x,y)$ which can be acquired at any time, and may be averaged over many noise cycles. Considering $m = 3$, $\phi_{R1} = 0^o$, $\Delta\phi_R = 90^o$, we have the three intensities

$$\begin{aligned} I_1(x,y) &= a(x,y) + b(x,y)\cos\Delta\phi(x,y) \\ I_2(x,y) &= a(x,y) + b(x,y)\cos[\Delta\phi(x,y) + 90^o] \\ I_3(x,y) &= a(x,y) + b(x,y)\cos[\Delta\phi(x,y) + 180^o] \end{aligned} \tag{4.26}$$

Now one observes that $I_3'(x,y) = [I_1(x,y) + I_3(x,y)]/2$, so replacing $I_3(x,y)$ by $2I_3'(x,y) - I_1(x,y)$ in the evaluation formula, Table 4.1, we get

$$\Delta\phi(x,y) = \arctan\frac{I_1 - 2I_2 + I_3}{I_1 - I_3} = \arctan\frac{I_3' - I_2}{I_1 - I_3'} \tag{4.27}$$

It is not necessary to take $\Delta\phi_R = 90^o$, because for the system of equations

$$\begin{aligned} I_1(x,y) &= a(x,y) + b(x,y)\cos\Delta\phi(x,y) \\ I_2(x,y) &= a(x,y) + b(x,y)\cos[\Delta\phi(x,y) + \Delta\phi_R] \\ I_3(x,y) &= a(x,y) + b(x,y)\cos[\Delta\phi(x,y) + 180^o] \end{aligned} \tag{4.28}$$

we have the solution

$$\Delta\phi(x,y) = \arctan\frac{I_1(1+\cos\Delta\phi_R) - 2I_2 + I_3(1-\cos\Delta\phi_R)}{(I_1 - I_3)\sin\Delta\phi_R} \tag{4.29}$$

which now is replaced by

$$\Delta\phi(x,y) = \arctan\frac{I_1\cos\Delta\phi_R - 2I_2 + I_3'(1-\cos\Delta\phi_R)}{(I_1 - I_3')\sin\Delta\phi_R} \tag{4.30}$$

4.5.3 Solution of the Phase Sampling Equations with Unknown Phase Shifts

With an unknown but constant phase shift $\Delta\phi_R$ we have an additional unknown, so that at least $m = 4$ phase shifted intensities have to be recorded [215]. The first case to be treated in the following is for $m = 4$. For ease of computation without loss of generality a starting value of $\phi_{R1} = -3\,\Delta\phi_R/2$ is assumed. Then the four equations are

$$\begin{aligned} I_1 &= a + b\cos(\Delta\phi - \frac{3\Delta\phi_R}{2}) = a + b\cos\Delta\phi\cos\frac{3\Delta\phi_R}{2} + b\sin\Delta\phi\sin\frac{3\Delta\phi_R}{2} \\ I_2 &= a + b\cos(\Delta\phi - \frac{\Delta\phi_R}{2}) = a + b\cos\Delta\phi\cos\frac{\Delta\phi_R}{2} + b\sin\Delta\phi\sin\frac{\Delta\phi_R}{2} \\ I_3 &= a + b\cos(\Delta\phi + \frac{\Delta\phi_R}{2}) = a + b\cos\Delta\phi\cos\frac{\Delta\phi_R}{2} - b\sin\Delta\phi\sin\frac{\Delta\phi_R}{2} \\ I_4 &= a + b\cos(\Delta\phi + \frac{3\Delta\phi_R}{2}) = a + b\cos\Delta\phi\cos\frac{3\Delta\phi_R}{2} - b\sin\Delta\phi\sin\frac{3\Delta\phi_R}{2} \end{aligned} \tag{4.31}$$

With $S_1 = I_1 + I_4$, $S_2 = I_2 + I_3$, $S_3 = I_1 - I_4$, $S_4 = I_2 - I_3$, $u = 2b\cos\Delta\phi$, $v = 2b\sin\Delta\phi$, $w = \cos(\Delta\phi_R/2)$ (4.25) simplifies to

$$\begin{aligned} S_1 &= 2a + u(4w^3 - 3w) \\ S_2 &= 2a + uw \\ S_3 &= v\sqrt{1-w^2}(4w^2 - 1) \\ S_4 &= v\sqrt{1-w^2} \end{aligned} \tag{4.32}$$

One well known solution to this nonlinear system of equations is the *Carre-formula* [216]

$$\Delta\phi = \arctan\frac{v}{u} = \arctan\frac{\sqrt{I_1 + I_2 - I_3 - I_4}\sqrt{3I_2 - 3I_3 - I_1 + I_4}}{I_2 + I_3 - I_1 - I_4} \tag{4.33}$$

The Carre-formula directly calculates $\Delta\phi(x, y)$ from the 4 recorded intensities. But many experiments have shown that a big advantage lies in a primary pointwise calculation of the unknown phase shift $\Delta\phi_R(x, y)$ [217]. According to the above taken definitions it is $\cos(\Delta\phi_R) = 2w^2 - 1$, and if (4.32) is solved for $2w^2 - 1$ we get for each point

$$\Delta\phi_R(x, y) = \arccos\frac{I_1(x,y) - I_2(x,y) + I_3(x,y) - I_4(x,y)}{2[I_2(x,y) - I_3(x,y)]} \tag{4.34}$$

By assumption $\Delta\phi_R$ has to be constant for all points of the interference pattern, so we can take the average $\overline{\Delta\phi_R}$ over all pixels (x, y). In this way fluctuations in $\Delta\phi_R(x, y)$, which may have been caused by speckles, are averaged out. Outliers in $\Delta\phi_R(x, y)$ preferably occur where the denominator of (4.34) is zero or near zero, they may be detected by a test for outliers and discarded before the averaging is done.

Having calculated the average phase shift $\overline{\Delta\phi_R}$, now only three unknowns are left and the interference phase can be determined from the first three intensities of (4.25) by

$$\Delta\phi_1(x, y) = \arctan\frac{I_3 - I_2 + (I_1 - I_3)\cos\overline{\Delta\phi_R} + (I_2 - I_1)\cos 2\overline{\Delta\phi_R}}{(I_1 - I_3)\sin\overline{\Delta\phi_R} + (I_2 - I_1)\sin 2\overline{\Delta\phi_R}} + \frac{3\overline{\Delta\phi_R}}{2} \tag{4.35}$$

In an analogous way the interference phase may be calculated from the last three intensities of (4.25) by

$$\Delta\phi_2(x, y) = \arctan\frac{I_4 - I_3 + (I_2 - I_4)\cos\overline{\Delta\phi_R} + (I_3 - I_2)\cos 2\overline{\Delta\phi_R}}{(I_2 - I_4)\sin\overline{\Delta\phi_R} + (I_3 - I_2)\sin 2\overline{\Delta\phi_R}} + \frac{\overline{\Delta\phi_R}}{2} \tag{4.36}$$

The arguments (x, y) to the intensities in (4.34) and (4.35) have been omitted for convenience. The 2π-steps of the interference phase distributions $\Delta\phi_1(x, y)$ and $\Delta\phi_2(x, y)$ arise at different points. This additional information can be used in the subsequent demodulation by considering the continuous interference phase variation of the two for the decision of adding or subtracting a further 2π.

The next case to be treated is $m = 5$ recorded intensities with unknown but constant mutual phase shifts. Without loss of generality here the best starting value for easy mathematics is $\phi_{R1} = -2\,\Delta\phi_R$. The recorded intensities are

$$\begin{aligned}
I_1 &= a + b\cos(\Delta\phi - 2\Delta\phi_R) = a + b\cos\Delta\phi\cos 2\Delta\phi_R + b\sin\Delta\phi\sin 2\Delta\phi_R \\
I_2 &= a + b\cos(\Delta\phi - \Delta\phi_R) = a + b\cos\Delta\phi\cos\Delta\phi_R + b\sin\Delta\phi\sin\Delta\phi_R \\
I_3 &= a + b\cos\Delta\phi \\
I_4 &= a + b\cos(\Delta\phi + \Delta\phi_R) = a + b\cos\Delta\phi\cos\Delta\phi_R - b\sin\Delta\phi\sin\Delta\phi_R \\
I_5 &= a + b\cos(\Delta\phi + 2\Delta\phi_R) = a + b\cos\Delta\phi\cos 2\Delta\phi_R - b\sin\Delta\phi\sin 2\Delta\phi_R
\end{aligned} \tag{4.37}$$

Again we define $u = 2b\cos\Delta\phi$ and $v = 2b\sin\Delta\phi$, but now $w = \cos\Delta\phi_R$. This gives the system of equations

$$\begin{aligned}
I_1 + I_5 &= 2a + u(2w^2 - 1) \\
I_2 + I_4 &= 2a + uw \\
2I_3 &= 2a + u \\
I_1 - I_5 &= 2vw\sqrt{1 - w^2} \\
I_2 - I_4 &= v\sqrt{1 - w^2}
\end{aligned} \tag{4.38}$$

A direct solution for the interference phase is

$$\Delta\phi(x,y) = \arctan\frac{v}{u} = \arctan\frac{\sqrt{4(I_2 - I_4)^2 - (I_1 - I_5)^2}}{2I_3 - I_1 - I_5} \tag{4.39}$$

As with $m = 4$ the intermediate step over the calculation of $\Delta\phi_R(x,y)$ and its averaging is recommended. For the calculation of the phase shift a number of formulas can be derived from (4.38). Perhaps the shortest is

$$\Delta\phi_R(x,y) = \arccos\frac{I_1 - I_5}{2(I_2 - I_4)} \tag{4.40}$$

More solutions can be designed by convex combinations of two solutions of the $m = 4$-case, e. g. with identical weights

$$\begin{aligned}
\Delta\phi_R(x,y) &= \arccos\frac{1}{2}\left[\frac{I_1 - I_2 + I_3 - I_4}{2(I_2 - I_3)} + \frac{I_2 - I_3 + I_4 - I_5}{2(I_3 - I_4)}\right] \\
&= \arccos\frac{I_1(I_3 - I_4) + (I_2 - I_3 + I_4)(I_2 - 2I_3 + I_4) - I_5(I_2 - I_3)}{4(I_2 - I_3)(I_3 - I_4)}
\end{aligned} \tag{4.41}$$

A further solution also recognizing all 5 measured intensities is

$$\Delta\phi_R(x,y) = \arccos\frac{(I_2 - I_4)(I_1 - 2I_2 + 2I_3 - 2I_4 + I_5)}{(I_2 - I_4)(I_1 - 2I_3 + I_5) - (I_1 - I_5)(I_2 - 2I_3 + I_4)} \tag{4.42}$$

Now after averaging over the $\Delta\phi_R(x,y)$ the interference phase can be determined by any of the formulas

$$\begin{aligned}
\Delta\phi(x,y) &= \arctan\frac{I_{i+2} - I_{i+1} + (I_i - I_{i+2})\cos\overline{\Delta\phi_R} + (I_{i-1} - I_i)\cos 2\overline{\Delta\phi_R}}{(I_i - I_{i+2})\sin\overline{\Delta\phi_R} + (I_{i+1} - I_i)\sin 2\overline{\Delta\phi_R}} \\
&\quad -(i+1)\overline{\Delta\phi_R} \qquad\qquad i = 1, 2, 3
\end{aligned} \tag{4.43}$$

A multiple calculation of $\Delta\phi$ again facilitates a noise suppression by averaging and an easier and more reliable demodulation.

A further phase shift algorithm may become important for the evaluation of speckle interferograms which are produced by the spatial carrier method employing a tilted reference wave, see Chap. 6. Let us assume an unknown constant phase shift of $\Delta\alpha$ in the horizontal direction and of $\Delta\beta$ in the vertical direction. Then the recorded intensities are

$$\begin{aligned} I_1 &= a + b\cos\Delta\phi \\ I_2 &= a + b\cos(\Delta\phi - \Delta\alpha) \\ I_3 &= a + b\cos(\Delta\phi + \Delta\alpha) \\ I_4 &= a + b\cos(\Delta\phi - \Delta\beta) \\ I_5 &= a + b\cos(\Delta\phi + \Delta\beta) \end{aligned} \tag{4.44}$$

These are the five intensities at least necessary for determining the five unknowns. With the definitions $S_1 = I_2 - I_3$, $S_2 = 2I_1 - I_2 - I_3$, $S_3 = I_4 - I_5$, $S_4 = 2I_1 - I_4 - I_5$, $u = b\cos\Delta\phi$, $v = b\sin\Delta\phi$ we get

$$\left(\frac{v}{u}\right)^2 = \frac{S_1^2 S_4 - S_3^2 S_2}{S_4^2 S_2 - S_2^2 S_4} \tag{4.45}$$

so that one formula for determining the interference phase is

$$\Delta\phi(x,y) = \arctan \frac{\sqrt{(I_2 - I_3)^2(2I_1 - I_4 - I_5) - (I_4 - I_5)^2(2I_1 - I_2 - I_3)}}{\sqrt{(2I_1 - I_2 - I_3)(2I_1 - I_4 - I_5)(I_2 + I_3 - I_4 - I_5)}} \tag{4.46}$$

If in phase shifting applications with unknown phase shifts even the condition of constant phase shift is not fulfilled, the unknown mutual phase shifts can be determined by the Fourier transform method described in Sec. 4.6, and the system of equations (4.22) has to be calculated with these determined reference phases. Another approach to automated reduction of phase shifted interferograms involves the recording of the individual beams giving rise to the interference. This enables the determination of the noise terms and the subsequent calculation of the phase shifts between two holographic interferograms [218].

4.5.4 Application of Phase Shift Evaluation Methods

The determination of pixels which promise to yield reliable interference phase values proceeds over checking the depth of the *intensity modulation* [219]. This modulation is contained in the $b(x,y)$ of (4.16). It can be calculated from the $u(x,y)$ and $v(x,y)$ of (4.19), which are at hand during the interference phase determination, by

$$b(x,y) = \sqrt{[u(x,y)]^2 + [v(x,y)]^2} \tag{4.47}$$

Where this $b(x,y)$ is greater than a prescribed threshold, the pixel (x,y) will render a reliable phase value.

For $m = 4$ interferograms and a phase shift roughly in the vicinity of 90° from (4.24) we get the quantity

$$b' = \sqrt{[I_1(x,y) - I_3(x,y)]^2 + [I_4(x,y) - I_2(x,y)]^2} \tag{4.48}$$

to be tested. The factor 2 can be recognized in the threshold. Experiments have shown that nearly the same results for $m = 4$ and $\Delta\phi_R$ near 90° can be achieved even by the simpler quantity

$$b'' = |I_1(x,y) - I_3(x,y)| + |I_4(x,y) - I_2(x,y)| \tag{4.49}$$

An example is given in Fig. 4.16. Fig. 4.16a shows one of the four recorded holographic

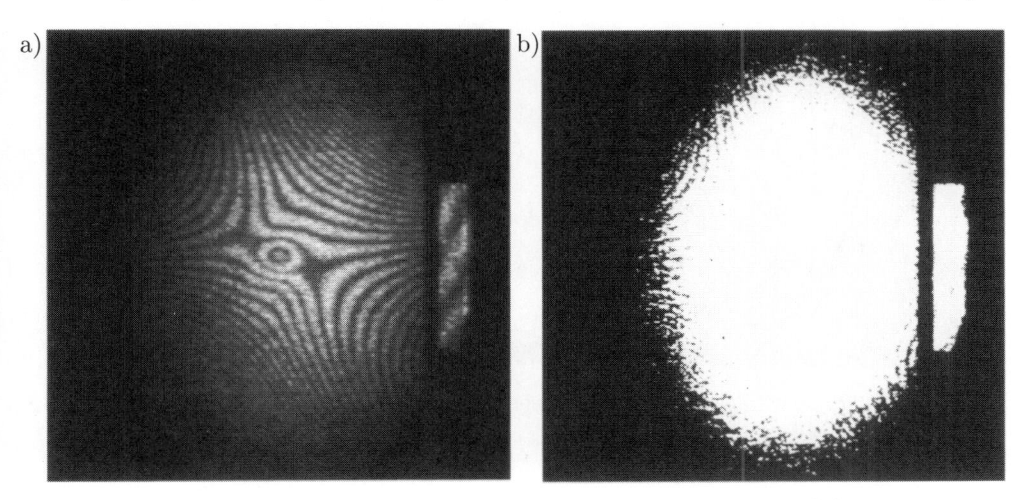

Figure 4.16: Determination of pixels with sufficient intensity modulation

interference patterns, Fig. 4.16b displays all pixels in white where the quantity of (4.49) exceeds the threshold. By applying a modified digital median filter isolated good pixels may be discarded, the interference phase at bad pixels surrounded by good ones may be determined by taking the average of the phases of the surrounding pixels. In this way sharp borders of the region of reliable pixels can be fixed.

The described method not only fits to find pixels of sufficient contrast but also enables one to detect the borders of the imaged object surface at all, if the object does not fill the whole frame. Altogether a mask is generated, only in its interior the phase shift evaluation has to be performed.

The big advantage of the methods with unknown constant phase shift and a calculation of this phase shift pointwisely out of the recorded intensities by (4.34), (4.40), (4.41), or (4.42) is the possibility first to check the calculated phase shift distribution for feasibility and second to discard outliers and to gain the most likely phase shift by averaging. Since a constant phase shift is assumed, strong variations would indicate non-constant phase steps. A linear trend along the pixels indicates a systematic error in the phase shifter, e. g. a tilt of the phase shifting piezo mounted mirror. Such an error can be compensated by fitting a plane to the phase shift values $\Delta\phi_R(x,y)$ and taking the values of this plane function as the $\overline{\Delta\phi_R(x,y)}$ in the interference phase calculation by (4.35), (4.36), or (4.43).

Further advantages are offered by the two or more formulas to calculate the interference phase (4.35) and (4.36) or (4.43), which give redundant information, e. g. for the demodulation process. If the interferograms are recorded in sequence with very short time gaps, e. g. with video-frequency, spurious vibrations can be assumed to contribute linearly to the shifted phase and are compensated inherently by the evaluation process.

A one-dimensional example of phase stepping with unknown but constant phase steps is shown in Figs. 4.17 and 4.18. Fig. 4.17 represents the four phase stepped holographic

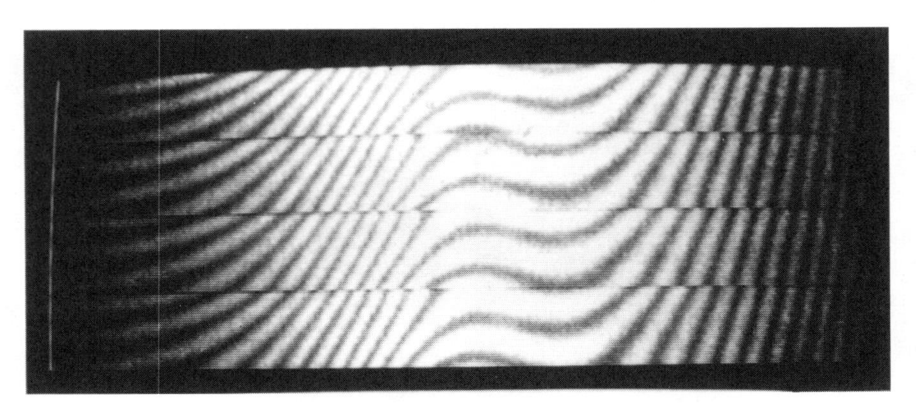

Figure 4.17: Four phase stepped holographic interferograms of tensile test specimen

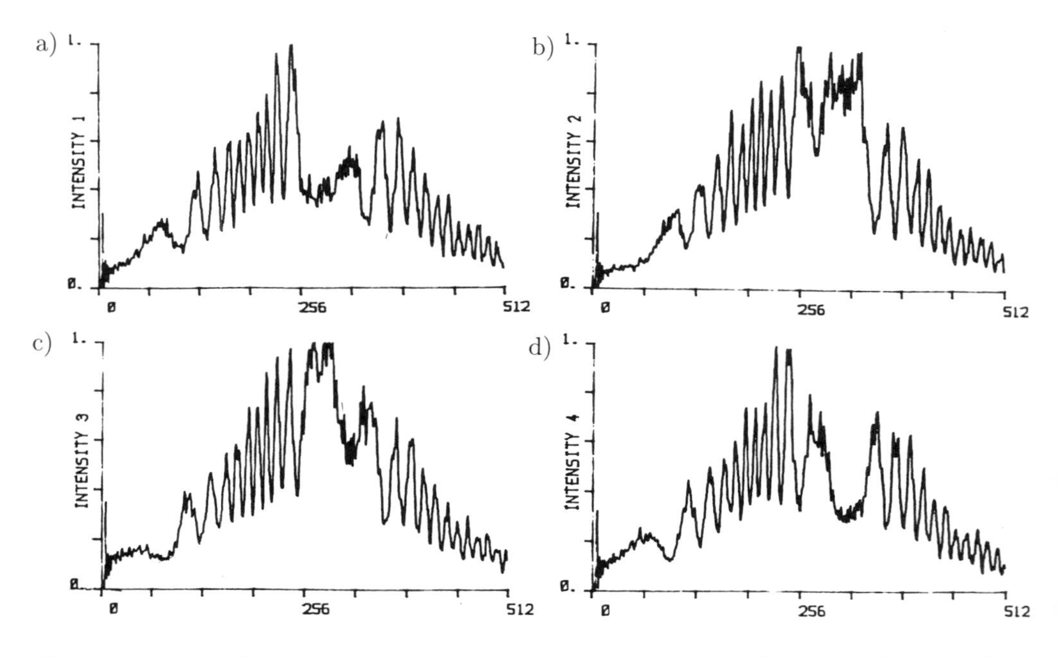

Figure 4.18: (a - d) Evaluation of phase stepped holographic interferograms along one line, intensity distributions

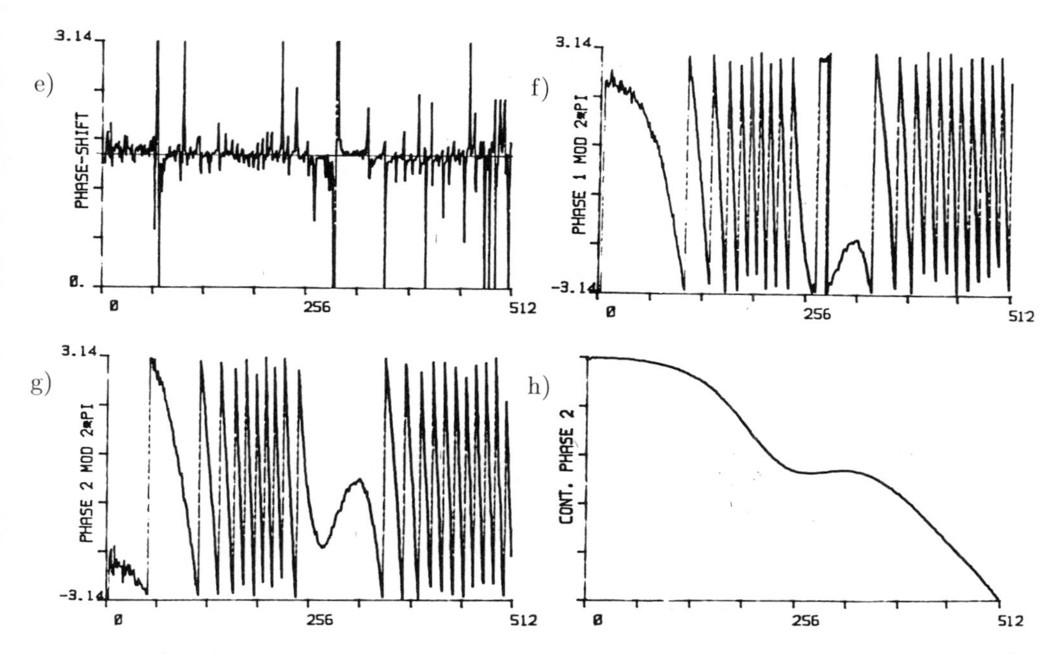

Figure 4.18 (e - h) Evaluation of phase stepped holographic interferograms along one line, (e) phase shift, (f) interference phase 1, (g) interference phase 2, (h) demodulated interference phase

interferograms of a tensile test specimen with an internal crack produced by the real-time method. The evaluation along one line is shown in several steps in Fig. 4.18. The four recorded intensity distributions along the common line are displayed in Figs. 4.18a to d. In Fig. 4.18e the phase step $\Delta\phi_R(x, y)$, calculated by (4.34), is given together with the straight line which corresponds to the averaged $\overline{\Delta\phi_R}$. Figs. 4.18f and g display the interference phase distributions modulo 2π determined by (4.35) and (4.36). The continuous interference phase, after unwrapping the 2π-discontinuities, is displayed in Fig. 4.18h. Although a varying background intensity, varying contrast, and even local saturation in the recorded intensities is present, the calculated interference phase distribution is clean and smooth. The changes in slope, decreasing to increasing and vice versa, are uniquely detected.

A two-dimensional evaluation by the same procedure of phase sampling is demonstrated at the example of a thermally loaded panel, consisting of an internal aluminum honeycomb structure with surface layers of carbon fiber reinforced plastic. Figs. 4.19a to d exhibit the four phase stepped interferograms arising from a temperature difference of 2° C. The size of the panel was about 80×80 cm. The resulting interference phase distribution, which can be interpreted as proportional to the normal displacement field due to the optimized sensitivity vectors, is presented in Fig. 4.20. Several debonds of the surface layer from the internal structure are clearly detectable.

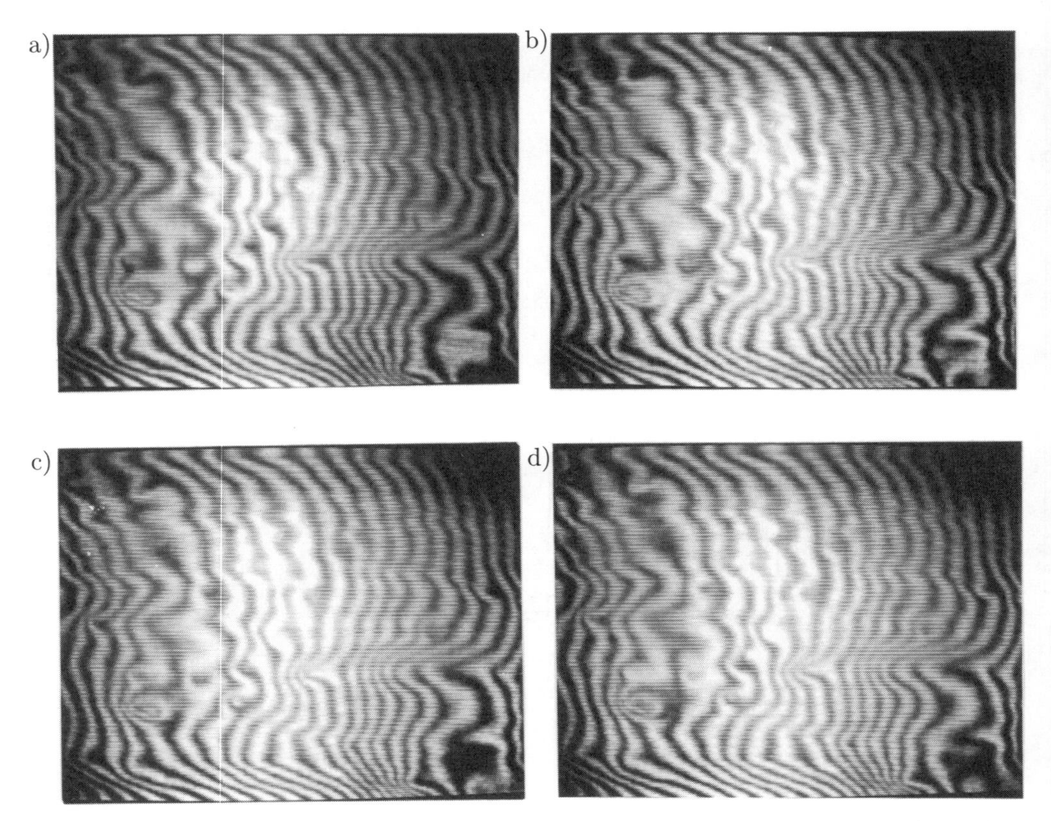

Figure 4.19: Four phase stepped holographic interferograms of a thermally loaded panel

4.5.5 Discussion of Phase Shift Evaluation Methods

The phase shift methods for determining the interference phase distributions in holographic interferometry offer a number of advantages. The price one has to pay for it is the additional technical effort to perform the phase shifts, the increased requirements necessary for stability as for the *two reference beam holography* with double exposure or for the *real-time method*, as well as the additional storage capacity of the computer, all compared to the recording of a single interference pattern. But the benefits are manifold [220, 221]:

- The evaluation procedure can be fully automated.
- The interference phase is calculated at all pixels, not only at the fringe centers. Thus we get the best *spatial resolution* possible with the available electronics. No interpolation between skeleton lines is necessary.
- Due to the multiple recorded interferograms the *sign ambiguity* is resolved automatically. The evaluated interference phase increases and decreases in the same manner as the original one.

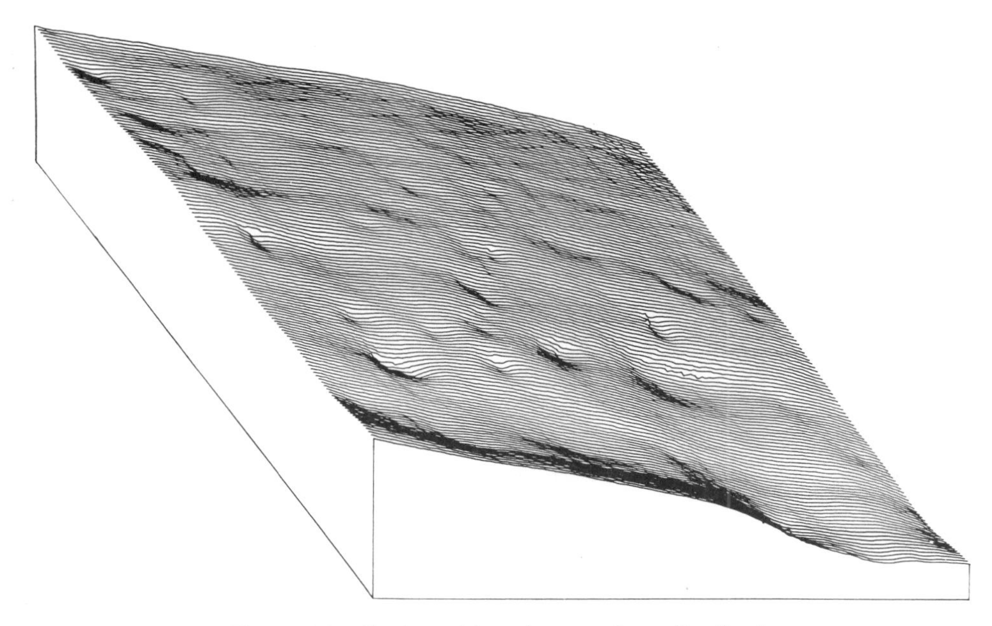

Figure 4.20: Evaluated interference phase distribution

- The additive and multiplicative noise components are inherently recognized and compensated for in the automatic evaluation process.
- There are many different algorithms which fit to different circumstances, e. g. environmental distortions, phase shifter miscalibration, unknown phase shifts, etc.
- The intermediate results of the evaluation procedure allow a detection, whether the whole measurement was erroneous, e. g. zero phase shifts, whether there are points with contrast too low to guarantee a reliable evaluation, or whether there are points where no object surface was existent at all, e. g. at holes or outside the object margins.
- A phase shift which is varying over the pattern can be detected by some algorithms and sometimes can be fitted by a low degree polynomial, especially when the phase of the reference wave is shifted by a piezo mounted mirror which undergoes an additional tilt.
- The resolution and accuracy of the determined interference phase is better than $1/20$ of 2π and with some experimental skill reaches $1/100$ of 2π.

Although the evaluation scheme, (4.22) and (4.25), applies for all sets of reference phases ϕ_{Rn}, $n = 1, \ldots, m$ which do not produce a singular matrix in (4.22), the best choice for the phase shifts between successive interferograms is between 30° and 150°. This recommendation concerns to the methods with known as well as to these with unknown phase shifts. For the methods with known phase shifts it is not necessary that all phase shifts are constant, but this simplifies the evaluation procedure.

The influence of different error sources on the phase shift methods has been investigated by a large number of researchers [169, 222, 223, 224, 225, 226]. The results can be summarized by the statement that all these error sources like insufficient quantization, spurious diffraction or reflection patterns, aberrations of the optics, vibrations, air turbulence, inhomogeneity of the reference beam wavefront etc., which can be modeled as additive or multiplicative noise degrade the precision and accuracy independently of the choice of the evaluation algorithm. Exceptions are detector nonlinearities and false phase shifts, e. g. by phase shifter miscalibration, whose consequences depend on the specific evaluation algorithm. Their influence has been tested in a comparison of different algorithms [227].

These tests have shown that a linear phase shift error degrades the calculated interference phase for all algorithms which assume a known constant phase shift. The degradation is reduced with the increase of the number of interferograms, especially for $m = 5$ or $m = 7$, [228, 229, 230]. The algorithms which calculate the unknown phase shift from the recorded interference patterns compensate for a linear phase shift error inherently and remain exact.

The same trend holds for nonlinear, quadratic phase shift errors. Although their influence remains present when applying the algorithms with unknown phase shifts, it is least then, compared to the algorithms assuming a known constant phase shift and employing the same number of frames.

A higher number of frames as well as making use of the algorithms with unknown phase shifts also is favorable if detector nonlinearities have to be conceived.

4.6 Fourier Transform Evaluation

4.6.1 Principle of the Fourier Transform Evaluation Method

In *Fourier transform evaluation* [231, 232, 233, 234, 235, 236, 237, 238, 239] essentially a linear combination of *harmonic spatial functions* is fitted to the recorded and stored intensity distribution $I(x, y)$, given by (4.1). The admissible spatial frequencies of these harmonic functions are defined by the user via the *cutoff frequencies* of a *bandpass filter* in the spatial frequency domain.

To do this the intensity function is expressed with the help of the complex exponential. Introducing

$$c(x, y) = \frac{1}{2} b(x, y) \, \mathrm{e}^{\mathrm{i}\Delta\phi(x, y)} \tag{4.50}$$

the intensity $I(x, y)$ of (4.1) becomes

$$I(x, y) = a(x, y) + c(x, y) + c^*(x, y) \tag{4.51}$$

with $a(x, y)$, $b(x, y)$ as described in Sec. 4.1, i being the imaginary unit and * denoting complex conjugation. The discrete two-dimensional Fourier transform via the *FFT* *algorithm* applied to $I(x, y)$ yields

$$\mathcal{I}(u, v) = \mathcal{A}(u, v) + \mathcal{C}(u, v) + \mathcal{C}^*(u, v) \tag{4.52}$$

with (u, v) being the spatial frequency coordinates. Since $I(x, y)$ is a real distribution in the spatial domain, $\mathcal{I}(u, v)$ is a *Hermitean* distribution in the spatial frequency domain, which means

$$\mathcal{I}(u, v) = \mathcal{I}^*(-u, -v) \tag{4.53}$$

The real part of $\mathcal{I}(u, v)$ is even and the imaginary part is odd. The *amplitude spectrum* $|\mathcal{I}(u, v)|$ thus looks point-symmetric with respect to the dc-term $\mathcal{I}(0, 0)$. $\mathcal{A}(u, v)$ contains the zero-peak $\mathcal{I}(0, 0)$ and the low frequency variations of the background. $\mathcal{C}(u, v)$ and $\mathcal{C}^*(u, v)$ carry the same information as evident from (4.53).

By *bandpass filtering* in the *spatial frequency domain*, $\mathcal{A}(u, v)$ and one of the terms $\mathcal{C}(u, v)$ or $\mathcal{C}^*(u, v)$ are eliminated. The remaining spectrum, $\mathcal{C}^*(u, v)$ or $\mathcal{C}(u, v)$, is no longer Hermitean, so the *inverse Fourier transform* applied to, e.g. $\mathcal{C}(u, v)$, gives a complex $c(x, y)$ with non-vanishing real and imaginary parts. The interference phase can be calculated by

$$\Delta\phi(x, y) = \arctan \frac{\text{Im } c(x, y)}{\text{Re } c(x, y)} \tag{4.54}$$

The inverse transform of $\mathcal{C}^*(u, v)$ instead of $\mathcal{C}(u, v)$ would result in $-\Delta\phi(x, y)$. The uncertainty about which of the symmetric parts of the spectrum belongs to $\mathcal{C}(u, v)$ and which to $\mathcal{C}^*(u, v)$ is a manifestation of the *sign ambiguity*, (4.2).

A one-dimensional example for the Fourier transform evaluation is given in Fig. 4.21. The intensity distribution with varying contrast and varying background of Fig. 4.21a is Fourier transformed, the amplitude spectrum is shown in Fig. 4.21b. After filtering, the spectrum of which the amplitude is given in Fig. 4.21c remains. The application of the inverse transform and (4.54) result in the phase modulo 2π shown in Fig. 4.21d.

Although the Fourier transform calculated by the FFT algorithm normally has its zero frequency at the left and the two-dimensional Fourier transform has its zero-frequency in the upper left corner when displayed, in the examples demonstrated here and in the following, we use the reordered display of the amplitude spectra, as determined by (A.63).

4.6.2 Noise Reduction by Spatial Filtering

The bandpass filtering in the spatial frequency domain not only makes the spectrum non-Hermitean, but also enables a reasonable *image enhancement* [234]. Low frequency *background variations*, e. g. a Gaussian illumination, lead to spectral components centered around the zero-component. Their influence is minimized by a bandpass filter which eliminates all spectral components up to a certain lower *cutoff frequency*. High frequency components, like *speckle noise*, are suppressed to a reasonable amount if the filter stops all frequencies higher than an upper cutoff frequency.

This capability is shown in Fig. 4.22. The intensity, Fig. 4.22a, is degraded by a Gaussian background, varying contrast, speckle noise, reduced *quantization* into 16 gray-levels,

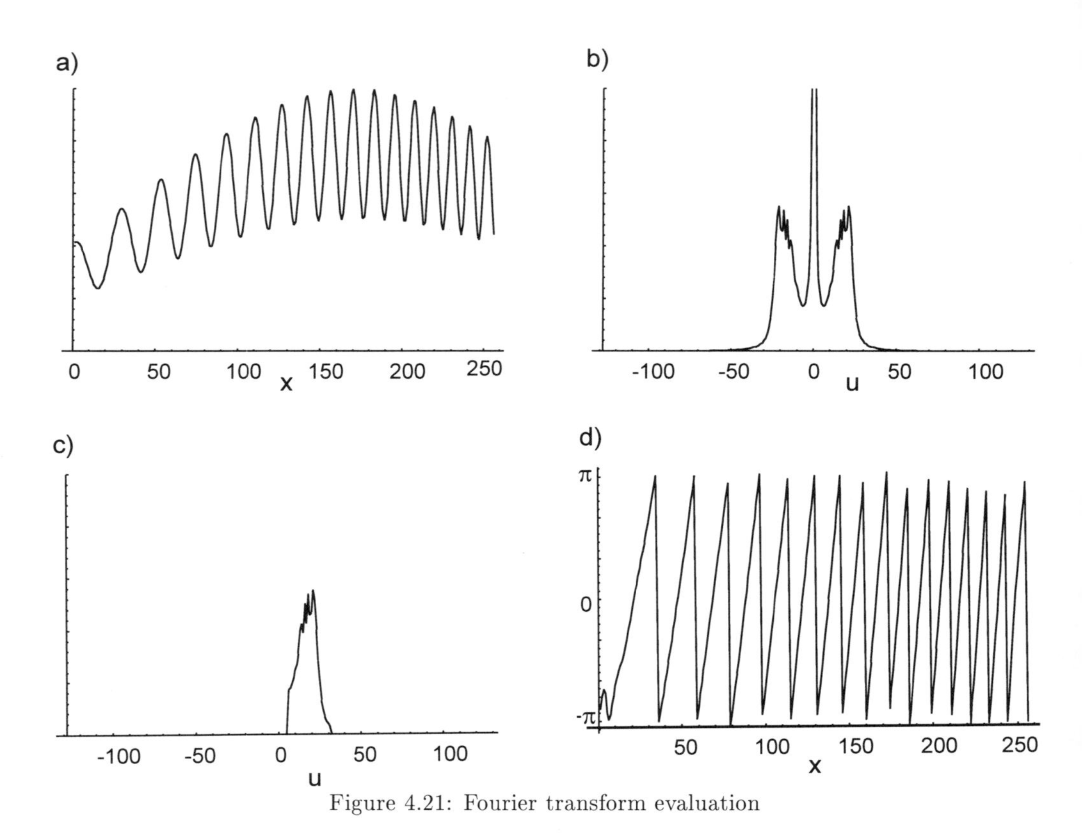

Figure 4.21: Fourier transform evaluation

and non-linear response with *saturation*. Nevertheless the finally evaluated interference phase distribution, Fig. 4.22d, is clean and fully modulated.

In eliminating high frequencies, one has to consider the fact that the *finite discrete Fourier transform* assumes a periodic input signal, whereas, in practice, data consists of one non-periodic stretch of finite length. The discontinuities from the right to the left or from the lower to the upper edge of the image lead to high frequency components in the spectrum. If these are filtered away, the resulting phase distribution suffers from a *wrap-around pollution* by a forced smooth continuation at the edges of the frame. Thus the marginal pixels at the edges of the frame get no reliable phase values; the number of these pixels depend on the choice of the upper cutoff frequency. The run-out at the left edge in Fig. 4.22d is caused by this effect.

Making the spectrum non-Hermitean means setting to zero the spectral value at each spatial frequency or at its symmetric counterpart. Since there is no general way to decide whether $\mathcal{I}(u,v)$ at a certain (u,v) belongs to $\mathcal{C}(u,v)$ or to $\mathcal{C}^*(u,v)$, or is a combination of contributions belonging to both of them, the easiest way is to eliminate one half-plane of the spatial frequency plane. Some of these halfplanes are displayed in Fig. 4.23. In Fig. 4.23a the passband is the $+u$-halfplane, displayed in white, and the shaded region is

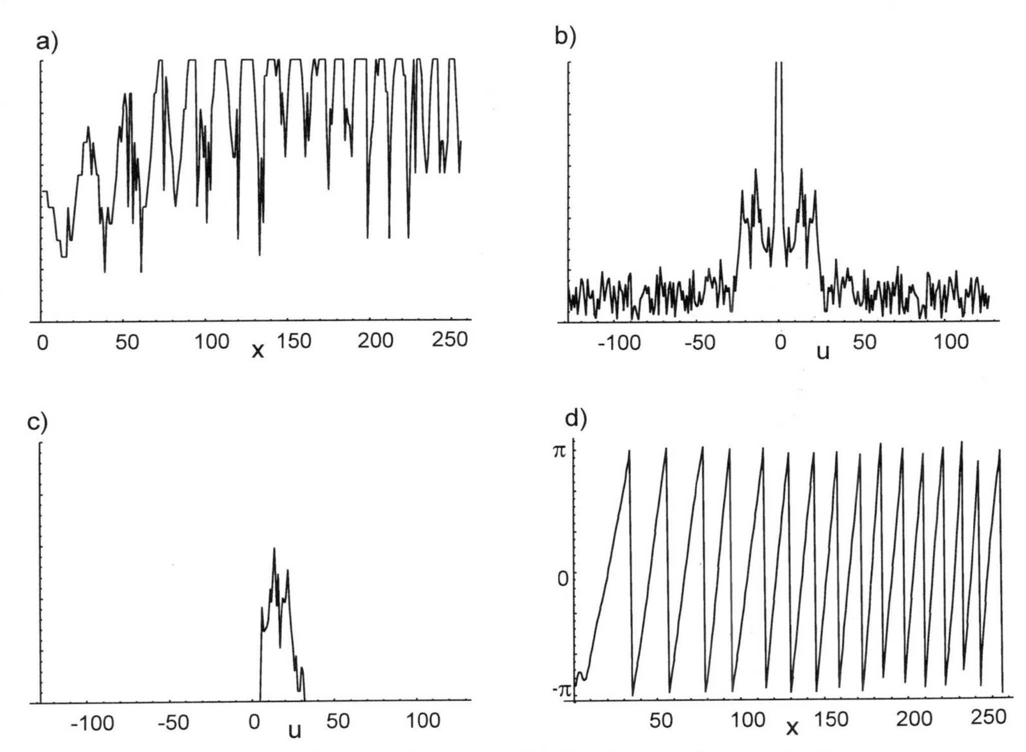

Figure 4.22: Image enhancement by Fourier transform evaluation

the stopband. In Fig. 4.23b the passband is the $-u$-halfplane, in Fig. 4.23c the passband is the $+v$-halfplane and in Fig. 4.23d we have the passband in the $-v$-halfplane. For Fourier transform evaluation the component at spatial frequency $(0, 0)$ is always set to zero. Filters which destroy the Hermitean property and eliminate low-frequency background as well as high-frequency speckle noise have the form shown in Fig. 4.24, with a passband in one halfplane and with defined lower and upper cutoff frequencies.

In many holographic applications it is possible to record and store the illuminated surface before the interference pattern is produced [233]. So in *real-time holographic interferometry* one may

- record the object surface illuminated only by the object wave with the reference wave blocked. The hologram plate may be present or absent.
- record a holographic reconstruction of the object with the object wave blocked. Only the reference wave illuminates the hologram plate.
- record the zero interference pattern, still without fringes, before starting the loading.

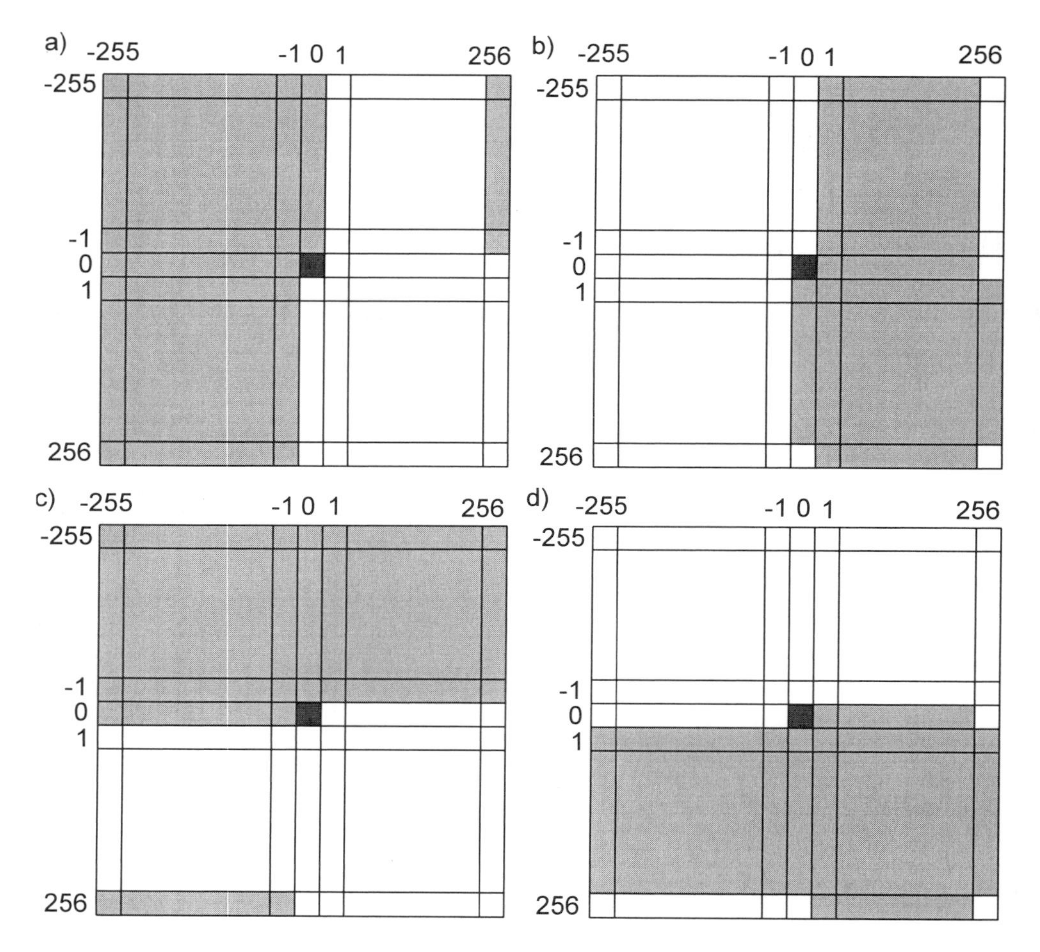

Figure 4.23: Halfplanes for filtering in the spatial frequency domain, reordered display

Let this recorded background be $a'(x, y)$ and its Fourier transform be $\mathcal{A}'(u, v)$. A normalized version of $\mathcal{A}'(u, v)$ is then subtracted from $\mathcal{I}(u, v)$

$$\mathcal{I}'(u, v) = \mathcal{I}(u, v) - \frac{\operatorname{Re} \mathcal{I}(0, 0)}{\mathcal{A}'(0, 0)} \mathcal{A}'(u, v) \tag{4.55}$$

Since the imaginary part of the dc-term of a Fourier transform is always zero, we have zeroed the dc-term and eliminated the background. Now from $\mathcal{I}'(u, v)$ one halfplane is eliminated and after the inverse transform has been obtained the interference phase is calculated as described above. This procedure is recommended if the interference pattern or the object does not cover the whole frame or if we have complicated background variations in the frequency range of the interference pattern.

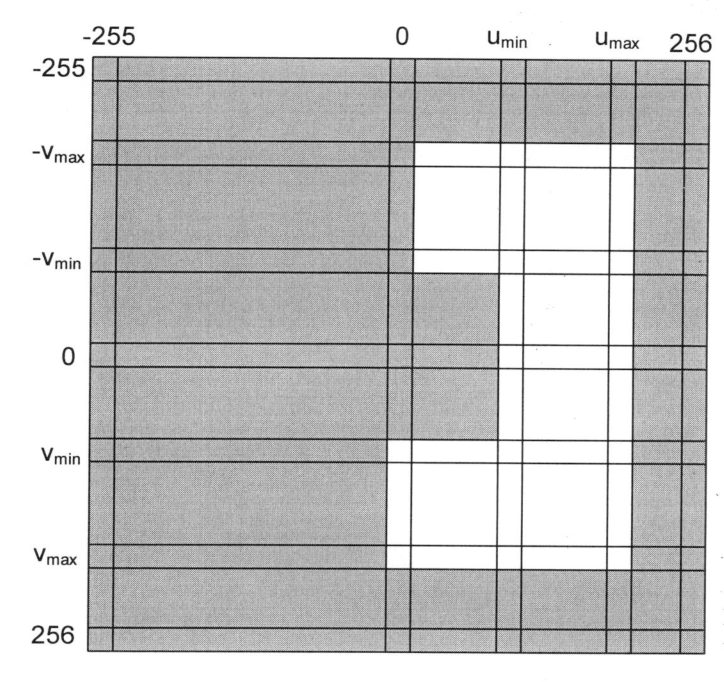

Figure 4.24: Bandpass filter, reordered display

4.6.3 Spatial Filtering and Sign Ambiguity

The filter of Fig. 4.23a, where only positive *spatial frequencies* in the horizontal, the u-direction, and both positive and negative frequencies in the vertical, the v-direction can pass, gives an interference phase distribution with an increasing interference phase in the horizontal direction, but increasing and decreasing interference phase in the vertical direction. For a passband in the $-u$-halfplane the phases are decreasing in the horizontal direction, Fig. 4.23b. In the cases of Figs. 4.23c and 4.23d, the roles of the directions interchange with respect to Figs. 4.23a and 4.23b.

These facts can be used for the detection of sign changes: two phase distributions determined from the same pattern, but with orthogonally oriented passbands are compared to resolve a local sign ambiguity. Only the global sign is left indefinite, this has to be decided from knowledge about the specific experiment. From a computational view the elimination of a halfplane oriented parallel to the u- or v-axes would look to be the easiest way to render the spectrum non-Hermitean. Other orientations of halfplanes, or even other filter strategies destroying the Hermitean property, are also feasible.

The Fourier transform evaluation of the holographically measured deformation of a plate thermally loaded in a microwave-oven is shown in Fig. 4.25. Overlayed on the interference pattern we see the metallic grid in the window protecting the environment from microwave radiation, Fig. 4.25a. The bright spot in the lower left is a lens in the oven's door transmitting the illumination wave. The amplitude spectrum of this pattern is displayed in Fig. 4.25b in a logarithmic gray-scale to compress the high dynamic range, especially that of the zero-peak. Two mutually orthogonally oriented bandpass filters

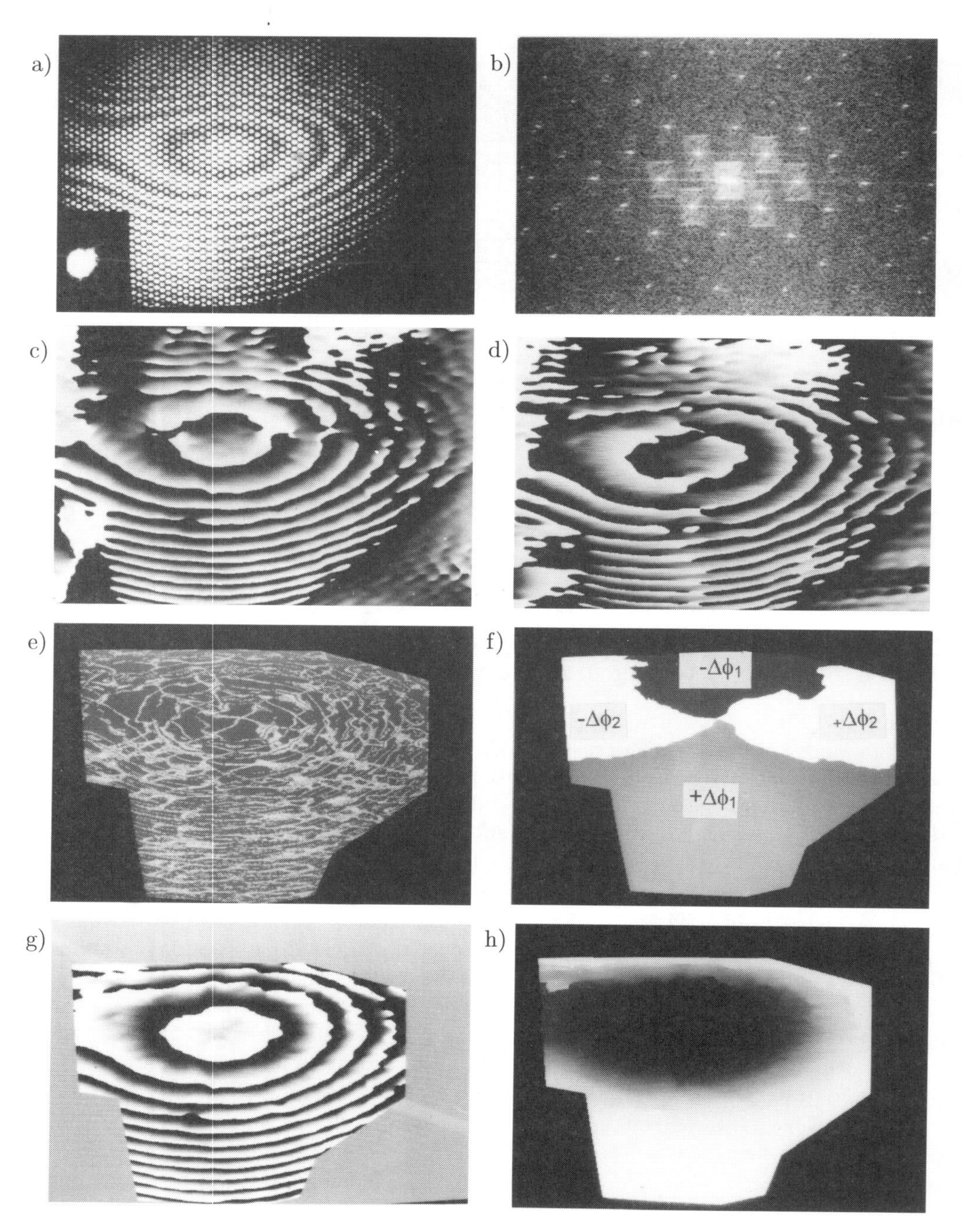

Figure 4.25: Fourier transform evaluation of thermally loaded plate

i)

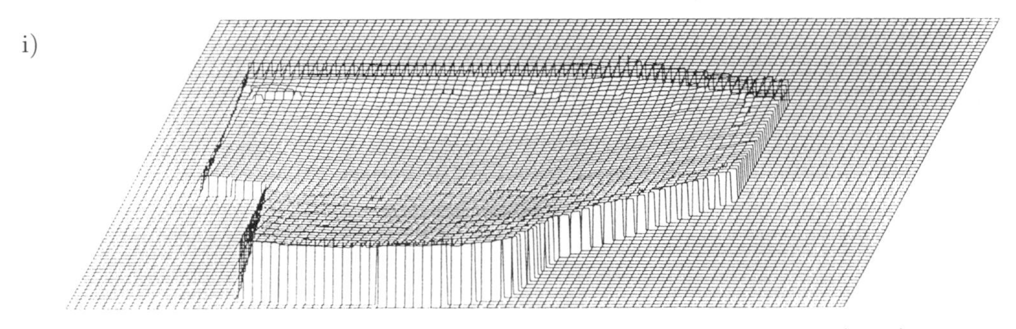

Figure 4.25: Fourier transform evaluation of thermally loaded plate (cont.)

with lower cutoff frequencies $u_{min} = v_{min} = 2$ and upper cutoff frequencies $u_{max} = 18$, $v_{max} = 34$ in both cases, are applied. The units are cycles per 512 pixels. The upper cutoff frequencies are chosen lower than the least frequency of the grid pattern. Therefore only fringe periods greater than the grid spacing can be evaluated.

The resulting interference phase distributions modulo 2π are given in gray-scale display in Figs. 4.25c and 4.25d. The phase $\Delta\phi_1$ of Fig. 4.25c is produced with the filter in the $+u$-halfplane, while the phase $\Delta\phi_2$ of Fig. 4.25d stems from the filter in the $+v$-halfplane. Artificial phases are given by the numerical process to pixels where no interference was present and these regions are masked out during further processing [240].

For the subsequent interactive sign correction, all pixels, where the two phase distributions differ less than a threshold, here chosen as 0.1, are marked.

$$||\Delta\phi_1(x,y)| - |\Delta\phi_2(x,y)|| < 0.1 \tag{4.56}$$

The borders of these regions are defined only along those lines, Fig. 4.25e, at which the phase takes the values $+\Delta\phi_1$, $-\Delta\phi_1$, $+\Delta\phi_2$, and $-\Delta\phi_2$. In this way a sufficiently continuous crossing of the borders is provided. The regions are shown in Fig. 4.25f, and the sign-corrected interference phase distribution modulo 2π is displayed in Fig. 4.25g. The demodulation, see Sec. 4.9, now leads to the continuous phase distribution of Fig. 4.25h, which as a pseudo-3D-display is shown in Fig. 4.25i.

4.6.4 Fourier Transform Evaluation of Phase Shifted Interferograms

Sign ambiguity is always present in the case of the evaluation of a single interferogram. After interactive sign-correction, as demonstrated above, there remains at least a global sign ambiguity, which means all phases have either the correct or the wrong sign. An exact determination of the sign-distribution is achieved if an additional phase stepped interferogram is produced with a mutual phase step ϕ_R [232, 241, 242]. Theoretically, ϕ_R must be in the range $0 < \phi_R < \pi$ but, in practice, values $\pi/3 < \phi_R < 2\pi/3$ are recommended. If this condition is fulfilled, the exact value of ϕ_R does not need to be known [232, 235].

The single phase stepping complicates the experimental procedure, but less than with the phase step methods, where several constant phase steps have to be provided. Let us now write the two intensity distributions

$$\begin{aligned} I_1(x,y) &= a(x,y) + b(x,y)\ \cos\left[\Delta\phi(x,y)\right] \\ I_2(x,y) &= a(x,y) + b(x,y)\ \cos\left[\Delta\phi(x,y) + \phi_R\right] \end{aligned} \tag{4.57}$$

Fourier transform processing of each intensity with the same bandpass filter parameters yields

$$\begin{aligned} c_1(x,y) &= \frac{1}{2}\, b(x,y) \exp[\mathrm{i}\Delta\phi(x,y)] \\ c_2(x,y) &= \frac{1}{2}\, b(x,y) \exp[\mathrm{i}\Delta\phi(x,y) + \mathrm{i}\phi_R(x,y)] \end{aligned} \tag{4.58}$$

From Eqs. (4.58) $\phi_R(x,y)$ is computed pointwise as

$$\phi_R(x,y) = \arctan \frac{\mathrm{Re}\ c_1(x,y)\ \mathrm{Im}\ c_2(x,y) - \mathrm{Im}\ c_1(x,y)\ \mathrm{Re}\ c_2(x,y)}{\mathrm{Re}\ c_1(x,y)\ \mathrm{Re}\ c_2(x,y) + \mathrm{Im}\ c_1(x,y)\ \mathrm{Im}\ c_2(x,y)} \tag{4.59}$$

The knowledge of $\phi_R(x,y)$ is used for determination of the sign-corrected interference phase distribution via

$$\Delta\phi(x,y) = \mathrm{sign}[\phi_R(x,y)] \arctan \frac{\mathrm{Im}\ c_1(x,y)}{\mathrm{Re}\ c_1(x,y)} \tag{4.60}$$

Figs. 4.26a and 4.26b show two phase stepped holographic interferograms. The sign of $\phi_R(x,y)$ as calculated by (4.59) is shown in Fig. 4.26c. Fig. 4.26d displays the phase distribution modulo 2π and Fig. 4.26e shows the sign-corrected phase distribution modulo 2π, calculated by (4.60). The demodulated interference phase distribution in a 3D-display is shown in Fig. 4.26f, respectively.

The Fourier transform calculation of the phase step by (4.59) allows a generalization of the phase step method of Sec. 4.5. Having recorded three or more phase stepped interference patterns with arbitrary phase steps, provided that these are $< \pi$, the phase steps ϕ_{Rn} can be evaluated by the Fourier transform procedure described above [243]. Taking the average of the absolute values $|\phi_{Rn}|$, we get the phase steps which are used to set up a system of equations similar to (4.22). This system is solved and the interference phase calculated by (4.23). This *generalized phase shifting interferometry*, using additional parallel Fizeau fringes which are evaluated by the Fourier transform method with spatial carrier, is presented in [244].

4.6.5 Spatial Heterodyning

In *spatial heterodyning* an additional *carrier frequency* is added to the interference pattern [239, 245, 246, 247, 248, 249, 250, 251, 252, 253, 254, 255, 256, 257]. A common way of generating the spatial carrier frequency in interferometry is to give a relatively large tilt to the reference mirror [254]. To produce linear fringes in holographic interferometry, the

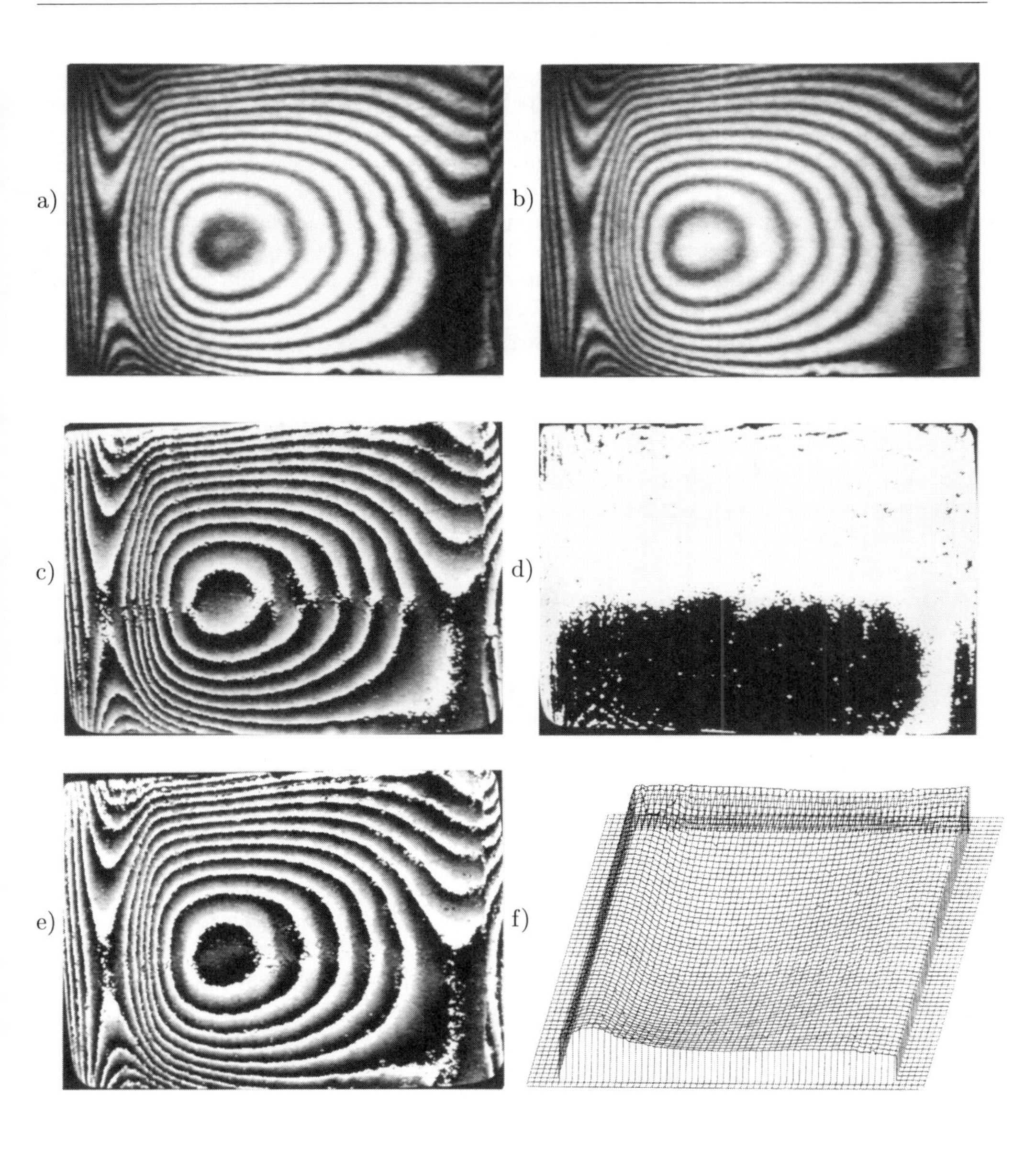

Figure 4.26: Fourier transform evaluation of phase shifted interferograms

object beam may be swung between the two exposures [258], the object may be tilted [259], or in a two reference wave arrangement the hologram position or the reference wave angles during reconstruction are properly controlled [260]. Great care must be taken to have a sensitivity vector not varying too much, to get equidistant carrier fringes. To fulfill the sampling theorem the employed detector array must have a spatial resolution high enough to resolve the spatial carrier and the spatial distribution of the detector sensitivity must be uniform over the array. For holographic measurements of phase objects, in [261] the use of Young's fringes as carriers, produced by a laterally shifted diffuser, is suggested. Cylindrical fringe carriers are treated in [262].

Now we let us assume that the spatial carrier has the fixed frequency f_0. Without loss of generality we further assume that the carrier fringes are parallel to the y-axis, meaning the carrier frequency has only an x-component. The recorded intensity pattern then is

$$I(x, y) = a(x, y) + b(x, y)\cos[\Delta\phi(x, y) + 2\pi f_0 x] \tag{4.61}$$

The formal analogy of (4.61) to temporal heterodyning (4.11) should be noticed. As for the Fourier transform evaluation now using (4.50) this intensity is written

$$I(x, y) = a(x, y) + c(x, y)\exp(2\pi \mathrm{i} f_0 x) + c^*(x, y)\exp(-2\pi \mathrm{i} f_0 x) \tag{4.62}$$

The Fourier transform of the intensity with respect to x then yields [263]

$$\mathcal{I}(u, y) = \mathcal{A}(u, y) + \mathcal{C}(u - f_0, y) + \mathcal{C}^*(u + f_0, y) \tag{4.63}$$

Since the chosen spatial carrier frequency f_0 is higher than the spatial variations of $a(x, y)$, $b(x, y)$, and $\Delta\phi(x, y)$, the partial spectra $\mathcal{A}$, $\mathcal{C}$, and $\mathcal{C}^*$ are well separated. $\mathcal{A}$ is concentrated around the dc-term at $u = 0$ and carries the low frequency background illumination. $\mathcal{C}$ and $\mathcal{C}^*$ are placed symmetrically to the dc-term and are centered around $u = f_0$ and $u = -f_0$. If by an adequate bandpass-filter first $\mathcal{A}$ and $\mathcal{C}^*$ are eliminated, and then $\mathcal{C}(u - f_0, y)$ is shifted by f_0 toward the origin, the carrier is removed and we obtain $\mathcal{C}(u, y)$. Taking the inverse Fourier transform of $\mathcal{C}(u, y)$ with respect to u yields $c(x, y)$ defined by (4.50). From this $c(x, y)$ the interference phase is calculated by (4.54) with phase values lying between $-\pi$ and $+\pi$. An alternative to the Fourier transform evaluation is the fitting of modified sinusoids to the carrier fringes [264].

A combination of spatial and temporal heterodyning is used for the simultaneous recording of multiple phase objects on a single *space-time interferogram* [59]. Although the method is proposed originally for Michelson- and Mach-Zehnder type interferometers, it may be used in holographic interferometry as well. An interferogram of the form

$$I(x, y, t) = a(x, y, t) + b(x, y, t)\cos[\Delta\phi(x, y, t) + 2\pi(f_{0X}x + f_{0Y}y + f_{0T}t)] \tag{4.64}$$

with the spatial carrier frequencies (f_{0X}, f_{0Y}) and the temporal carrier frequency f_{0T}, can be evaluated by the three-dimensional form of the above evaluation scheme. Now phase variations with wider spatio-temporal bandwidths can be determined, than would be possible by using only a single carrier frequency. Partial spectra $c_n(u, v, \omega)$ not separated by using either one of the carrier frequencies, are separable when using both spatial and temporal carrier frequencies. Furthermore the spatio-temporal frequency bandwidth available for the system can be effectively utilized. Several images having bandwidths less than the image detection system can be recorded multiplexed on a single interferogram.

4.6.6 Spatial Synchronous Detection

Spatial heterodyning consists of filtering and shifting of frequency components in the spatial frequency domain. The analogue in the spatial domain is the *spatial synchronous detection* method of [265, 266]. This method assumes a holographic interference pattern with a spatial carrier as described by (4.61). The spatial carrier may be artificially introduced by tilting one of the interfering wavefronts or is a constituent of the interference pattern.

In spatial synchronous detection the recorded and stored intensity, (4.61), is multiplied by R_1 and R_2 given by

$$\begin{aligned} R_1(x,y) &= \cos\, 2\pi f_0 x \\ R_2(x,y) &= \sin\, 2\pi f_0 x \end{aligned} \tag{4.65}$$

The multiplication results in

$$\begin{aligned} &I(x,y)R_1(x,y) \\ &= [a(x,y) + b(x,y)\cos(\Delta\phi(x,y) + 2\pi f_0 x)]\,[\cos\, 2\pi f_0 x] \\ &= a(x,y)\cos\, 2\pi f_0 x + \frac{b(x,y)}{2}\cos(\Delta\phi(x,y) + 4\pi f_0 x) + \frac{b(x,y)}{2}\cos(\Delta\phi(x,y)) \end{aligned} \tag{4.66}$$

and

$$\begin{aligned} &I(x,y)R_2(x,y) \\ &= [a(x,y) + b(x,y)\cos(\Delta\phi(x,y) + 2\pi f_0 x)]\,[\sin\, 2\pi f_0 x] \\ &= a(x,y)\sin\, 2\pi f_0 x + \frac{b(x,y)}{2}\sin(\Delta\phi(x,y) + 4\pi f_0 x) - \frac{b(x,y)}{2}\sin(\Delta\phi(x,y)) \end{aligned} \tag{4.67}$$

Of the three terms in (4.66) and (4.67) the first is of high frequency, just the carrier frequency. The second term has an even higher frequency, but the third term represents a low spatial frequency component which can be separated from the remaining terms by low-pass filtering. Afterwards the interference phase is calculated from the isolated cosine- and sine-terms by (4.54).

The filtering in the spatial domain, intended to extract the third terms, is done by convolving the products IR_1 and IR_2 with a window function, which must be several fringe periods wide. This window function may be a rectangular window, or with better results, a Hanning window or one of the numerous other digital filters.

4.7 Digital Holography

In the preceding sections the interference phase has been determined from temporal or spatial intensity variations. If one has numerical access to the whole optical field, amplitude and phase, the interference phase can be calculated by simple subtraction of the phases of the fields before and after deformation. This has become feasible by *digital holography* [267].

A second motivation for digital holography are the inefficient recording media. Silver halides provide the best resolution with high sensitivity and thus good quality holographic reconstructions, but need the wet chemical processing. Photothermoplasts require special electronics, are limited in size, resolution, and diffraction efficiency. Therefore holographers have waited for the possibility of fast electronic recording with sufficient resolution to record and evaluate holograms and holographic interferograms in nearly real-time.

A first response to this demand was the development of the ESPI- and DSPI-techniques of speckle metrology, Chap. 6, where analog TV-cameras or digital CCD-targets are employed. To achieve speckles of resolvable size and to prevent high frequency microinterferences, which are no longer resolvable by the targets, in these methods the reference wave is co-linear to the object wave. The object surface is imaged onto the target, the superposition of the co-linear reference gives a speckle field. The speckle fields of two states of the object surface, produced and recorded in this way, are then subtracted pointwisely one from the other, the resulting correlation fringes indicate the interference phase distribution belonging to the surface deformation.

Contrary to these techniques the digital holography approach directly investigates the microinterference pattern produced by reference and object wave, which is recorded on the CCD-target. The CCD-targets in the last years have got more and smaller pixels, the hope is that this trend continues and will relieve us of the limitations still set today.

4.7.1 Principle of Digital Holography

The fundamentals of digital reconstruction of wavefields from sampled holograms are treated by Yaroslavskii and Merzlyakov [268], the trace has been taken up recently by Schnars [269]. In digital holography a *Fresnel hologram* is generated on a CCD-target, which acts as the recording medium, is digitized and quantized, and is stored in the image processing system memory. To keep the following derivations easy, the reference wave is assumed as a collimated plane wave impinging normally onto the CCD-target, Fig. 4.27.

The optical reconstruction of the real image is performed by illuminating a hologram with the reference wave alone, Sec. 2.6.2. The diffraction of the plane reference wave at the hologram is described by the *Fresnel-Kirchhoff integral* (2.69). If it is assumed that the distance z between the hologram and the object, and thus the distance between hologram and real image, is much greater than the maximum dimensions of the CCD-chip, precisely

$$z^3 \gg \max \left\{ \frac{\pi}{4\lambda} \left[(x-\xi)^2 + (y-\eta)^2 \right]^2 \right\} \tag{4.68}$$

then the near-field or Fresnel-approximation (2.71) is valid. So we may use the *discrete finite Fresnel transform* (2.86) to calculate the complex amplitude of the wave that would be diffracted if the digitally stored hologram were a physical one. Let $\tau(k,l)$ be the stored hologram transmission sampled on a rectangular raster of $N \times M$ points, Fig. 4.27. Δx and Δy are given by the pixel distances of the CCD-target. Then the real image in the plane at z is

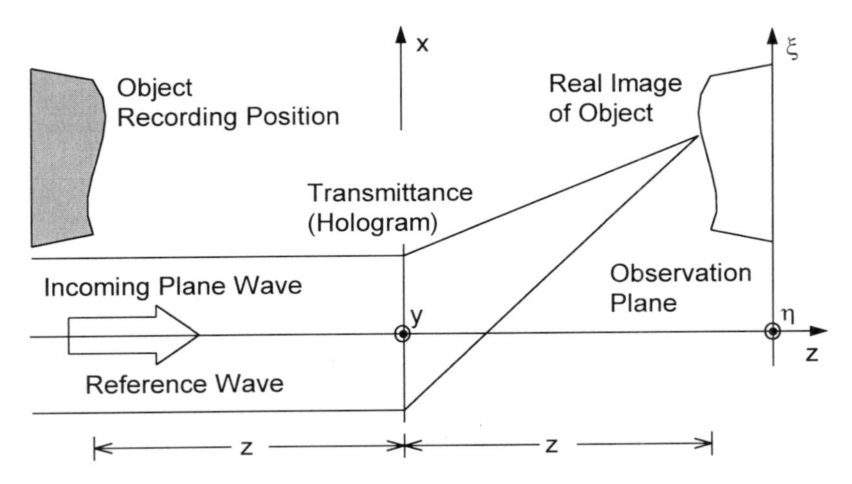

Figure 4.27: Geometry for numerical reconstruction of a real image from a digitally stored hologram

$$
\begin{aligned}
E(n,m,z) &= \frac{\mathrm{i}E_0}{\lambda z}\mathrm{e}^{-\mathrm{i}\pi\lambda z\left(\frac{n^2}{N^2\Delta x^2}+\frac{m^2}{M^2\Delta y^2}\right)} \\
&\times \sum_{k=0}^{N-1}\sum_{l=0}^{M-1}\tau(k,l)\mathrm{e}^{-\frac{\mathrm{i}\pi}{\lambda z}(k^2\Delta x^2+l^2\Delta y^2)}\mathrm{e}^{\mathrm{i}2\pi\left(\frac{kn}{N}+\frac{lm}{M}\right)} \qquad (4.69)
\end{aligned}
$$

$E_0 \neq 0$ represents the real amplitude of the reconstructing reference wave and can be set arbitrarily. As shown in Sec. 2.4.4 (4.69) is the *inverse Fourier transform* of the product consisting of the hologram $\tau(k,l)$ multiplied with an exponential factor which may be regarded as a Fresnel zone plate. The Fourier transform is performed by the *FFT algorithm* providing an effective calculation, Sec. A.6.

The reconstructed real image $E(n,m,z)$ is a complex function, so both the intensity and the phase can be calculated pointwisely. This is in contrast to the optical reconstruction making only the intensity visible. The intensity is determined by

$$I(n,m,z) = |E(n,m,z)|^2 = \{\mathrm{Re}[E(n,m,z)]\}^2 + \{\mathrm{Im}[E(n,m,z)]\}^2 \qquad (4.70)$$

and the phase distribution is calculated by

$$\phi(n,m,z) = \arctan\frac{\mathrm{Im}[E(n,m,z)]}{\mathrm{Re}[E(n,m,z)]} \qquad (4.71)$$

with values in $]-\pi,+\pi]$, if the arctan-function takes into account the numerator and the denominator separately, Sec. 4.1.1. As a consequence of the surface roughness of the object under test, the phase varies randomly.

Since only the intensity ratio between different points of the image is of interest, the constant intensity factor $\mathrm{i}E_0/(\lambda z)$ in (4.69) does not need to be calculated. In the same way, the phase at one pixel has no meaning, only the difference of the phases at the same point of the wave field corresponding to different states of the same object surface will be of interest in holographic interferometric metrology. But this difference does not depend on the phase factor $\exp\{-i\pi\lambda z[n^2/(N^2\Delta x^2)+m^2/(M^2\Delta y^2)]\}$, so this phase factor can be neglected in the calculation, too.

A first experimental verification of digital holography is performed with a die of side length 11 mm as the object, positioned in a distance 1.054 m from the target. The CCD-array consists of 1024 × 1024 pixels, each 6.8 μm × 6.8 μm large. The gray-values are quantized into 256 discrete grey levels. A 30 mW He-Ne-laser with wavelength 632.8 nm renders the coherent light. Fig. 4.28a displays the digitally sampled hologram, while the

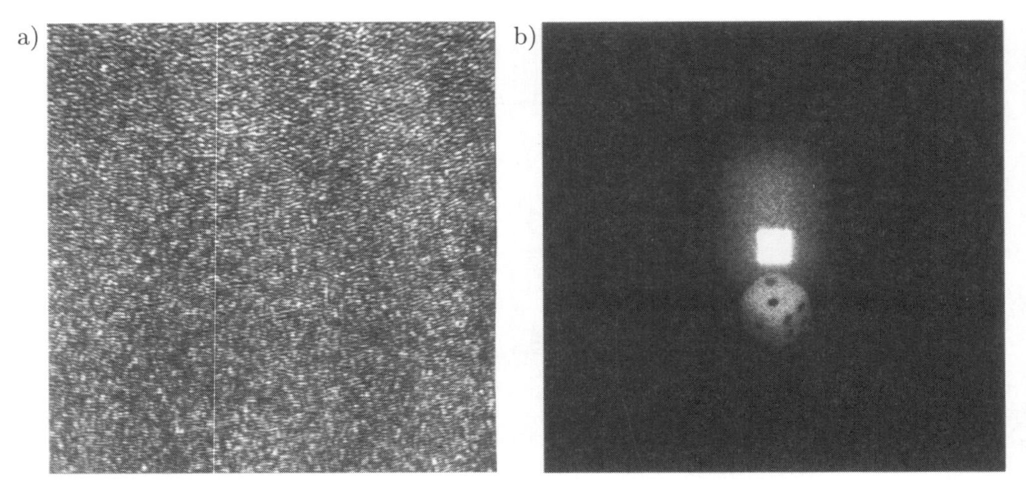

Figure 4.28: Digitally sampled (a) and reconstructed (b) off-axis hologram

numerically reconstructed intensity field is shown in Fig. 4.28b. The undiffracted reference wave is noticeable in the center of the reconstruction. One has to remember the fact, that the reconstructing reference wave was introduced only numerically. Due to the off-axis geometry during recording, the undiffracted reference wave and the real image of the object are well separated. Furthermore speckles are present in the reconstructed intensity field. They exhibit the same average size and distribution as would have occurred in an optical reconstruction with the same numerical aperture as the CCD-dimensions.

The three-dimensional nature of the real image, which is typical for holography, can be demonstrated by reconstructing in different planes. Fig. 4.29 shows the in-line holography of two transparencies placed 10 cm apart. One of the transparencies has a cross of opaque dots, while the other has a circle of opaque dots, Fig. 4.29a. The digitally recorded in-line hologram is displayed in Fig. 4.29b. The reconstructions of the real images with $d = 46$ cm and $d = 56$ cm, corresponding to the distances of the transparencies during recording, are given in Figs. 4.29c and d. The effect of parallax is hardly to demonstrate, due to the small dimensions of the hologram, but it exists too.

In (4.69) it is easily recognized that for $z \to \infty$, which would imply a Fourier transform hologram, the reconstruction algorithm reduces to a mere Fourier transform without prior multiplication. But in this case all depth information is lost.

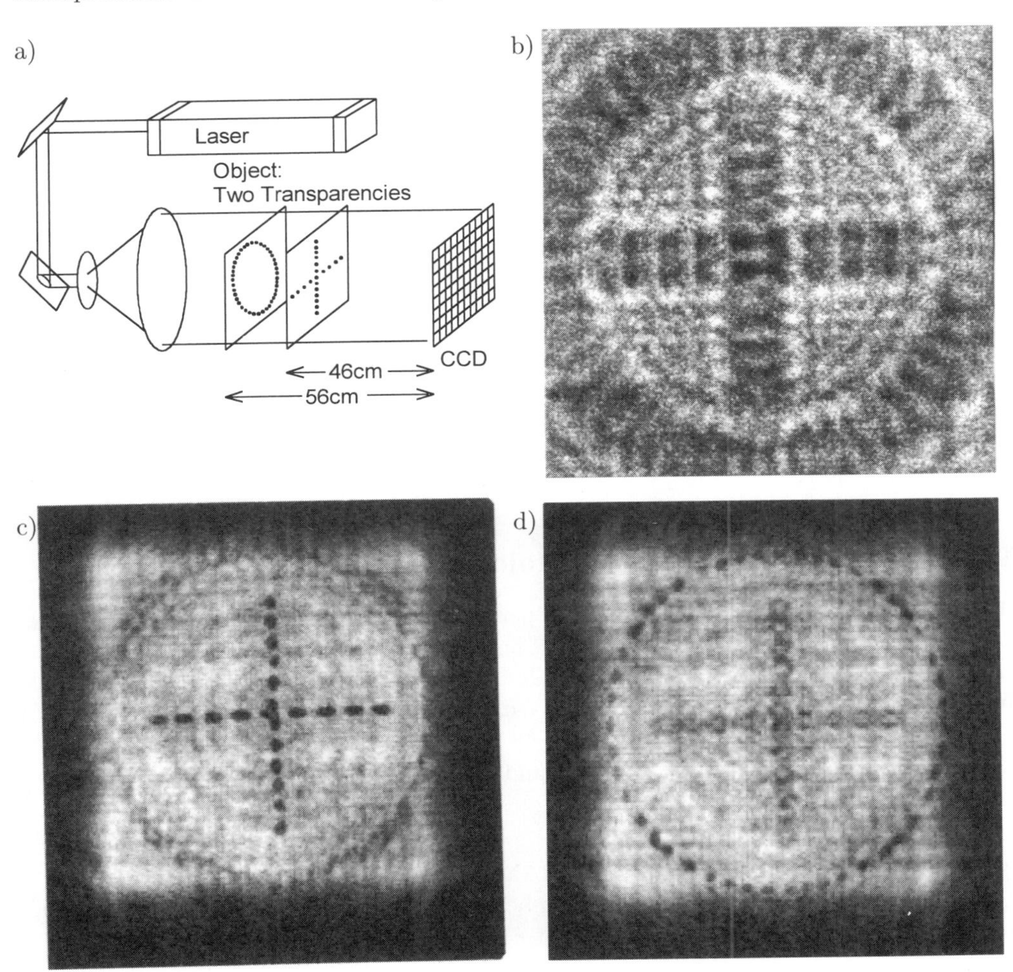

Figure 4.29: Digital in-line holography, (a) Set-up, (b) digital hologram, (c and d) numerical reconstruction with d = 46 cm (c), d = 56 cm (d) (Courtesy of U. Schnars, BIAS)

4.7.2 Digital Holographic Interferometry

In the same way as holography can be extended to double exposure holographic interferometry, digital holography can be employed to perform *digital holographic interferometry*. In a straightforward translation two holograms with the amplitude transmittances $\tau_1(k,l)$ and $\tau_2(k,l)$, corresponding to different states of the object are recorded and stored subsequently and then added pointwisely, [267]. Since the Fresnel transform (4.69) is essentially a Fourier transform and thus is linear, (A.28) and Table A.3, it reconstructs the sum of the two wave fields. The resulting intensity exhibits the cosine-shaped interference pattern according to (3.3). This interference pattern may be evaluated by one of the various interference phase determination methods.

But in digital holography there is a more effective way of interference phase determination: If the two holograms $\tau_1(k,l)$ and $\tau_2(k,l)$ are not added but reconstructed separately, their phase distributions $\phi_1(n,m)$ and $\phi_2(n,m)$ can be calculated separately by (4.71). The interference phase, which is the phase difference between the wave field of the object before and after a change of the loading, is then calculated by

$$\Delta\phi(n,m) = \begin{cases} \phi_2(n,m) - \phi_1(n,m) & \text{if} \quad \phi_2(n,m) \geq \phi_1(n,m) \\ \phi_2(n,m) - \phi_1(n,m) + 2\pi & \text{if} \quad \phi_2(n,m) < \phi_1(n,m) \end{cases} \tag{4.72}$$

Although the individual phases $\phi_1(n,m)$ and $\phi_2(n,m)$ are randomly fluctuating from point to point, the difference modulo 2π is deterministic. This difference is the interference phase distribution governed by the deformation of the object surface or the variation of the refractive index in the transparent object. The two possible approaches to achieve the interference phase distribution in digital holographic interferometry are summarized in Fig. 4.30.

4.7.3 Application of Digital Holography

Digital holography is limited mainly by the parameters of the CCD-target. Let us assume for simplicity that the pixels have no spacing between and no overlap, then the pixel center distances Δx and Δy agree with the pixel dimensions. The *sampling theorem* states that only spatial frequencies less than $f_{x,max}$ in x-direction and $f_{y,max}$ in y-direction with

$$f_{x,max} = \frac{1}{2\Delta x} \qquad \text{and} \qquad f_{y,max} = \frac{1}{2\Delta y} \tag{4.73}$$

can be reliably reconstructed.

The frequency is the inverse distance of the fringes, which is $\pi/|\boldsymbol{k}''|$ as shown in Sec. 2.2.1. Let the *maximum admissible angle* between object wave and reference wave be $\gamma_{x,max}$, then by (2.30) we get

$$\gamma_{x,max} = 2 \arcsin\left(\frac{\lambda}{2} f_{x,max}\right) \tag{4.74}$$

The analogous relations hold in the y-direction.

As a numerical example take a sensor of 1024×1024 pixels, each $6.8\ \mu m \times 6.8\ \mu m$ large and use a wavelength of $0.633\ \mu m$. Then γ_{max} is 2.67^o and for the *off-axis configuration* of Fig. 4.31 the object 1 m away from the target must be smaller than 4.6 cm.

Larger objects have to be placed further apart from the CCD-target or their apparent size has to be reduced by a lens. This latter option is feasible since there is no difference, whether the object wavefront is coming directly or imaged by an optical system to the CCD-chip where it interferes with the reference wave. A typical arrangement employing

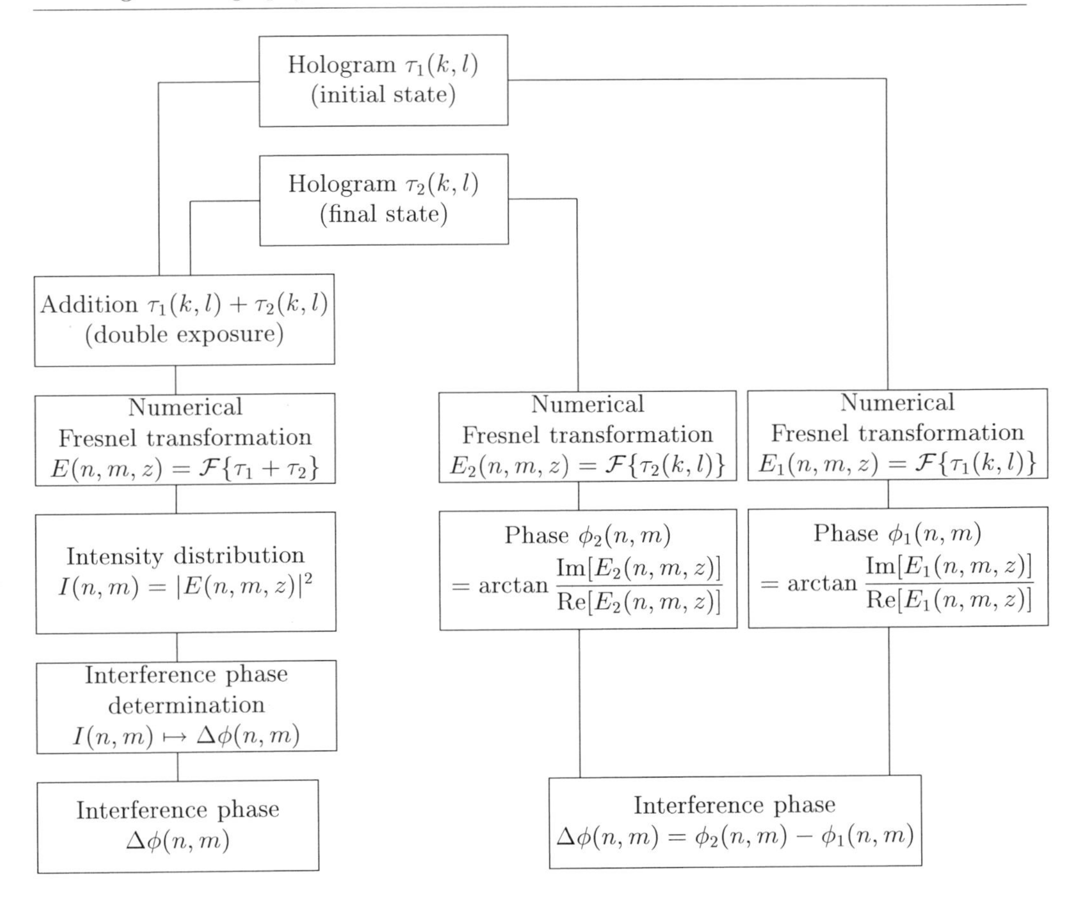

Figure 4.30: Digital holographic interferometry

a concave lens to reduce the angle of the object wave is shown in Fig. 4.32. Instead of the large angle wavefronts (α) from the original object now the wavefronts from the virtual image of the object (β) are impinging onto the target. So all angles between rays from

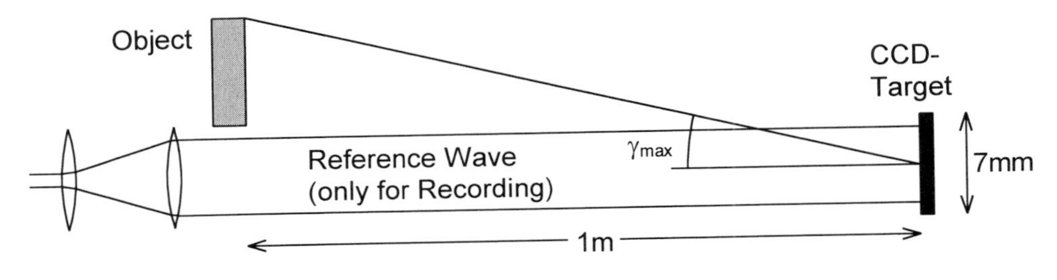

Figure 4.31: Off-axis configuration for digital holography

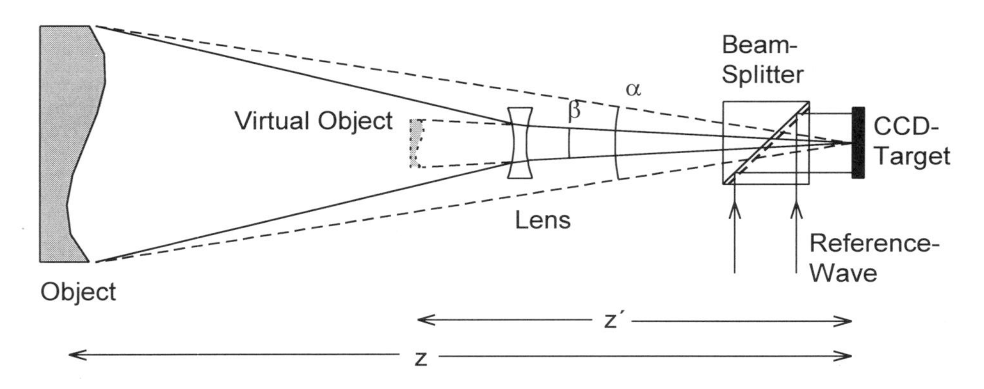

Figure 4.32: Digital holography arrangement for objects of large dimensions

the virtual object and the reference wave remain below the threshold defined in (4.74). For numerical reconstruction, the distance z has to be replaced by the distance z' of the virtual object to the target [270]. With this arrangement the transient deformation of an impact loaded plate consisting of fiber reinforced plastic was measured [271]. The plate was clamped at three sides in a rigid frame. A pneumatically accelerated steel ball hit the plate at the open side and caused the bending waves propagating along the plate. Two holograms were digitally recorded: The first one with the plate at rest before the impact, the second one 5 μs after the impact. A ruby laser in single-pulse operation with a pulse length of about 30 ns was used. For recording the second hologram, first the CCD-shutter was opened, still in darkness. Then the ball was released and intercepted the light beam of a photoelectric barrier, which generated the start signal for the laser. The time delay between impact and laser pulse was electronically adjusted. Fig. 4.33a shows

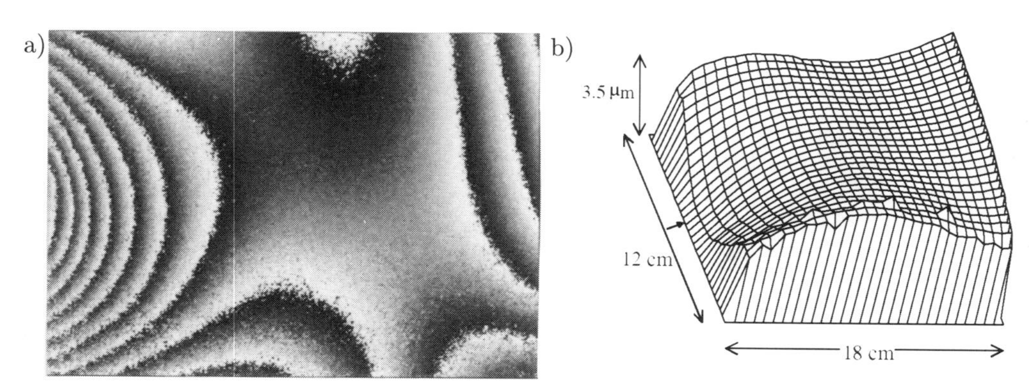

Figure 4.33: Bending of an impact-loaded plate measured by digital holography

the interference phase distribution modulo 2π after subtraction of the two reconstructed phase fields and Fig. 4.33b displays the unwrapped interference phase which corresponds to the deformation field. The open side of the plate, hit by the ball, in Fig. 4.33a is the left vertical side.

Up to now we have only considered the numerical reconstruction of the *real image*. Of course the *virtual image* can be made visible also by simulating numerically an imaging lens, Fig. 4.34.

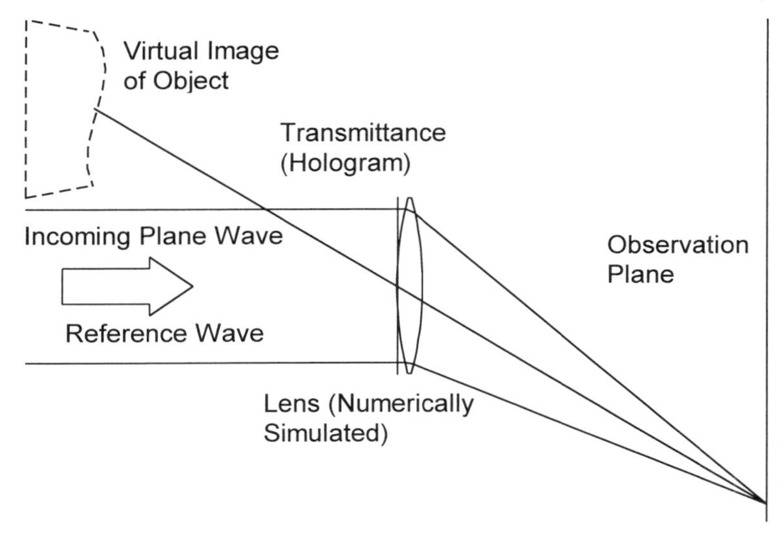

Figure 4.34: Reconstruction of the virtual image

The lens with focal length f is given by the complex transmittance (2.74)

$$T_L(x,y) = \mathrm{e}^{\dfrac{\mathrm{i}\pi}{\lambda f}(x^2+y^2)} \tag{4.75}$$

If we aspire at a magnification of 1 we must set

$$\frac{1}{f} = \frac{2}{z} \tag{4.76}$$

Replacing $\tau(x,y)$ by $\tau(x,y)T_L(x,y)$ in the Fresnel-Kirchhoff integral (2.72), we get

$$E(\xi,\eta,z) = \frac{\mathrm{i}E_0}{\lambda z}\mathrm{e}^{-\dfrac{\mathrm{i}\pi}{\lambda z}(\xi^2+\eta^2)}\int_{-\infty}^{\infty}\int_{-\infty}^{\infty}\tau(x,y)\mathrm{e}^{+\dfrac{\mathrm{i}\pi}{\lambda z}(x^2+y^2)}\mathrm{e}^{\dfrac{\mathrm{i}2\pi}{\lambda z}(x\xi+y\eta)}dxdy \tag{4.77}$$

differing from (2.72) only by the sign in the exponential the transmittance is multiplied with. This can be carried over to the numerical evaluation by changing the sign in the corresponding exponential of (4.69).

As we have seen, digital holography not only can be used in conjunction with *off-axis holography* for measuring deformations of opaque objects, but it can be applied without further difficulties with *in-line configurations*, to *phase objects* or *scattering particles* in transparent media. *Holographic contouring* as described in Sec. 5.6 also is feasible with digital holography.

In the case of additional *rigid body translations* digital holography reveals an advantage over holographic interferometry with optical reconstruction. Let the rigid body translations be larger than the speckle size so that no correlation of the microinterference patterns takes place in optical holographic interferometry or in the ESPI/DSPI-methods. Having stored the digitized holograms of the two object states, this translation can be determined by a numerical correlation of the two reconstructed complex amplitude distributions or by a numerical realization of the speckle-photography method [272]. Then the determined translation can be compensated by shifting the reconstructed fields by a proper number of pixels in x- and y-direction, permitting a subsequent interference phase determination.

The digital recording of the holograms without changing plates or transporting thermoplastic film and without chemical or electric processing now enables the analysis of *dynamic events*. A sequence of holograms can be recorded with video frequency. The numerical evaluation has to be performed later on, since the numerical Fresnel transform today still requires some time. Fast computation may be performed by special purpose hardware processors or parallel computers, but a general purpose PC normally is sufficient for these tasks.

4.8 Dynamic Evaluation

Up to now in this chapter we have considered methods evaluating the interference fringes which appear superimposed over the reconstructed virtual image of the object. The fringes are observed from a fixed point of observation. These approaches to the evaluation of holographic interference patterns can be summarized as the *static evaluation methods* [273].

On the other hand the holographic interference fringes are generally localized in space, Sec. 3.3, so different observation points generate different interference patterns. A continuous variation of the observation point thus will induce a continuous change of the observed interferogram during reconstruction. This is the basic idea behind the so called *dynamic evaluation methods*, which have been proposed very early after the invention of holographic interferometry [63, 274, 275]. But despite of some conceptual advantages the dynamic methods have by far not found the wide-spread applications like the static methods.

4.8.1 Principles of Dynamic Evaluation

In the dynamic evaluation methods the interference orders are counted which are moving over a single observed object point while the observation point is continuously changed. Let us consider the case of a deformation of an opaque surface. An arbitrary point P of the surface may be displaced by the vector $\boldsymbol{d}(P)$ between the two exposures of a *double exposure hologram*, Fig. 4.35. Let this point be observed from an observation point which continuously moves from B_1 to B_2, then according to (3.20) and (3.21) the interference phase changes from $\Delta\phi_1(P) = \frac{2\pi}{\lambda}\boldsymbol{d}(P)\cdot[\boldsymbol{b}_1(P)-\boldsymbol{s}(P)]$ to $\Delta\phi_2(P) = \frac{2\pi}{\lambda}\boldsymbol{d}(P)\cdot[\boldsymbol{b}_2(P)-\boldsymbol{s}(P)]$. What has to be measured is the interference phase difference $\Delta\Delta\phi_{1,2}(P)$ between the two

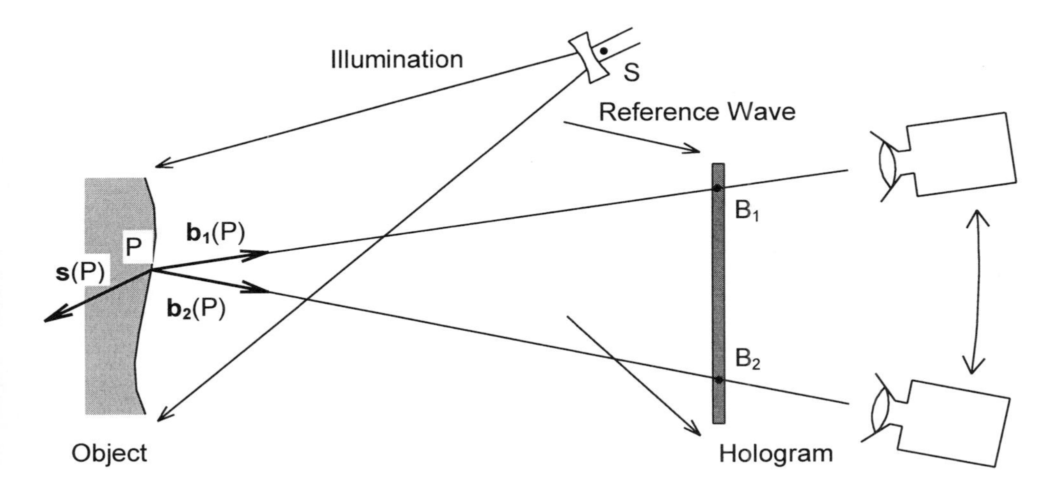

Figure 4.35: Geometry for dynamic evaluation

observation points B_1 and B_2, which is

$$\begin{aligned}\Delta\Delta\phi_{1,2}(P) &= \Delta\phi_2(P) - \Delta\phi_1(P) \\ &= \frac{2\pi}{\lambda}\boldsymbol{d}(P)\cdot[\boldsymbol{b}_2(P) - \boldsymbol{b}_1(P)]\end{aligned} \tag{4.78}$$

This is the central equation for the dynamic evaluation methods.

Two advantages of the dynamic methods now become obvious: (1) the interference phase difference does not depend on the illumination vector $\boldsymbol{s}(P)$, and (2) there is no absolute phase problem, since the additive phase term which is identical in $\Delta\phi_1(P)$ and $\Delta\phi_2(P)$, is canceled in the subtraction leading to $\Delta\Delta\phi_{1,2}(P)$. What remains is the necessity to determine $\Delta\Delta\phi_{1,2}(P)$ with the proper sign distribution, especially sign changes which may occur for complicated curves of localization, must be recognized exactly.

The main sensitivity of the dynamic evaluation methods is in direction $\boldsymbol{b}_2(P) - \boldsymbol{b}_1(P)$, which is roughly parallel to the hologram plane and thus in most applications nearly tangential to the object surface. This complements the static methods, where the main sensitivity is for displacements in direction $\boldsymbol{b}(P) - \boldsymbol{s}(P)$, which is nearly normal to the surface.

If a one-dimensional displacement parallel to the hologram plane has to be evaluated, a single measurement of $\Delta\Delta\phi_{1,2}(P)$ suffices and a scalar form of (4.78) has to be solved for $d(P)$. For a two-dimensional displacement two measurements $\Delta\Delta\phi_{1,2}(P)$ and $\Delta\Delta\phi_{3,4}(P)$ are required, which may be evaluated from a single hologram if both displacement components are parallel to the hologram. In the general three-dimensional case three observations are necessary and the three-dimensional system of equations (4.78) has to be solved. The three observations can be taken through two holograms whose normals should form a significantly large angle, $> 30^o$, to the best a 90°-angle. An alternative to the second hologram can be a mirror close to the object, so that the investigated object point can be observed in two directions employing a single hologram [275].

4.8.2 Dynamic Evaluation by a Scanning Reference Beam

A straightforward approach to dynamic evaluation is to record many static reconstructions during a shift of the camera whose optical axis is kept intersecting a fixed object surface point in the virtual image. The angular separation between the frames must be small enough that the fringe count cannot be lost during data analysis. In [273] this was performed with thirty-nine data frames, which have been recorded and evaluated with the image processing facilities of those days. The experimental error could be hold down to about 5 %.

A more direct fringe counting is possible, if the reconstruction of the holographic interferogram is performed by a thin reconstruction beam conjugate to the original reference beam [276, 277]. In this way, a real object image is obtained at the position of the original object, Fig. 4.36. The position of the reconstruction beam on the hologram plate defines the actual observation point. When the position of the reconstruction beam on the hologram is changed, the projected fringes move across the real image of the object surface. The varying intensity is recorded by a photodetector placed at the object point, whose displacement is to be evaluated. To get the displacements at many points the detector position must be scanned over the real image, the same way as in temporal heterodyning, Sec. 4.4.

Due to the small beam diameter, the aperture at the hologram is low and thus fulfills the requirements of the *observer projection theorem*, Sec. 3.3.5. That means the fringes are projected onto the real image plane independently of the localization in space.

The hologram may be scanned along several lines by a conjugate reference beam, which is produced by a spherical mirror or a converging lens corrected for spherical aberrations. Since more than three measurements now are taken, (4.78) is evaluated by the least squares method [277].

A thin conjugate reference beam can scan along closed loops. A repeated closed loop scan may filter temporal fluctuations by averaging. Arbitrary pairs of observation points B_i now can be chosen from the loop, Fig. 4.36 [276, 278, 279, 280].

Nothing is found in literature about how the fringes, especially fractions of the fringes, are counted, or about the translation of the recorded intensity to interference phase. A change from increasing to decreasing interference order, or vice versa, must be detected. In this context the closed loop scanning offers some advantages: The phase difference along one whole turnaround must be zero. If the interference phase is calculated by one-dimensional Fourier transform evaluation, Sec. 4.6, due to the continuous junction at the ends of a whole-turn sample, there is no wrap-around error. Turning points of the fringe order, which must be existent, should be easily detected in the primarily monotonic phase distribution after Fourier transform evaluation. On the other hand phase shifting combined with dynamic evaluation should be principally feasible.

A major drawback of the dynamic evaluation is the existence of moving parts in the evaluation configuration and the necessary scanning of the real image, comparable to temporal heterodyning.

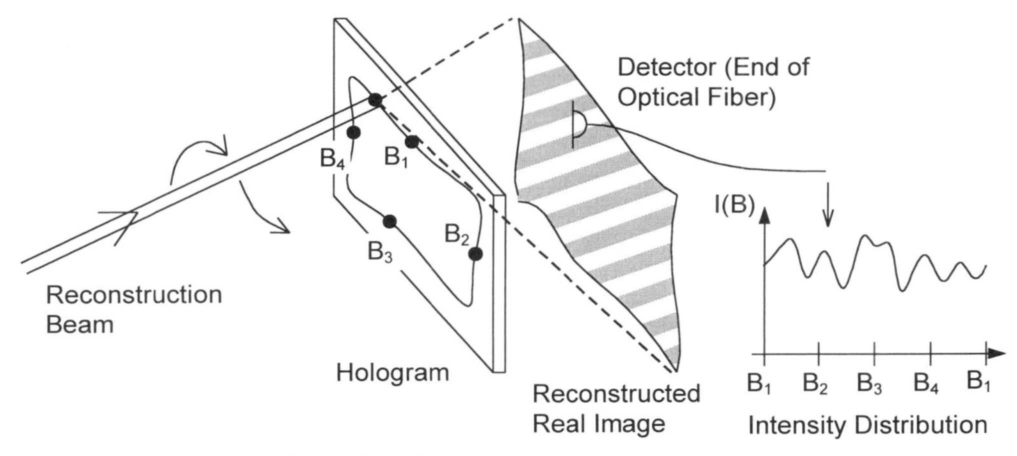

Figure 4.36: Dynamic evaluation by a scanning reconstruction beam

4.9 Interference Phase Demodulation

As outlined in Sec. 4.1.1, the evaluated interference phase distributions are ambiguous in having only values between $-\pi$ and $+\pi$. They are said to be wrapped modulo 2π [281, 282]. Although in most practical applications a continuous interference phase distribution is expected, we get a sawtooth-like phase as the one shown in Fig. 4.37a. The process of resolution of the 2π-discontinuities by adding a step function consisting only of 2π-steps, Fig. 4.37b, is called *continuation*, *phase unwrapping* or *demodulation*.

4.9.1 Prerequisites for Interference Phase Demodulation

There are a number of requirements to be fulfilled in order of a reliable demodulation: The first is that continuous phase data are adequately sampled. Each demodulation procedure checks interference phase differences of neighboring pixels. If due to a violation of the sampling theorem this difference exceeds π, the demodulation must fail because of the introduction of unnecessary phase jumps. On the other hand undersampled high frequency fringe data may result in small phase differences between two pixels and a phase jump between these pixels is missed.

Statistical noise like speckle noise is a common cause for false identification of phase jumps. As soon as the amplitude of the noise approaches π, the actual phase jumps become obscured. A low pass filter for smoothing the interference phase data is not recommendable, since it will wash out the sharp 2π-steps. Instead median filtering is a good choice, since it is 2π-step preserving [174].

Each demodulation procedure assumes a continuous interference phase distribution which is wrapped into an interval of length 2π. So discontinuities resulting from large height steps or holes or edges of objects not filling the full frame have to be avoided. The region in which the demodulation should be performed has to be defined by masking. If

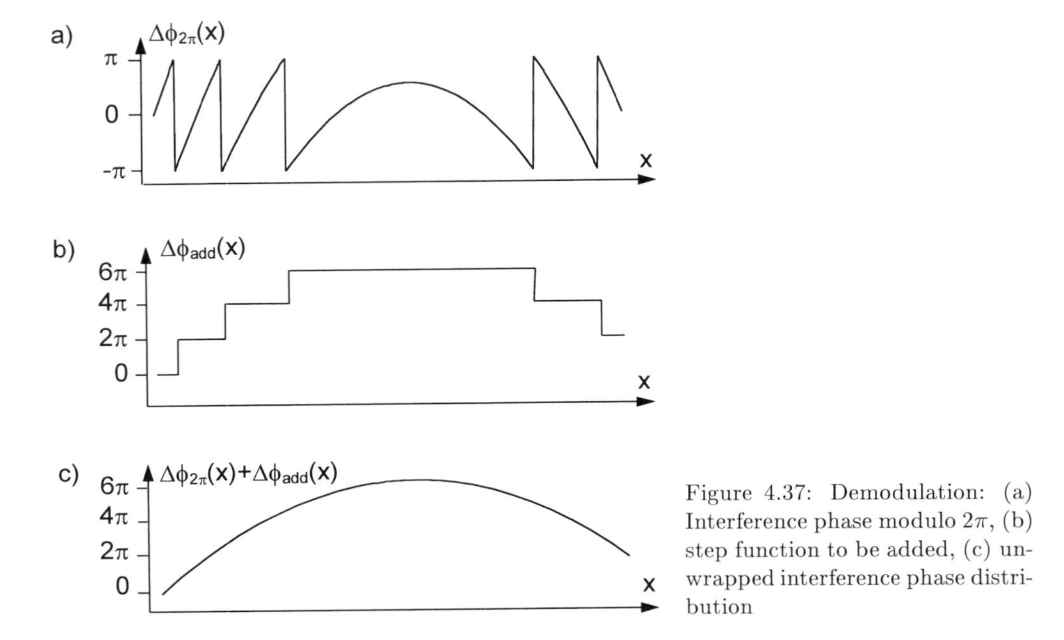

Figure 4.37: Demodulation: (a) Interference phase modulo 2π, (b) step function to be added, (c) unwrapped interference phase distribution

the region for demodulation is not topologically connected, then no correct relations of the interference phase in the unconnected parts can be derived without additional side information, e. g. particular points in each part where the exact displacement is known. An approach to automatic phase unwrapping in the presence of surface discontinuities is described in [283].

Especially if the interference phase distribution has been determined by the Fourier transform method without additionally introduced carrier fringes, 4.6, then sign-errors may occur. The sign-error must be corrected, e. g. as described in Sec. 4.6.3, before demodulation. Each demodulation requires sign-correct interference phase distributions modulo 2π [284], only a global sign-change may be tolerable, leading to a global sign-change in the unwrapped continuous phase.

The demodulation procedures can roughly be categorized into *path dependent demodulation* techniques, where the order in which the pixels are investigated and unwrapped, is predetermined by the process, and into *path independent demodulation* techniques. In this latter case the order of investigated pixels is determined by the phase values at the pixels.

4.9.2 Path-Dependent Interference Phase Demodulation

A straightforward attempt to demodulation of a one-dimensional interference phase distribution $\Delta\phi(x)$ is done by checking the phase differences of adjacent pixels $\Delta\phi(x+1) - \Delta\phi(x)$. If this difference is less than $-\pi$, an additional 2π is added to $\Delta\phi$ from $x+1$ onwards; if the difference is greater than $+\pi$, one more 2π is subtracted from $\Delta\phi$ starting at pixel $x + 1$. Several of these 2π-terms may cumulate, Fig. 4.37b, so that at each pixel an integer multiple of 2π, $\phi_{add}(x) = 2\pi n(x)$ with $n(x) \in \{\ldots, -1, 0, 1, \ldots\}$ must be added for unwrapping. The starting point need not necessarily lie at the leftmost pixel $x = 1$. If a central starting pixel x_0 is chosen, differences to the right $\Delta\phi(x+1) - \Delta\phi(x)$ and to the left $\Delta\phi(x-1) - \Delta\phi(x)$ have to be calculated.

Demodulation algorithms based on this approach can be elegantly defined [285, 286], but these algorithms strongly depend on numerical differentiation, which will amplify the influence of noise in the phase data. If, due to noise, a wrong difference occurs, leading to an erroneous 2π-term added or subtracted or missing a necessary 2π-term, the resulting phase error spreads up to the outmost pixel (if it is not neutralized by another error in the opposite direction). In many practical applications an object is illuminated by an expanded Gaussian laser beam. Then at the margins of the recorded interference pattern the contrast decreases, resulting in a higher probability for erroneous interference phase values and thus false phase differences between adjacent pixels. Therefore a central starting point in the pattern is most often recommendable.

A final check for plausibility of the unwrapped interference phase distribution may be performed by fitting a line through several adjacent pixels of the unwrapped phase and comparing the phase values of the next pixels to this line. This may detect false 2π-steps due to noise, as long as a sufficiently smooth phase distribution can be assumed.

The described one-dimensional demodulation procedure can be transferred to two dimensions along rows and columns of pixels. Let us start, e.g., with one row. Once this row is demodulated, the pixels of this row act as starting pixels for column demodulation. There are some trivial variations of this scheme: One may process one row, then go down one pixel, process the next row in reverse direction and so on. The two-dimensional phase data are now treated like a folded one-dimensional data set [287]. Or each row is unwrapped independently from the other rows and then the rows are adjusted by scanning through a center column [288] or by comparing the phase differences averaged along each row.

Another approach makes use of the two dimensions [289]. At each pixel the phase differences to the pixel above in the previous row and the one to the left in the same row are checked. If both differences indicate the same $n(x)$, this is taken for unwrapping. If the differences differ, the pixel is masked and its unwrapping is postponed until more neighbors of it are unwrapped. This concept may be improved to spiral scanning with starting at a central pixel [290]. Instead of checking the pixels above and to the left, by application of a spin filter the direction of the highest slope of the phase distribution can be calculated, and the pixels in this direction are tested for unwrapping [291].

Demodulation along predetermined paths gets difficulties if coming to a masked region, e. g. at holes in the surface. The spiral scanning then can be modified to scan along the boundaries of the masked area [290].

4.9.3 Path-Independent Interference Phase Demodulation

To avoid the difficulties with a possible spreading of erroneous phase, or a failing at holes, *path independent demodulation* procedures are recommended [235, 292, 293, 294, 295, 296]. The following algorithm interprets the interference phase distribution modulo 2π as a graph, where the points are the nodes, and the arcs are the connections between neighboring points. 4-neighborhoods or 8-neighborhoods may be used. With each arc a value $d_{2\pi}(\Delta\phi_1, \Delta\phi_2)$ is associated, defined by the phase values $\Delta\phi_1$ and $\Delta\phi_2$ of the two points it connects:

$$d_{2\pi}(\Delta\phi_1, \Delta\phi_2) = \min\{|\Delta\phi_1 - \Delta\phi_2|, |\Delta\phi_1 - \Delta\phi_2 + 2\pi|, |\Delta\phi_1 - \Delta\phi_2 - 2\pi|\}. \tag{4.79}$$

The values $d_{2\pi}$ may be interpreted as a distance modulo 2π.

The demodulation now proceeds along paths where these distances are least. Along these paths the probability of an erroneous demodulation is least. Points with wrong phase are surrounded this way and the same is true for regions without interference phase at all, indicated by a mask. If a point possesses an erroneous phase that cannot be reached correctly along any path, the incorrectly demodulated point in the vast majority of cases remains isolated in the finally resulting interference phase distribution.

The algorithm proceeds as follows:

1. For a starting point all emanating free arcs are recorded in a list together with their values $d_{2\pi}$.
2. The minimal value in the list is searched for. The demodulation term 0, -2π, or $+2\pi$ for this arc is stored in an extra file. The arc, together with its value, is discarded from the list and marked to avoid repeated consideration.
3. The final node of the just considered arc acts as a new starting point in step 1. Only those arcs which have not been stored formerly in the extra file in step 2, are considered as free. If no free arcs emanate from this node, proceed to step 2 directly.
4. If the capacity of the list is exhausted, the list is checked for arcs which may have already entered the extra file along another path. These arcs are deleted.
5. Steps 1 to 4 are repeated until all points have been an end node of an arc in the extra file.
6. The interference phase distribution is demodulated by using the values in the extra file. There is one and only one path to each point. The -2π- or $+2\pi$-terms of course may cumulate along these paths.

The demodulation of step 6 may be performed during step 2 in an extra image of already demodulated points. Another modification of this algorithm, which shortens the computational effort, allows only arcs with values less than a prescribed threshold to be recorded in the list. In this way the number of comparisons for searching the minimum in step 2 can be drastically reduced.

This demodulating algorithm is demonstrated by the numerical example of Fig. 4.38a. Fig. 4.38b shows, how the algorithm detects the best track and circumvents the one bad spot. Fig. 4.38c displays the resulting unwrapped phase distributions after demodulation along predetermined horizontal paths and after path independent phase demodulation. The interference phase distributions of Figs. 4.25j and k, and 4.26 f are unwrapped by this path independent algorithm.

A demodulation algorithm which is claimed to be noise-immune [297, 298] first checks the interference phase distribution for inconsistent points. This check, which is described in more detail in Sec. 4.9.4, is based on the observation that for a continuous phase distribution the integral along any closed loop must be zero. In the discrete case this integral is calculated by the sum of the differences between the phases of all adjacent pixels along the loop. By testing the 4-point loops in all 2×2-arrays, inconsistent points which act as sources for discontinuities in the unwrapping, are identified. Each such source is paired with a source of opposite sign or with the boundary of the phase-map by the shortest possible cut-line. The unwrapping process then will not cross these cut-

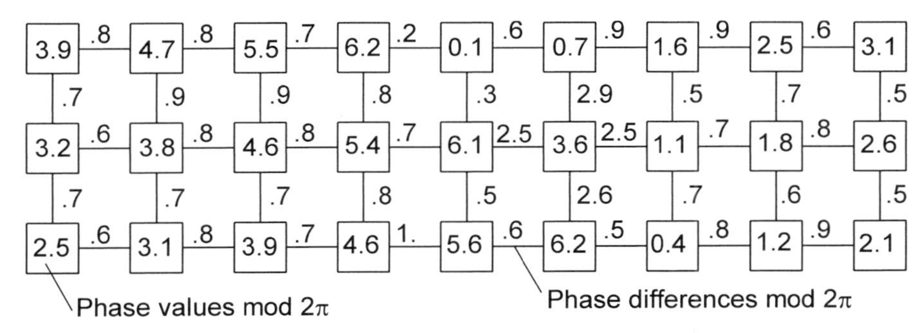

Figure 4.38: (a) Path independent interference phase demodulation, interference phase values

lines but follows the pixels neighboring the cut-line. A similar approach but based on an identification of probable discontinuities by searching for regions in which the phase curvature exceeds some threshold is presented in [284]. For that purpose second derivatives of the phase data are calculated and investigated.

A *regional processing* algorithm where the phase map modulo 2π first is segmented into regions containing no phase ambiguities, is described in [299, 300, 301]. After in these regions the demodulation is completed, adjacent edges of the regions are investigated for discontinuities. The regions then are mutually phase shifted in order to reduce the discontinuities between them. Now the regions can merge until at the end one region fills the whole data array.

A combination of the concepts of regional processing, pixel-processing along predetermined paths, and processing in order of minimal likelihood for erroneous demodulation is known as *tile processing* [302, 303, 304]. The whole data array is divided into rectangular sub-arrays, the tiles. In each tile the interference phase is unwrapped by a path dependent method. Then the phase data at the edges of adjacent tiles are compared and the phase is adjusted by the same 2π-multiple for the whole tile to fit to the adjacent tile. Tiles with a

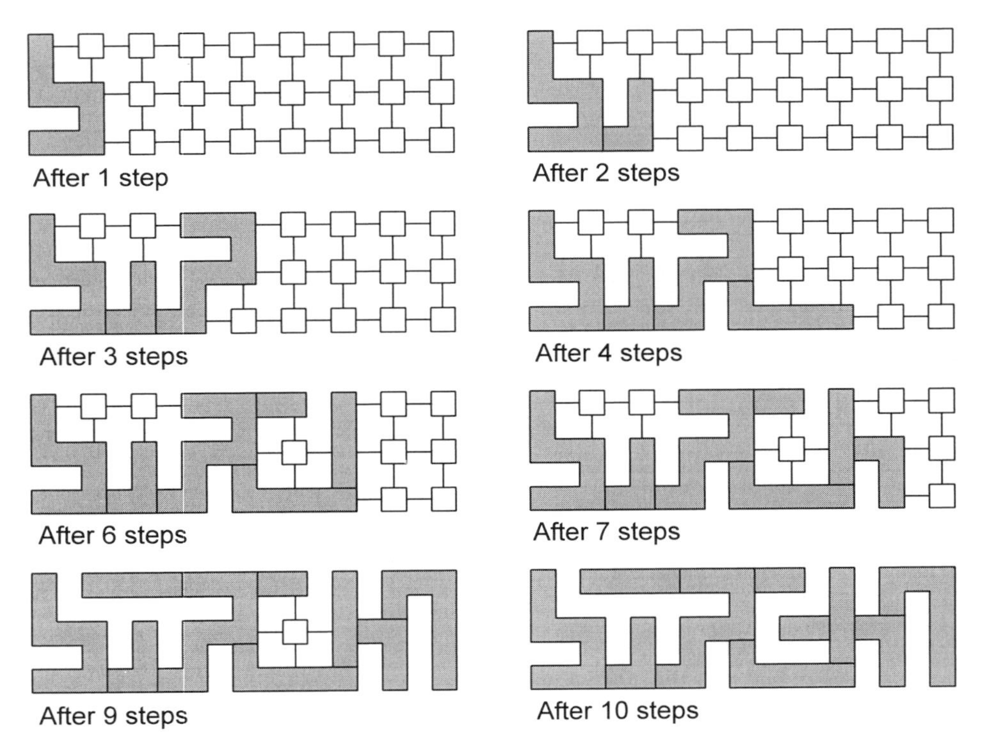

Figure 4.38: (b) Path independent interference phase demodulation, demodulation steps

large percentage of edge pixels agreeing to the edge pixels of adjacent formerly processed tiles get a high degree of confidence and thus are preferred in the order of processing. Any discontinuity will be restricted to a region within one tile. The main parameter affecting this method is the size of the tiles which should be chosen such that on the average there are one or two fringes crossing each tile [287].

4.9.4 Interference Phase Demodulation by Cellular Automata

The use of *cellular automata* for interference phase demodulation has first been proposed in [305]. A modification of the algorithm of [305] is given in [306]. Cellular automata are simple, discrete mathematical systems, whose global behaviour results from collective effects of a large number of cells. The state of each cell evolves in discrete time steps according to rather simple local neighborhood rules.

For phase unwrapping the decision at each pixel whether 2π are added or subtracted or not, depends on the phase differences to the neighboring points. In [305] these are the 4 nearest neighbors in horizontal and vertical direction, in [306] all 8 neighbors to the central point of a 3×3-neighborhood are considered. Each phase difference of the investigated pixel to one of its neighbors gives a vote for addition of 2π if the difference is greater than π, a vote for subtraction of 2π if it is less than $-\pi$, and no vote if the absolute

3.9	4.7	5.5	6.2	6.4	7.0	7.9	8.8	9.4
3.2	3.8	4.6	5.4	6.1	3.6	1.1	1.8	2.6
2.5	3.1	3.9	4.6	5.6	6.2	6.7	7.5	8.4

Path-dependent demodulation (horizontal direction)

3.9	4.7	5.5	6.2	6.4	7.0	7.9	8.8	9.4
3.2	3.8	4.6	5.4	6.1	9.9	7.4	8.1	8.9
2.5	3.1	3.9	4.6	5.6	6.2	6.7	7.5	8.4

Path-independent demodulation

Figure 4.38: (c) Path independent interference phase demodulation, comparison of results

difference is less than π. Then 2π are added or subtracted according to the majority of votes. In the case of a tie, addition or subtraction is chosen arbitrarily, only if no vote at all was given, the actual value is retained. This procedure applied to all pixels is called a *local iteration*.

After a number of local iterations, - a number, which mainly depends on the maximum distance of 2π-steps in the original phase-map, - oscillation between successive patterns occurs. When this oscillatory state is reached, the arithmetic average of the two states is computed at each pixel and the process starts anew. The averaging is called *global iteration*. If a steady state is reached after global iteration, the process stops.

The process of cellular automata demodulation is shown in Fig. 4.39 at the example of a simulated 32×32-pixel interference phase modulo 2π, Fig. 4.39a. The first oscillation takes place after 6 local iterations, Fig. 4.39d. After 3 global iterations, the phase is successfully demodulated, Fig. 4.39i.

The algorithm is robust against distortions and noise, but there are possible points or regions, which would lead to new 2π-discontinuities after each global iteration, so that the algorithm would never terminate. Therefore a *check for consistency* has to be performed. This consists of summing the four phase differences along the closed path connecting the four pixels of a 2×2-neighborhood. Here each phase difference is taken modulo 2π. If the sum is zero, the neighborhood is said to be consistent, otherwise it is interpreted as being inconsistent. Fig. 4.40a gives a consistent 2×2-array, since

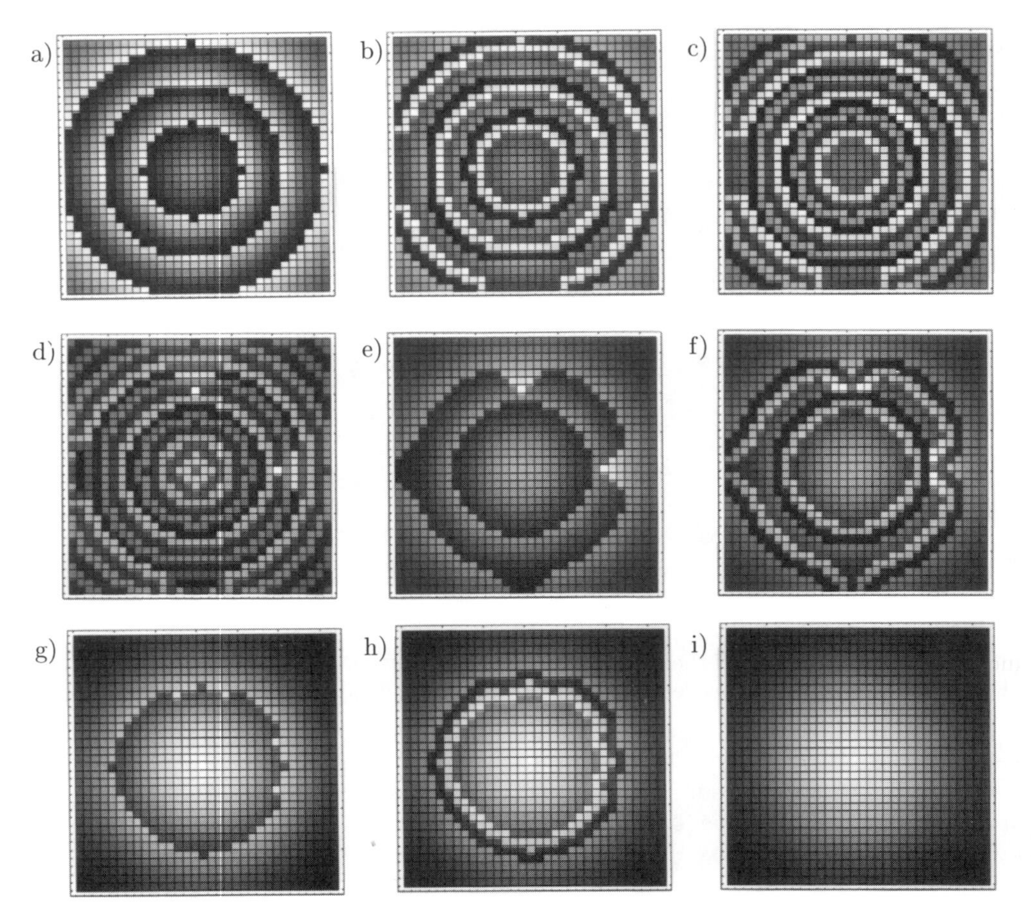

Figure 4.39: Cellular automata demodulation: (a) Interference phase modulo 2π, (b) phase after 1 local iteration, (c) phase after 2 local iterations, (d) phase after 6 local iterations, (e) phase after 1 global iteration, (f) phase after 1 global and 1 local iterations, (g) phase after 2 global iterations, (h) phase after 2 global and 1 local iterations, (i) phase after 3 global iterations

a)

2.67	1.94
-2.77	-2.04

b)

2.45	1.08
-2.53	-1.15

Figure 4.40: Consistent (a) and inconsistent (b) 2×2-arrays of interference phase values modulo 2π

$[-2.77-2.67]+[-2.04+2.77]+[1.94+2.04]+[2.67-1.94] = 0.84+0.73-2.30+0.73 = 0$. Here $[.]$ denotes the value modulo 2π in $]-\pi, \pi]$. Fig. 4.40b is an example of an inconsistent array, for $[-2.53 - 2.45] + [-1.15 + 2.53] + [1.08 + 1.15] + [2.45 - 1.08] = 1.30 + 1.38 + 2.23 + 1.37 = 6.28$. The inconsistent regions are masked and discarded from being tested in the local iterations. The inconsistency test may be performed once before the first local iteration [305] or additionally after each global iteration [306]. Nevertheless there remain the problems that the inconsistency check does not guarantee convergence to the steady state for all possible phase-maps, especially those with phase dislocations induced by aliasing [305], or that the inconsistency regions will increase in size [306].

4.9.5 Further Approaches to Interference Phase Demodulation

The demodulated interference phase distribution is assumed to have no sharp discontinuities which would give rise to high frequency components in the Fourier spectrum. A check of the bandwidth of a one-dimensional phase distribution by the Fourier transform is the base for the *bandlimit demodulation* [307, 308]. Here all possible step functions with one $+2\pi$- or -2π-step are tested, the step function that minimizes the bandwidth is selected as one that unwraps at one position. This process is repeated until no further reduction of the bandwidth is obtained. The method has been shown to be robust in the presence of signal-independent random additive noise, provided that the correct interference phase function is slowly varying. The method will fail if the number of phase steps increases. The extension of the bandlimit approach to two-dimensional phase-maps on a more effective basis than mere repeated row- or column-processing has still to be done.

An actual approach to demodulation is the application of a *neural network* of the Hopfield type [309, 310]. In *Hopfield networks* all possible connections between all neurons are existent. The synaptic weights only take on the inhibitory value 0 or the excitatory value 1. Hopfield networks are used to minimize a so-called Hopfield energy function. For demodulation purposes with each pixel a number of neurons is associated. A merit function $E(\phi_{add})$ is defined, which measures the deviation from smoothness of the phase distribution unwrapped by ϕ_{add}. The $n(x)$ in $\phi_{add}(x) = 2\pi n(x)$ is represented by the number of the excited ones of all neurons associated with pixel x. This merit function is related to the Hopfield energy function in a discrete-time Hopfield network. According to the equations of the network dynamics, the neurons change their states in such a manner that the value of the energy function decreases with the temporal evolution of the network until a stable state is reached.

As with the cellular automata approach in defining the merit function here we may use 4- or 8-pixel neighborhoods. Furthermore inconsistencies have to be excluded. This is done as described for the cellular automata demodulation, Sec. 4.9.4. But now no cut-lines as in [298] are defined, instead the synaptic connections across the cut-lines are prohibited.

All demodulation techniques presented so far spatially unwrap a single interference phase distribution. In *temporal phase unwrapping* as proposed in [311, 312] a number of measurements with increasing load is recorded, where the time increments are chosen small enough to fulfill the sampling theorem requirements. Now unwrapping is performed along the time axis for each pixel independently from its neighboring pixels. So objects not filling the full frame, unconnected regions, or regions with poor signal-to-noise ratios will not corrupt the demodulation of good data points. The requirement of sign-correct phase data is fulfilled by using the phase shift method with four interferograms and a mutual phase shift of 90°, see (4.25). By a proper combination of the equation for phase shift evaluation (4.25) and the calculation of the phase difference between successive time instants, this temporal phase difference can be calculated with values in $]-\pi, +\pi]$, as long as the sampling theorem is fulfilled. The overall interference phase without any 2π-discontinuity then for each time instant is determined by summation of the temporal phase differences.

A demodulation by two-dimensional fitting of unwrapped phase maps in a least squares sense is proposed in [293]. The problem of two-dimensional *least squares phase unwrapping* is shown to be equivalent to the solution of Poisson's equation on a discrete rectangular grid with Neumann boundary conditions. It is solved by application of a fast discrete cosine transform [313]. If less reliable pixels or regions are known, resulting from regionally varying noise, aliasing, phase inconsistencies, measurement errors, shadows, or no phase data at all, these parts of the pattern can be appropriately weighted. Weighting may be binary into 0 and 1 or continuously inverse proportional to the probability of noise or error. An algorithm based on Picard iteration is given for weighted least squares phase unwrapping. An algorithm which provides faster convergence is developed on the basis of the method of preconditioned conjugate gradients. Several examples in [293] show, how phase noise, data inconsistencies and other degradations are automatically accommodated by this least squares approach.

5

Processing of the Interference Phase

The calculation of the interference phase distribution as described in Chap. 4 normally is not the final goal of a quantitative evaluation of holographic interference patterns, but an intermediate step in the determination of the physical quantity of interest: components of the displacement vector field, strains, stresses, vibration amplitudes, contours, refractive index distributions etc. In Chap. 2 the formation of holographic interference patterns dependent on the geometry of the holographic arrangement and a given change of the optical wave field has been described. Based on these foundations in this chapter the inverse problem is treated, namely to conclude from an already evaluated holographic interference phase distribution and a given geometry of the holographic arrangement to the values and directions of the parameters describing the underlying physical process. The basics of computer-aided methods determining a variety of different physical quantities are presented. Since the topics considered in this chapter are under current research, there still remain some open problems which should be identified here and which may be solved in the near future.

5.1 Displacement Determination

In order to determine the components of a *displacement vector field* from the already evaluated interference phase distribution some assumptions have to be met. First a continuous displacement variation at least in defined areas of the interferogram is assumed, and second there should only remain a global uncertainty about the sign. These assumptions allow a continuous counting of the interference order along uniterrupted paths through the interferogram. Normally these assumptions are used in the demodulation of the interference phase, Sec. 4.9, to achieve a continuous phase distribution. Nevertheless there remains the problem of the overall sign ambiguity, Sec. 4.1.1, and the absolute phase problem, Sec. 4.1.2. There are applications, where the displacement at a reference point in the interferogram is known, so one can calculate the absolute phase at this point. This case needs a separate treatment from the case where no zero order or reference displacement is known. Moreover we have to discriminate between a sensitivity vector that varies over the interferogram and a constant sensitivity vector, as was pointed out in Sec. 4.1.2.

5.1.1 Displacement Determination with Known Reference Displacement

The basis of the displacement determination is the equation (3.21)

$$\Delta\phi(P) = \boldsymbol{d}(P) \cdot \boldsymbol{e}(P) \tag{5.1}$$

We assume that we have evaluated the interference phase distribution $\Delta\phi(P)$ without sign ambiguity by e. g. a phase shift method of Sec. 4.5. With a *known reference displacement* $\boldsymbol{d}(P_0)$ at a point P_0 the interference phase $\Delta\phi(P)$ at P_0 and at all P which are accessible by an uninterrupted path from P_0 can be calculated. The sensitivity vector for each P of interest has to be calculated according to (3.18) - (3.20), which has to be performed once for a representative P, if the sensitivity vector is constant, or separately for each P, if it is not constant. Then the displacement is determined by inversion of (5.1).

The easiest case is the measurement of only the normal component using an arrangement with a constant sensitivity vector $\boldsymbol{e}(P)$ having a non-zero component only in the normal direction. Then $d_z(P)$ is determined by

$$d_z(P) = \frac{\Delta\phi(P)}{e_z(P)} \tag{5.2}$$

as outlined in (3.25) and (3.26).

For a varying sensitivity vector this approach is only feasible if it is guaranteed that $d_x(P) = d_y(P) = 0$ for all P. Otherwise errors may occur, since the components $e_x(P)$ or $e_y(P)$ do not vanish for all P. Sometimes this error can be estimated by assuming upper and lower bounds for $d_x(P)$ and $d_z(P)$, taking the extremal $e_x(P)$ and $e_y(P)$ for the given geometry, and putting these values into (5.1). If the error remains small, especially for an optimized holographic arrangement, 3.2.1, one can live with it.

If all three components of the displacement vector have to be determined, one has to solve the whole three-dimensional *system of equations* defined by (5.1). Therefore three measurements have to be performed to achieve the phase values $\Delta\phi^1(P)$, $\Delta\phi^2(P)$ and $\Delta\phi^3(P)$ for each point P. These interference phase values correspond to the same object point and thus to the same displacement vector, but to different sensitivity vectors [314]. The sensitivity vectors should span a three-dimensional vector space, Sec. 5.2.3. The three interference patterns leading to the three phases may be recorded from the same observation direction, but with different illumination directions. This allows an easy identification of corresponding pixels in each recorded pattern, but requires repeatable or at least stable loading of the object to take the multiple holograms, each after switching between the illumination directions. Simultaneous recording on a single plate is possible with three object and three reference beams. Crosstalk then is avoided by noncorrelation between beams not belonging together, which is realized by optical path length differences exceeding the coherence length of the used laser [315]. If the fringes are moving, as is often the case with thermal loading, one has to trigger a simultaneous recording of the patterns along different observation directions [316]. But then the interferograms suffer under different *perspective distortion*, corresponding points P have to be identified, Sec. 5.2.2.

Let the three sensitivity vectors at P be $\boldsymbol{e}^1(P)$, $\boldsymbol{e}^2(P)$ and $\boldsymbol{e}^3(P)$, respectively. In the following raised indices denote different sensitivity vectors while lower indices stand for different object points. The lower indices x, y, z denote the Cartesian coordinates. Then at each point P we have to solve the system of linear equations

$$\begin{pmatrix} \Delta\phi^1(P) \\ \Delta\phi^2(P) \\ \Delta\phi^3(P) \end{pmatrix} = \begin{pmatrix} e_x^1(P) & e_y^1(P) & e_z^1(P) \\ e_x^2(P) & e_y^2(P) & e_z^2(P) \\ e_x^3(P) & e_y^3(P) & e_z^3(P) \end{pmatrix} \begin{pmatrix} d_x(P) \\ d_y(P) \\ d_z(P) \end{pmatrix} \tag{5.3}$$

to obtain $\boldsymbol{d}(P)$. This may be written in vector notation by $\boldsymbol{\Delta\phi}(P) = \boldsymbol{E}(P) \cdot \boldsymbol{d}(P)$ with $\boldsymbol{E}(P)$ representing the so called *sensitivity matrix*. The solution is

$$\boldsymbol{d}(P) = \boldsymbol{E}^{-1}(P) \cdot \boldsymbol{\Delta\phi}(P) \tag{5.4}$$

If more than three interferograms with more than three sensitivity vectors are recorded and the corresponding interference phases $\Delta\phi^1(P), \ldots, \Delta\phi^n(P)$ have been evaluated, the solution is found according to the *least squares method* by

$$\boldsymbol{d}(P) = \left(\boldsymbol{E}^T(P)\boldsymbol{E}(P)\right)^{-1} \left(\boldsymbol{E}^T(P) \cdot \boldsymbol{\Delta\phi}(P)\right) \tag{5.5}$$

where $\boldsymbol{E}(P)$ now is the corresponding $n \times 3$-matrix of the n sensitivity vectors [163, 317]. But still it is required that these n sensitivity vectors span a three-dimensional vector space.

5.1.2 Displacement Determination with Unknown Reference Displacement

For a displacement determination with *unknown reference displacement* let us first assume that the *sensitivity vector* is constant across the investigated surface of the object. The evaluated interference phase has the general form

$$\Delta\phi(P) = \boldsymbol{d}(P) \cdot \boldsymbol{e}(P) + \phi_0(P) \tag{5.6}$$

where ϕ_0 is an unknown constant. If we consider a point P_0 on the object as a reference, we can count the phase to each other point P_1 by

$$\begin{aligned} \Delta'\phi(P_1) &= \Delta\phi(P_1) - \Delta\phi(P_0) \\ &= [\boldsymbol{d}(P_1) - \boldsymbol{d}(P_0)] \cdot \boldsymbol{e}(P_1) = \Delta\boldsymbol{d}(P_1) \cdot \boldsymbol{e}(P_1) \end{aligned} \tag{5.7}$$

Now we can proceed as in Sec. 5.1.1 with the only restriction that not the absolute displacement but the displacement $\Delta\boldsymbol{d}$ relative to the reference point P_0 is measured.

Relaxing now the assumption that the sensitivity vector is constant across the surface, we may write $\boldsymbol{e}(P_1) = \boldsymbol{e}(P_0) + \Delta\boldsymbol{e}(P_1)$ Then (5.7) becomes

$$\begin{aligned} \Delta'\phi(P_1) &= \boldsymbol{d}(P_1) \cdot \boldsymbol{e}(P_1) - \boldsymbol{d}(P_0) \cdot \boldsymbol{e}(P_0) \\ &= \boldsymbol{d}(P_1) \cdot [\boldsymbol{e}(P_0) + \Delta\boldsymbol{e}(P_1)] - \boldsymbol{d}(P_0) \cdot [\boldsymbol{e}(P_0) + \Delta\boldsymbol{e}(P_1)] + \boldsymbol{d}(P_0) \cdot \Delta\boldsymbol{e}(P_1) \\ &= \Delta\boldsymbol{d}(P_1) \cdot \boldsymbol{e}(P_1) - \boldsymbol{d}(P_0) \cdot \Delta\boldsymbol{e}(P_1) \end{aligned} \tag{5.8}$$

Now it is no longer possible to eliminate the displacement $\boldsymbol{d}(P_0)$ of the reference point because of the variation $\Delta \boldsymbol{e}$ of the sensitivity vector [318], as shown already in the example in Sec. 4.1.2, Fig. 4.3.

Equations (5.7) and (5.8) allow the following interpretation: A *rigid body translation*, which implies a constant displacement vector $\boldsymbol{d}$ for all P will yield a constant interference phase if also the sensitivity vector is constant, (5.7), thus no fringes can be observed. On the other hand for a varying sensitivity vector the term $\boldsymbol{d}(P_0) \cdot \Delta \boldsymbol{e}(P_1)$ in (5.8) will not vanish for all P_1 and fringes are produced although $\boldsymbol{d}(P_1)$ is constant over all P_1. This observation gives rise to a method for measuring absolute vectorial displacements from fringe patterns produced with the help of *varying sensitivity vectors* [318].

Let us assume that interference phase values $\Delta\phi_m^n$ are determined at M points P_m for N sensitivity vectors $\boldsymbol{e}^n$:

$$\Delta\phi_m^n = \boldsymbol{d}(P_m) \cdot \boldsymbol{e}^n(P_m) + \phi_0^n(P_1) \qquad n = 1, \dots, N; m = 1, \dots, M \tag{5.9}$$

Since the phase is assumed to be continuously counted between the P_m, there are only N unknown constants $\phi_0^n(P_1)$, namely one for each sensitivity vector, but for a single arbitrary reference point, here chosen as P_1. Altogether we have NM equations with $3M + N$ unknowns. This system of equations may be solved only if $NM \geq 3M + N$. One sees that no solutions exist for $N \leq 3$ or $M = 1$. $N = 4$ implies $M \geq 4$, $N = 5$ requires $M \geq 3$, and all $N \geq 6$ yield solutions provided that $M \geq 2$. We recognize that we need at least $N = 4$ sensitivity vectors, which means at least four holographic recordings with different illumination directions and/or observation directions.

The case $N = M = 4$ should be considered in more detail. The sixteen equations (5.9) are organized into the matrix equation

$$\begin{pmatrix} \Delta\phi_1^1 \\ \Delta\phi_1^2 \\ \Delta\phi_1^3 \\ \Delta\phi_1^4 \\ \Delta\phi_2^1 \\ \Delta\phi_2^2 \\ \Delta\phi_2^3 \\ \Delta\phi_2^4 \\ \Delta\phi_3^1 \\ \Delta\phi_3^2 \\ \Delta\phi_3^3 \\ \Delta\phi_3^4 \\ \Delta\phi_4^1 \\ \Delta\phi_4^2 \\ \Delta\phi_4^3 \\ \Delta\phi_4^4 \end{pmatrix} = \begin{pmatrix}
e_{x1}^1 & e_{y1}^1 & e_{z1}^1 & 0 & 0 & 0 & 0 & 0 & 0 & 0 & 0 & 0 & 1 & 0 & 0 & 0 \\
e_{x1}^2 & e_{y1}^2 & e_{z1}^2 & 0 & 0 & 0 & 0 & 0 & 0 & 0 & 0 & 0 & 0 & 1 & 0 & 0 \\
e_{x1}^3 & e_{y1}^3 & e_{z1}^3 & 0 & 0 & 0 & 0 & 0 & 0 & 0 & 0 & 0 & 0 & 0 & 1 & 0 \\
e_{x1}^4 & e_{y1}^4 & e_{z1}^4 & 0 & 0 & 0 & 0 & 0 & 0 & 0 & 0 & 0 & 0 & 0 & 0 & 1 \\
0 & 0 & 0 & e_{x2}^1 & e_{y2}^1 & e_{z2}^1 & 0 & 0 & 0 & 0 & 0 & 0 & 1 & 0 & 0 & 0 \\
0 & 0 & 0 & e_{x2}^2 & e_{y2}^2 & e_{z2}^2 & 0 & 0 & 0 & 0 & 0 & 0 & 0 & 1 & 0 & 0 \\
0 & 0 & 0 & e_{x2}^3 & e_{y2}^3 & e_{z2}^3 & 0 & 0 & 0 & 0 & 0 & 0 & 0 & 0 & 1 & 0 \\
0 & 0 & 0 & e_{x2}^4 & e_{y2}^4 & e_{z2}^4 & 0 & 0 & 0 & 0 & 0 & 0 & 0 & 0 & 0 & 1 \\
0 & 0 & 0 & 0 & 0 & 0 & e_{x3}^1 & e_{y3}^1 & e_{z3}^1 & 0 & 0 & 0 & 1 & 0 & 0 & 0 \\
0 & 0 & 0 & 0 & 0 & 0 & e_{x3}^2 & e_{y3}^2 & e_{z3}^2 & 0 & 0 & 0 & 0 & 1 & 0 & 0 \\
0 & 0 & 0 & 0 & 0 & 0 & e_{x3}^3 & e_{y3}^3 & e_{z3}^3 & 0 & 0 & 0 & 0 & 0 & 1 & 0 \\
0 & 0 & 0 & 0 & 0 & 0 & e_{x3}^4 & e_{y3}^4 & e_{z3}^4 & 0 & 0 & 0 & 0 & 0 & 0 & 1 \\
0 & 0 & 0 & 0 & 0 & 0 & 0 & 0 & 0 & e_{x4}^1 & e_{y4}^1 & e_{z4}^1 & 1 & 0 & 0 & 0 \\
0 & 0 & 0 & 0 & 0 & 0 & 0 & 0 & 0 & e_{x4}^2 & e_{y4}^2 & e_{z4}^2 & 0 & 1 & 0 & 0 \\
0 & 0 & 0 & 0 & 0 & 0 & 0 & 0 & 0 & e_{x4}^3 & e_{y4}^3 & e_{z4}^3 & 0 & 0 & 1 & 0 \\
0 & 0 & 0 & 0 & 0 & 0 & 0 & 0 & 0 & e_{x4}^4 & e_{y4}^4 & e_{z4}^4 & 0 & 0 & 0 & 1
\end{pmatrix} \begin{pmatrix} d_{x1} \\ d_{y1} \\ d_{z1} \\ d_{x2} \\ d_{y2} \\ d_{z2} \\ d_{x3} \\ d_{y3} \\ d_{z3} \\ d_{x4} \\ d_{y4} \\ d_{z4} \\ \phi_{01} \\ \phi_{02} \\ \phi_{03} \\ \phi_{04} \end{pmatrix} \tag{5.10}$$

which again is written shortly as $\boldsymbol{\Delta\phi} = \boldsymbol{E} \cdot \boldsymbol{d}$. The solution is found in analogue to (5.4) or (5.5). It is easily shown that the matrix $\boldsymbol{E}$ becomes singular when there is no variation of the sensitivity vectors across the surface of the object. As soon as the displacements have been calculated for four points by the presented method, the absolute phase can be determined for all points and the displacement of other object points now may be calculated by the methods of Sec. 5.1.1. The above mentioned method is equivalent to the method presented in [319, 320, 321], which also uses four sensitivity vectors and four object points but reduces the rank of the matrix to twelve by employing fringe order differences corresponding to interference phase differences. Therefore only displacements relative to a reference point are calculated in this latter case.

If the interference phase differences are determined by one of the *dynamic evaluation* methods of Sec. 4.8, also no reference displacement has to be known. For a three-dimensional evaluation of the displacement vector field at each object point of interest P a system of at least three equations of the form of (4.78) has to be solved. Let three interference phase differences $\Delta\Delta\phi^{1,2}(P)$, $\Delta\Delta\phi^{3,4}(P)$ and $\Delta\Delta\phi^{5,6}(P)$ be evaluated at P, then from (4.78) we get

$$\begin{pmatrix} \Delta\Delta\phi^{1,2}(P) \\ \Delta\Delta\phi^{3,4}(P) \\ \Delta\Delta\phi^{5,6}(P) \end{pmatrix} = \frac{2\pi}{\lambda} \begin{pmatrix} b_x^2(P) - b_x^1(P) & b_y^2(P) - b_y^1(P) & b_z^2(P) - b_z^1(P) \\ b_x^4(P) - b_x^3(P) & b_y^4(P) - b_y^3(P) & b_z^4(P) - b_z^3(P) \\ b_x^6(P) - b_x^5(P) & b_y^6(P) - b_y^5(P) & b_z^6(P) - b_z^5(P) \end{pmatrix} \begin{pmatrix} d_x(P) \\ d_y(P) \\ d_z(P) \end{pmatrix} \tag{5.11}$$

which has to be inverted. For more than three measurements the corresponding least squares system analogous to (5.5) has to be solved.

The two-dimensional case [63], when one component of the displacement vector is known, may lead to matrices of lower rank. But it is not treated separately here since the underlying theory is equivalent to the three-dimensional case.

It must be noted that the matrices employed in the evaluation methods of this section are prone to bad condition. So they have to be checked by a measure of condition as will be introduced in Sec. 5.2.3.

5.1.3 Elimination of Overall Displacement

In a number of applications the interesting displacement field is overlayed by another displacement, e. g. a defect induced deformation is embedded in rigid body motions and homogeneous deformations. To validate the risk by this defect it may be helpful to discriminate the defect induced deformation from the overall displacement field. This can be done quantitatively in the displacement domain after the determination of the displacement field. But for a number of problems it is sufficient to consider the interference phase distribution to achieve good qualitative statements.

The overall displacement field is determined by the method of *Gaussian least squares*, which in the following is demonstrated at the interference phase distribution of a tensile test specimen with an internal crack [322, 323]. Experimental observations and theoretical considerations show that the deformation is a combination of

- a constant translation t_x in longitudinal direction,
- a linearly increasing translation $\varepsilon(x - x_0)$, caused by the strain ε,
- a constant transversal translation t_z,

- a linearly increasing translation in transversal direction $\tan\gamma(x - x_0)$ due to a tilt γ,
- the displacement $d_z(x)$ created by the defect.

If an arrangement with no sensitivity in y-direction is used, the interference phase outside the defect range can be written according to Eq. (5.1) as

$$\begin{aligned}\Delta\phi(x) &= d_x(x)e_x(x) + d_z(x)e_z(x) \\ &= t_x e_x(x) + \varepsilon(x - x_0)e_x(x) + t_z e_z(x) + \tan\gamma(x - x_0)e_z(x) \end{aligned} \tag{5.12}$$

The parameters t_x, t_z, ε, and $\tan\gamma$ are now determined by the least squares method from the measured interference phase values $\Delta\phi(x_i)$ outside the defect area. The system of equations to be solved is

$$\begin{pmatrix} \sum e_{xi}^2 & \sum(x_i - x_0)e_{xi}^2 & \sum e_{xi}e_{zi} & \sum(x_i - x_0)e_{xi}e_{zi} \\ \sum(x_i - x_0)e_{xi}^2 & \sum(x_i - x_0)^2 e_{xi}^2 & \sum(x_i - x_0)e_{xi}e_{zi} & \sum(x_i - x_0)^2 e_{xi}e_{zi} \\ \sum e_{xi}e_{zi} & \sum(x_i - x_0)e_{xi}e_{zi} & \sum e_{zi}^2 & \sum(x_i - x_0)e_{zi}^2 \\ \sum(x_i - x_0)e_{xi}e_{zi} & \sum(x_i - x_0)^2 e_{xi}e_{zi} & \sum(x_i - x_0)e_{zi}^2 & \sum(x_i - x_0)^2 e_{zi}^2 \end{pmatrix} \begin{pmatrix} t_x \\ \varepsilon \\ t_z \\ \tan\gamma \end{pmatrix} = \begin{pmatrix} \sum \Delta\phi_i e_{xi} \\ \sum \Delta\phi_i (x_i - x_0)e_{xi} \\ \sum \Delta\phi_i e_{zi} \\ \sum \Delta\phi_i (x_i - x_0)e_{zi} \end{pmatrix} \tag{5.13}$$

where $e_{xi} = e_x(x_i)$, $e_{zi} = e_z(x_i)$, $\Delta\phi_i = \Delta\phi(x_i)$. $e_x(x_i)$, $e_z(x_i)$ and the coordinate of the clamping x_0 are determined from the holographic arrangement. After solving (5.13), the normal displacement $d_z(x)$ in the defect range can be expressed as

$$d_z(x) = \frac{\Delta\phi(x)}{e_z(x)} - [t_x + \varepsilon(x - x_0)]\frac{e_x(x)}{e_z(x)} - t_z - \tan\gamma(x - x_0) \tag{5.14}$$

Fig. 5.1a shows an evaluated interference phase distribution and Fig. 5.1b displays the deformation induced by the defect alone.

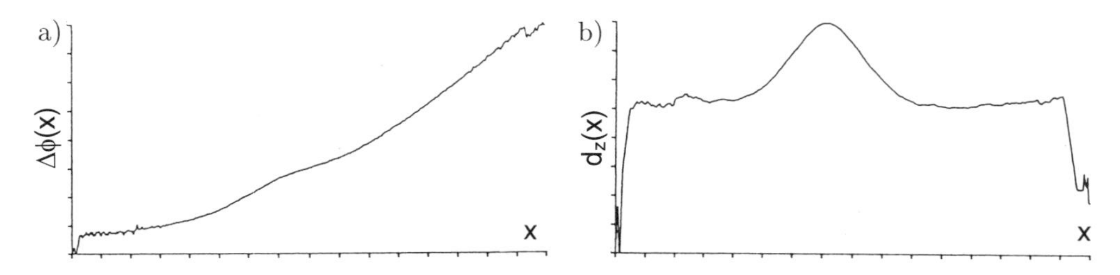

Figure 5.1: Defect induced deformation

The same method applied in two dimensions is presented in Figs. 5.2 (a) to (f). Fig. 5.2a to d show four phase stepped holographic interferograms of a pressurized vessel.

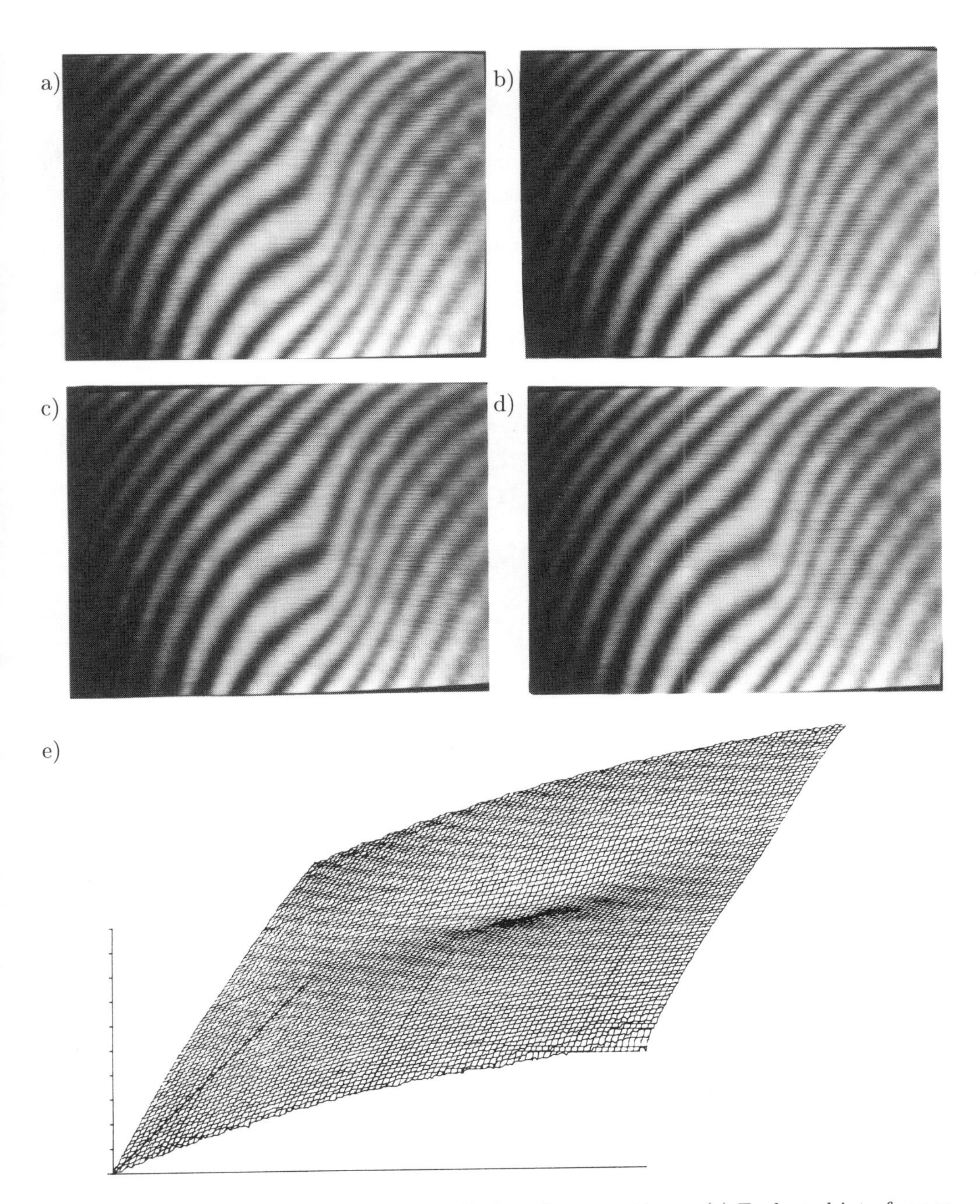

Figure 5.2: (a) - (d) Phase shifted holographic interference patterns, (e) Evaluated interference phase

The interference phase, Fig. 5.2e is determined by phase sampling evaluation. The surface deformation caused by the internal defect is given in Fig. 5.2f. The displayed distribution is proportional to a normal displacement with a peak to valley difference of about 700 nanometers.

The fitting of quadratic forms to the interference phase is described in [324], while fitting of polynomials is presented in [265]. Of special interest here is the use of *Zernike polynomials* which generally describe wavefronts in optical testing. [138] reports on a nonlinear least squares fitting program to model the intensity of the holographic interferogram.

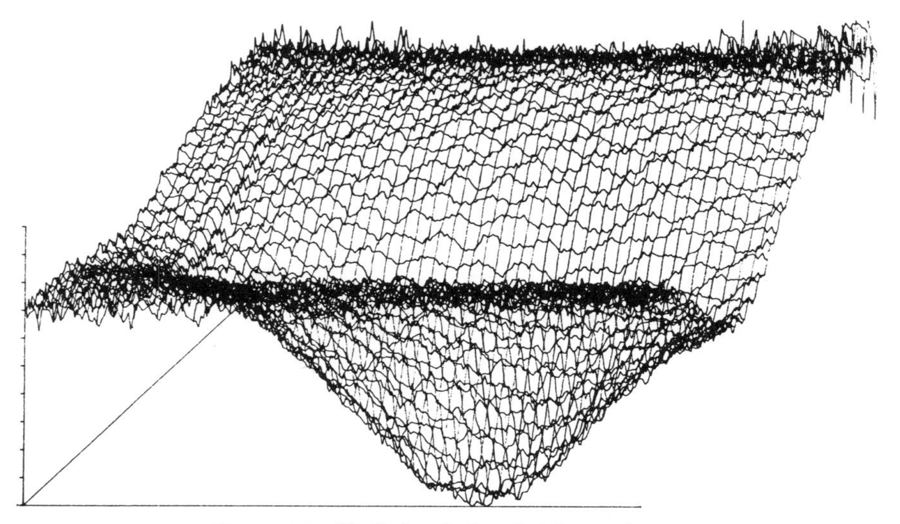

Figure 5.2: (f) Defect induced deformation

5.1.4 Non-Vibration Isolated Objects

In *holographic interferometric metrology* and in *holographic nondestructive testing* the displacement of a surface relative to the illumination source and to the observer is measured. Spurious motions of the object may corrupt the measurement result and thus have to be avoided. The common method is to place all optical components as well as the object onto a vibration isolated table in a laboratory environment. But especially when dealing with large scale technical objects in industrial environments it is not generally possible to avoid stochastic vibrations of the object and the holographic arrangement. Stochastic motions and vibrations of the object during and between the two recordings of a double-exposure hologram normally destroy any stationary wave field, thus no hologram nor interference pattern will be gained. In the following some approaches to circumvent these problems are reported.

To avoid the smearing of the microinterference due to motion during the exposure it must be guaranteed that no significant relative motion between the components occurs.

Generally the tendency to unwanted motions can be diminished by keeping the holographic arrangement compact, and by avoiding complicated folded beam paths with many mirrors or beam splitters. If object motions cannot be avoided, *pulsed lasers* instead of *CW-lasers* have to be employed [325]. Although good quality holograms of *non-vibration isolated objects* can be generated with a pulsed laser, there remains the uncertainty about parasitic object motion between the two exposures of a double exposure holographic interferogram.

The problem can be solved by the *object related triggering* of the laser [326]. To achieve this a CW-laser with an intra-cavity *acoustooptic modulator* is used. External modulation with a Pockels cell is principally possible but the acoustooptic modulator offers the advantage that the laser loses no energy during the closing time of the modulator. The modulator is controlled by the motion of the object, which is measured at one surface point. This measurement can be performed optically or mechanically by a contacting method. In [326] an acceleration detector, amplifier, and comparator is used. A temporal modulation of the laser is accomplished, exposing the hologram plate only when the chosen object point passes an interval of about $\lambda/10$ around its resting position. The method can be viewed as the transient counterpart of *stroboscopic holography* applied to periodically vibrating structures.

Another way to stabilize the microinterference pattern is the shifting of one of the interfering waves, normally the reference wave, by *feedback control* [327]. The feedback system consists of a phase detector, the electronics, which yield the control signal and an optical phase modulator. The phase detector for generating the electrical signal proportional to the relative phase of the interfering beams may consist of a magnifying microscope objective, a slit and a photomultiplier [327]. The phase modulator driven by the control signal for changing the relative phase of the two wavefields is a mirror fixed on a piezoelectric crystal. Using such a feedback control not only compensates for spurious motions of the object itself which may be caused by acoustic noise, but also for stochastic variations in the interfering wavefields caused by e. g. thermal drift or air turbulence [327].

A fringe stabilization method related to both the aforementioned methods is the stabilization by *frequency modulation*, where laser frequency shifts are generated by changing the laser cavity length [328]. The holographic interferometer in this case has different optical pathlengths for object and reference wave. Because of the difference in transit times of beams traversing the unequal paths, a shift in laser frequency results in a relative phase shift of the interfering waves and thus in a lateral shift of the generated fringes at the recording medium. The control signal generated by the photomultiplier and the electronics like in the aforementioned method now is applied to a piezoelectric transducer upon which one of the laser cavity reflectors is mounted. With such a system undesired fringe motion is compensated for, whether due to laser frequency drift or interferometer path difference perturbations [328].

The easiest way to perform a compensation of unwanted object motions by modulating the phase of the reference wave is to direct the reference beam over a small mirror fixed to the object [329, 330, 331]. The reference wave thus is reflected out of the object illumination wave by wavefront division. The mirror undergoes the same motions as the point of the object it is fixed to. A mathematical analysis is based on the fact that the mirror is specularly reflecting while the object surface is diffusely reflecting [329]. The analysis shows that with such a reference mirror not only the rigid body motions of the object are compensated but additionally we have one point in the fringe pattern where the interference order is known to be zero, namely the point of the mirror reflecting the illumination source point to the observation point.

The consequent continuation of these ideas leads to the fixation of the whole holographic arrangement onto the object whose surface deformations should be measured. Especially for large scale objects this is feasible. In such a way the deformation of a 4.5 m high *pressure vessel* with steel walls of 45 mm thickness has been measured holographically [322]. Some interferograms produced in this experiment have been shown in Fig. 5.2. The whole holographic arrangement including the laser was attached to the vessel by magnetic feet. Moreover, good results are already obtained, if the object is sprayed with a retroreflective paint and only the hologram plate is rigidly attached to it [332].

Although there exist effective methods to stabilize the microinterference fringes to be recorded in the hologram there still may remain an unknown rigid body motion between the two exposures of a double-exposure hologram. A compensation for this motion is possible during the reconstruction when the holographic interference pattern is generated. If the two states are recorded on separate hologram plates, these two plates after development may be tilted and shifted in the reference wave, a procedure called *sandwich hologram interferometry* [333, 334, 335, 336, 337, 338, 339, 340]. The relative motion of the plates leads to a relative motion of the reconstructed wavefields, with the consequence that an object translation can be compensated. It is helpful if an area on the object, on the frame, or else in the arrangement, is visible through the holograms, where it is known that no displacement must have occurred. While in sandwich holography the two plates are close together in a common reference wave, - thus the name of the method, - the two plates can be fully separated with individual reference waves acting on them [341, 342]. Then one of the two plates is fixed to a hologram holder that has all degrees of freedom to move with a precision better than 1 μm. This method is suitable for compensating stress-induced translations in non-destructive testing leading to defect fringes on an infinite fringe background [343]. A detailed theoretical analysis of fringe modifications in the holographic interferometric measurement of large deformations is given in [344, 345].

In holographic measurements of phase objects the idea of dividing the reference wave out of the object wave can be realized not only by using a reference mirror at or nearby the object but also by extracting the zero frequency out of the whole field in the focal plane of a lens, while the object wave is contained in the first diffraction order [346]. The phase object for this approach is illuminated by a plane wave that has passed a Ronchi grating. To separate the reference and the object wave a spatial filter with two holes must be placed in the focal plane of a lens. A second lens images these two fields onto a plane where they interfere. Since object beam and reference beam take a common path, this interferometer is proof against external vibrations.

5.2 The Sensitivity Matrix

In the preceding section it has been shown how the displacement vectors are determined from the evaluated interference phase distributions. Connecting elements are the sensitivity vectors which in all evaluations of more than one dimension are combined in the *sensitivity matrix*. This section now will discuss how to obtain the components of the sensitivity matrix and how its condition can be quantified, which is the main property defining the achievable accuracy of the measurement results.

5.2.1 Determination of the Sensitivity Vectors

The *sensitivity vectors* $\mathbf{e}(P)$ are calculated by (3.20) with the help of the unit vectors $\boldsymbol{s}(P)$ in illumination direction and $\boldsymbol{b}(P)$ in observation direction as defined in (3.18) and (3.19). To find out the sensitivity vectors of a specific experiment in an arbitrary but fixed Cartesian coordinate system one needs the coordinates of the *illumination point* S, which is the focal point of the optics used for expanding the illuminating laser beam, the coordinates of the *observation point* B, which is the center of the entrance pupil of the observing optical system, and the coordinates of the object points P of interest. For collimated illumination or observation the unit vectors are colinear with the illumination or observation directions and are identical for all object surface points.

Although the choice of the origin of the coordinate system is arbitrary, two options yield easier mathematics, since some components then will be zero: First, if the surface to be measured is plane, an origin at the surface with two axes lying in the surface plane should be chosen; second, for arbitrarily shaped object surfaces, a coordinate system with one axis colinear to the line connecting the illumination with the observation point and the origin halfway between these points is advantageous. Nevertheless the determination of the sensitivity vectors remains involved, especially for objects of complicated three-dimensional shape, for wide-angle or multiple illumination directions, or when mirrors are used for back or side views of the tested objects.

The straightforward approach is to make simple yardstick measurements between representative points of the holographic arrangement and then to calculate the necessary coordinates. Often the geometry or the contour of the object surface is given by design data, otherwise it has to be measured. There are optical methods for *contour measurement*, like triangulation-, projected fringe-, moiré-, or holographic methods, which may be applied [347, 18]. In the following only the holographic interferometric methods of these will be discussed.

In the preceding section methods were treated how to determine the unknown displacement vectors of surface points from known sensitivity vectors and measured interference phase values. The methods for the determination of the sensitivity vectors to be presented here employ known displacements or known rotations together with measured phases to determine the a priori unknown sensitivity vectors [348].

Consider three separate double exposure holograms of the object undergoing three different, but known, *rigid body displacements* $\boldsymbol{d}^1$, $\boldsymbol{d}^2$, and $\boldsymbol{d}^3$, which must not be effected in coplanar directions. Assume the same recording and observation geometry for the three holograms with respect to the object. This ensures an invariant sensitivity vector from one recording to the other. Now evaluate the interference phases $\Delta\phi^1(P)$, $\Delta\phi^2(P)$, and $\Delta\phi^3(P)$ corresponding to the three displacements and the sensitivity vector $\boldsymbol{e}(P)$ is obtained as the solution of

$$\begin{pmatrix} \Delta\phi_1(P) \\ \Delta\phi_2(P) \\ \Delta\phi_3(P) \end{pmatrix} = \begin{pmatrix} d_x^1 & d_y^1 & d_z^1 \\ d_x^2 & d_y^2 & d_z^2 \\ d_x^3 & d_y^3 & d_z^3 \end{pmatrix} \begin{pmatrix} e_x(P) \\ e_y(P) \\ e_z(P) \end{pmatrix} \tag{5.15}$$

The alternative approach is to use three rigid body rotations. Let the rotation vectors defining direction and rotation angles be $\boldsymbol{\theta}^1$, $\boldsymbol{\theta}^2$, and $\boldsymbol{\theta}^3$. We employ the so called *fringe-vector* $\boldsymbol{K}_f(P)$, whose magnitude is inversely proportional to the normal distance between the spatial fringe shells intersecting the object surface and whose direction coincides with the direction of this normal pointing to the fringe shell of higher interference order. The fringe vector for rotations is given by

$$\boldsymbol{K}_f(P) = -\boldsymbol{\theta} \times \boldsymbol{e}(P) \tag{5.16}$$

where $\times$ denotes the vectorial product. The threefold evaluation of $\boldsymbol{K}_f(P)$ for each P of interest with $\boldsymbol{\theta}^1$, $\boldsymbol{\theta}^2$, and $\boldsymbol{\theta}^3$ and the identical sensitivity vectors enables one to set up the system of equations

$$\begin{pmatrix} K_{fx}^1(P) & K_{fy}^1(P) & K_{fz}^1(P) \\ K_{fx}^2(P) & K_{fy}^2(P) & K_{fz}^2(P) \\ K_{fx}^3(P) & K_{fy}^3(P) & K_{fz}^3(P) \end{pmatrix} = \begin{pmatrix} \theta_x^1 & \theta_y^1 & \theta_z^1 \\ \theta_x^2 & \theta_y^2 & \theta_z^2 \\ \theta_x^3 & \theta_y^3 & \theta_z^3 \end{pmatrix} \begin{pmatrix} 0 & e_z(P) & -e_y(P) \\ -e_z(P) & 0 & e_x(P) \\ e_y(P) & -e_x(P) & 0 \end{pmatrix} \tag{5.17}$$

which is then solved to obtain $\boldsymbol{e}(P)$ for each P. With these methods the problem of the determination of the geometry data is transferred to the task of performing precisely the necessary displacements or rotations and to evaluate the interference phase or fringe-vector unambigously [349].

Another holographic approach is proposed in [350], where the object contours are measured by the holographic contouring using displaced illumination points, see Sec. 5.6.3. The main advantages are that the fringe pattern now is located on the surface and shows good visibility even for large illumination point displacements, and that the method works contactless with respect to the object.

5.2.2 Correction of Perspective Distortion

The determination of more than one component of the displacement vector necessitates the generation of several interference patterns which in turn are produced with different configuration geometries. If the observation direction varies between the recordings, the fringe patterns suffer from different *perspective distortions*. Especially when measuring at non-plane objects, errors induced by defocussing of the object surface and distortion due to perspective have to be taken into account [351]. For an oblique sight of the inspected surface, this surface appears deformed in the recorded image and often does not fill the full frame. A *spatial transform* has to map the pixels of the recorded interferogram to new pixels in an output interferogram, such that identical points of the object surface, which are imaged to different pixels in the various recorded images, are mapped to identical pixels in all output images.

In the following, an algorithm is given for the spatial transform which corrects perspective distortion in the case of plane surfaces. Without loss of generality, we assume a rectangular surface area to be evaluated, which by perspective distortion is deformed to an arbitrary convex quadrangle, Fig. 5.3. The spatial transform is performed by a *bilinear*

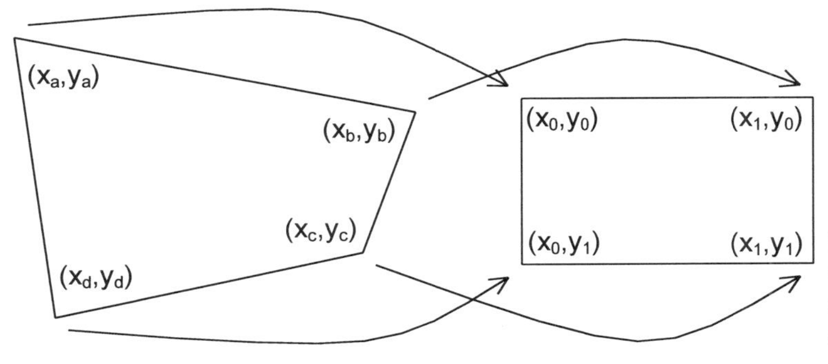

Figure 5.3: Bilinear mapping of a quadrangle onto a rectangle

interpolation, which on the one hand is fast and computationally simple, and on the other hand produces a smooth mapping that preserves continuity and connectivity.

Let $I(x, y)$ be the recorded and stored perspectively distorted interferogram with pixel coordinates (x, y). Then the corrected interference pattern is $I'(x, y)$:

$$I'(x, y) = I(x', y') = I(ax + by + cxy + d, ex + fy + gxy + h) \tag{5.18}$$

This bilinear transformation is defined by the values of the eight coefficients a through h. By specifying the mapping of the four vertices (x_a, y_a), (x_b, y_b), (x_c, y_c), (x_d, y_d) of the quadrangle to the four vertices (x_0, y_0), (x_1, y_0), (x_1, y_1), (x_0, y_1) of the output rectangle, we create a system of four equations

$$\begin{pmatrix} x_a \\ x_b \\ x_c \\ x_d \end{pmatrix} = \begin{pmatrix} x_0 & y_0 & x_0 y_0 & 1 \\ x_1 & y_0 & x_1 y_0 & 1 \\ x_1 & y_1 & x_1 y_1 & 1 \\ x_0 & y_1 & x_0 y_1 & 1 \end{pmatrix} \begin{pmatrix} a \\ b \\ c \\ d \end{pmatrix} \tag{5.19}$$

After inversion of the matrix we get the four coefficients a, b, c, and d as

$$\begin{pmatrix} a \\ b \\ c \\ d \end{pmatrix} = \frac{1}{(x_0 - x_1)(y_1 - y_0)} \begin{pmatrix} y_1 & -y_1 & y_0 & -y_0 \\ x_1 & -x_0 & x_0 & -x_1 \\ -1 & 1 & -1 & 1 \\ -x_1 y_1 & x_0 y_1 & -x_0 y_0 & x_1 y_0 \end{pmatrix} \begin{pmatrix} x_a \\ x_b \\ x_c \\ x_d \end{pmatrix} \tag{5.20}$$

The y_a through y_d are given by a system of equations similar to (5.19)

$$\begin{pmatrix} y_a \\ y_b \\ y_c \\ y_d \end{pmatrix} = \begin{pmatrix} x_0 & y_0 & x_0 y_0 & 1 \\ x_1 & y_0 & x_1 y_0 & 1 \\ x_1 & y_1 & x_1 y_1 & 1 \\ x_0 & y_1 & x_0 y_1 & 1 \end{pmatrix} \begin{pmatrix} e \\ f \\ g \\ h \end{pmatrix} \tag{5.21}$$

where the four coefficients e, f, g, and h are given by

$$\begin{pmatrix} e \\ f \\ g \\ h \end{pmatrix} = \frac{1}{(x_0 - x_1)(y_1 - y_0)} \begin{pmatrix} y_1 & -y_1 & y_0 & -y_0 \\ x_1 & -x_0 & x_0 & -x_1 \\ -1 & 1 & -1 & 1 \\ -x_1 y_1 & x_0 y_1 & -x_0 y_0 & x_1 y_0 \end{pmatrix} \begin{pmatrix} y_a \\ y_b \\ y_c \\ y_d \end{pmatrix} \tag{5.22}$$

Thus the coefficients of the bilinear transform are defined. The transform can be implemented by (5.18), but a computationally more efficient algorithm is based on a line by line processing of the output image. Each new pixel coordinate is calculated by an increment from the foregoing one. The increments are constant along one line and are raised by a constant increment from line to line.

The spatial transform for each pixel calculates the original position of this pixel in the input image. In most cases it stems from a fractional position in the input pattern, meaning that its origin is between four adjacent pixels. So an interpolation is necessary to determine the gray level of the output pixel. The simplest interpolation is the nearest neighbor interpolation. A better solution, especially if significant gray level changes occur over one unit of pixel spacing, consists of applying the *bilinear interpolation*: Let us look for the intensity $I(n+x, m+y)$ at the fractional position $(n+x, m+y)$ with integers n and m and $x, y \in (0., 1.)$, Fig. 5.4. The bilinear interpolation then gives

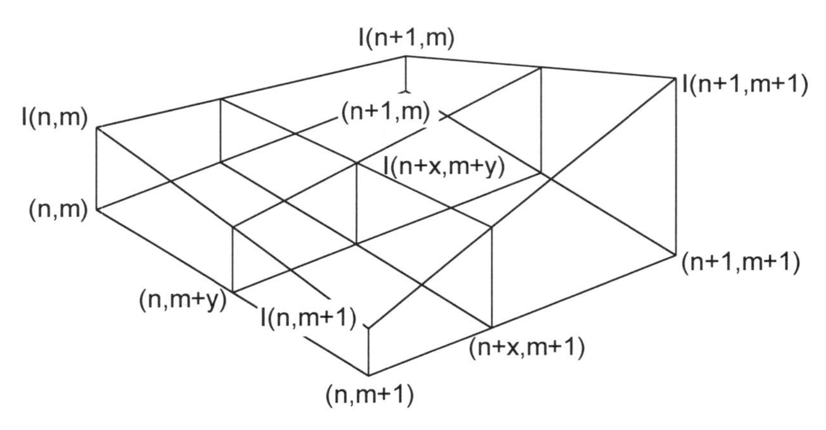

Figure 5.4: Bilinear mapping of the intensity

$$\begin{aligned} I(n+x, m+y) = I(n,m) + [I(n+1,m) - I(n,m)]x + [I(n,m+1) - I(n,m)]y \\ +[I(n+1,m+1) + I(n,m) - I(n,m+1) - I(n+1,m)]xy \end{aligned} \tag{5.23}$$

An example for a correction of perspective distortion of a plane rectangular plate is given in Fig. 5.5.

5.2.3 Condition of the Sensitivity Matrix

The determination of three-dimensional displacement fields requires the solution of linear systems of equations, as e. g. given in (5.3), (5.10), or (5.11). It is well known that if

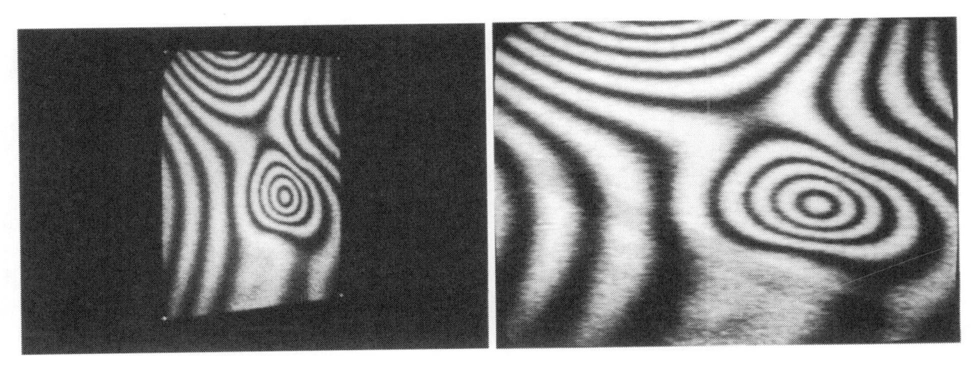

Figure 5.5: Correction of perspective distortion

the equations are linearly dependent, leading to a *singular matrix* whose determinant is zero, the system is not solvable. However, even if the matrices on the right-hand side of (5.3), (5.10), or (5.11) are not singular, the solutions may show errors of up to some 100 percent in magnitude and direction if we have an *ill-conditioned matrix*.

A figure to measure the *condition of a system of linear equations*, or equivalently that of the matrix $\boldsymbol{E}$ of this system, is the *Hadamard condition number* $K_H(\boldsymbol{E})$ defined as

$$K_H(\boldsymbol{E}) = \frac{|\det \boldsymbol{E}|}{\prod_{i=1}^{n} \alpha_i} \tag{5.24}$$

with

$$\alpha_i = \sqrt{\sum_{k=1}^{n} e_{ik}^2} \tag{5.25}$$

where n is the rank of matrix $\boldsymbol{E}$, whose elements are given by e_{ik}; det denotes the determinant. The system is ill-conditioned if $K_H(\boldsymbol{E}) \ll 1.0$, and is optimum for $K_H(\boldsymbol{E}) = 1.0$. A condition number of $K_H(\boldsymbol{E}) > 0.1$ is considered to be desirable.

The elements of the matrix are the components of the sensitivity vectors, these are derived from measurements of the geometry of the holographic arrangement. A good condition means that the directions of the sensitivity vectors are as different as possible and linearly independent. As a rule of thumb the sensitivity vectors should have as different directions in the three dimensions as possible.

Of course, the geometry has to be measured precisely. The precision of the geometry data influences the achievable accuracy of the measurement. However, even more important than a precise measurement of the geometry is a proper separation of the sensitivity vectors. With an ill-conditioned system even the smallest errors in the geometry values would cause severe measurement errors. The use of an overdetermined system of equations, which is solved by Gaussian least squares, yields on principle a higher accuracy due to the averaging property, [352, 353, 354, 355], but it will cause nearly the same errors if the sensitivity vectors are not separated far enough.

5.3 Holographic Strain and Stress Analysis

In experimental mechanics the deformation of test objects in response to a mechanical or thermal load is studied in order to determine strains, stresses, or bending moments. These quantities are of interest because they affect the strength, safety, and lifetime of mechanical structures or components [119, 24]. A structure most likely will fail where the strain or stress is maximal.

In this section the derivation of strains, stresses, and bending moments from holographically measured displacements of objects with opaque diffusely reflecting surfaces will be discussed. After the fundamental definitions, the most common structures, namely beams and plates are treated in more detail, followed by a more general approach using the fringe-vector theory.

5.3.1 Definition of Elastomechanical Parameters

Let $P = (x_P, y_P, z_P)$ be the Cartesian coordinate description of a point of a solid object which is displaced by $\boldsymbol{d}(P) = (d_x(P), d_y(P), d_z(P))$. This notation is preferred here for reasons of consistency, while in the standard literature the components of the displacement vector often are written $(u, v, w) = (d_x, d_y, d_z)$. If there is no confusion, the argument P will be omitted in the sequel.

Strain generally is a tensor completely specified by nine components, of which six are independent [356, 24]. At any point in a solid body the three components of *normal strain* are

$$\begin{aligned}
\varepsilon_x &= \frac{\partial d_x}{\partial x} \\
\varepsilon_y &= \frac{\partial d_y}{\partial y} \\
\varepsilon_z &= \frac{\partial d_z}{\partial z}
\end{aligned} \tag{5.26}$$

and the three independent *shear strains* are

$$\begin{aligned}
\gamma_{xy} &= \frac{\partial d_x}{\partial y} + \frac{\partial d_y}{\partial x} \\
\gamma_{yz} &= \frac{\partial d_y}{\partial z} + \frac{\partial d_z}{\partial y} \\
\gamma_{zx} &= \frac{\partial d_z}{\partial x} + \frac{\partial d_x}{\partial z}
\end{aligned} \tag{5.27}$$

While the normal strains describe the change of length per unit length in each coordinate direction, the shear strains measure the decrease in the angle between two line segments initially orthogonal and parallel to the coordinate axes. This is depicted in two dimensions for a small rectangle with sidelengths Δx and Δy in Fig. 5.6. Since the angles are assumed to be only small, the arctangent in the exact definition of the shear strain is approximated by its argument in the above definition.

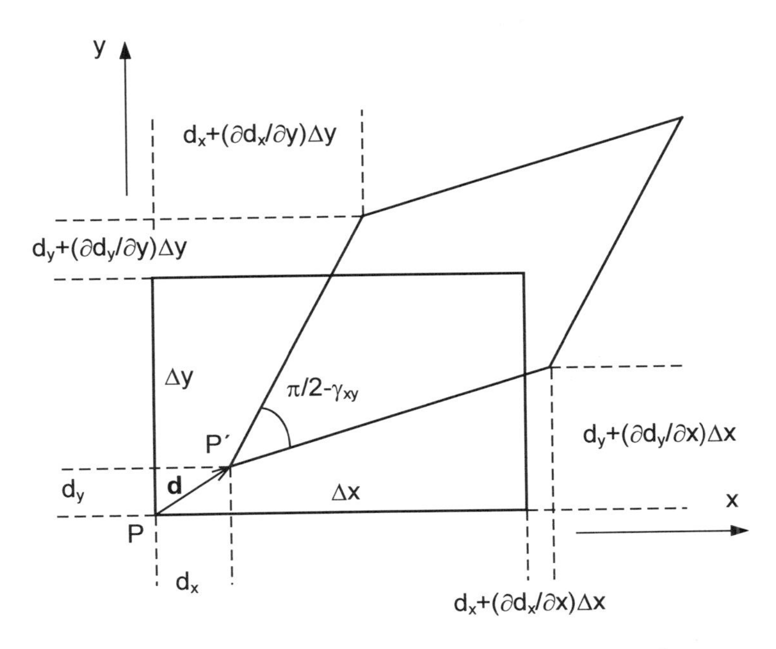

Figure 5.6: Elastic deformation of a small solid rectangle

The strains are expressed as derivatives of the displacement components. Also the *rotations* of an object can be written by these derivatives: The components ω_x, ω_y, and ω_z describing the rotations about the x-, y-, and z-axes, resp., are

$$\omega_x = \frac{1}{2}\left(\frac{\partial d_z}{\partial y} - \frac{\partial d_y}{\partial z}\right)$$
$$\omega_y = \frac{1}{2}\left(\frac{\partial d_x}{\partial z} - \frac{\partial d_z}{\partial x}\right)$$
$$\omega_z = \frac{1}{2}\left(\frac{\partial d_y}{\partial x} - \frac{\partial d_x}{\partial y}\right) \tag{5.28}$$

If external forces or moments affect a solid body, internal reactive forces maintain the equilibrium. As long as the masses are homogeneously distributed in the body, the reactive forces are distributed in planes. At each point of the solid an arbitrary cut dA can be defined. The *force* $\boldsymbol{F}$ related to the unit plane element dA is the *stress* $\boldsymbol{s}$

$$\boldsymbol{s} = \frac{d\boldsymbol{F}}{dA} \tag{5.29}$$

which is composed of the *normal stress* $\sigma = dF_n/dA$ and the *tangential* or *shear stress* $\tau = dF_t/dA$, Fig. 5.7. The description of the complete stress state at a point requires three planes or equivalently a cubic element to define the *stress tensor*, Fig. 5.8

$$\boldsymbol{S} = \begin{pmatrix} \sigma_{xx} & \tau_{xy} & \tau_{xz} \\ \tau_{yx} & \sigma_{yy} & \tau_{yz} \\ \tau_{zx} & \tau_{zy} & \sigma_{zz} \end{pmatrix} \tag{5.30}$$

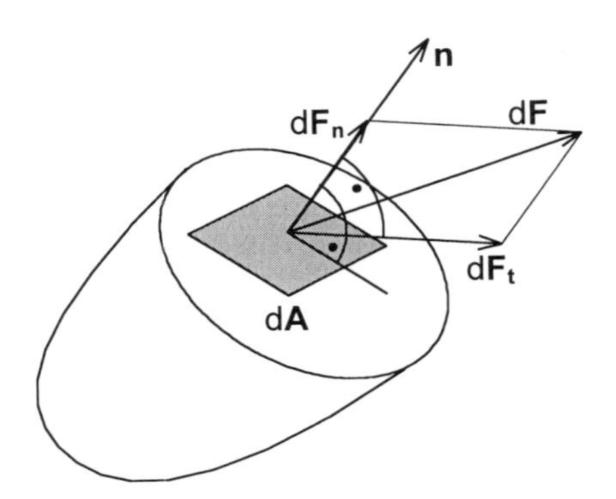

Figure 5.7: Normal and shear stress

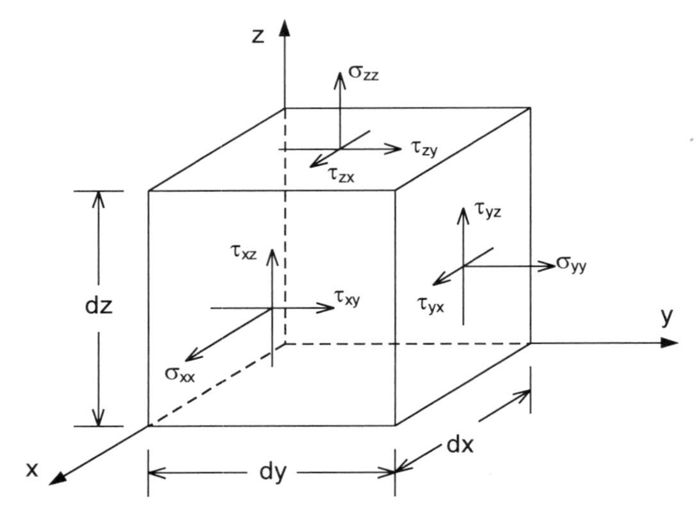

Figure 5.8: Elements of the stress tensor

Since $\tau_{xy} = \tau_{yx}$, $\tau_{xz} = \tau_{zx}$, and $\tau_{yz} = \tau_{zy}$, we need three normal stresses and three shear stresses to describe the stress state of a point of a solid body. Some relations between strains and stresses for special cases will be discussed in the next subsection.

The aim of holographic strain and stress analysis is the determination of the strain or stress state of a tested component, which normally is done by measuring the displacement vector field and then calculating the strains and stresses according to the above introduced equations. But holographic interferometry does not provide sufficient information to process all the derivatives of (5.26) to (5.28). In particular the displacement derivatives in the direction normal to the surface of the opaque body cannot be evaluated. By applying the methods of Sec. 5.1 for a plane surface in the x-y-plane we may evaluate the displacement vector components $d_x(x, y)$, $d_y(x, y)$, $d_z(x, y)$ or precisely $d_x(x, y, z = 0)$, $d_y(x, y, z = 0)$, $d_z(x, y, z = 0)$, but we have no access to the general $d_x(x, y, z)$, $d_y(x, y, z)$, $d_z(x, y, z)$ in the interior $z \neq 0$ of the body.

Nevertheless what we have is sufficient to calculate the in-plane strains ε_x, ε_y and γ_{xy} in the plane surface as well as the in-plane rotation ω_z about an axis normal to the surface.

We notice that the rotation components (5.28) are arithmetic averages of the rotations of two orthogonal faces of a cubic element, but this averaging is not required to evaluate the out-of-plane rotation of object surface points, $z = 0$ [24]. So the out of plane rotations at the surface are

$$\omega_x = \frac{\partial d_z}{\partial y} \qquad \omega_y = \frac{\partial d_z}{\partial x} \tag{5.31}$$

5.3.2 Beams and Plates

The holographically measurable displacements $\boldsymbol{d}(x, y, z = 0)$ are sufficient to describe the strains in the case of *plane stress*. We speak of a state of plane stress if a thin flat specimen is affected only by stresses parallel to its surface. Typical examples are the stretching of thin sheets or membranes and the tearing of flat tensile specimens commonly used to measure the mechanical properties of materials [24]. The in-plane strains for these objects are calculated from the measured $d_x(x, y)$- and $d_y(x, y)$-components of the displacement vector field.

The strains in another class of objects like beams, plates, or shells are determined from the measured displacement component $d_z(x, y)$ normal to the object surface.

A *beam* is a long slim solid component with a constant cross section subjected to transverse point-loads, distributed or area-loads, axial loads, and bending moments, Fig. 5.9. The length L of a beam generally is large compared with the width l and the thickness h: $L \gg l$, $L \gg h$. A beam is a common model for a number of technical components, like supporting beams, shafts, connecting rods, or turbine blades to name only a few.

A popular test object among holographers is the *cantilever beam* showing only deflections $d_z(x)$ in z-direction. The reasons for the popularity are that these deflections are easy to measure holographically - direction of highest sensitivity -, most often we have $d_z = 0$ at the base of the beam thus avoiding the absolute phase problem, and the theory of these beams is simple and well elaborated to allow a comparison of experimental and theoretical results.

If the beam consists of a material which behaves elastically for the applied load amplitudes, which means stress and strain are proportional, then for example in the x-direction we have

$$\sigma_{xx} = E\varepsilon_x \tag{5.32}$$

where E is the *modulus of elasticity*. Also the shear stresses and the shear strains are proportional

$$\tau_{zx} = G\gamma_{zx} \tag{5.33}$$

with G the *shear modulus of elasticity*. G and E are connected via

$$G = \frac{E}{2(1+\nu)} \tag{5.34}$$

where ν is the *Poissons ratio*, that in a specimen subjected to tensile or compressive load in x-direction describes the constant ratio of lateral strain to longitudinal strain

$$\varepsilon_z = \nu\varepsilon_x \tag{5.35}$$

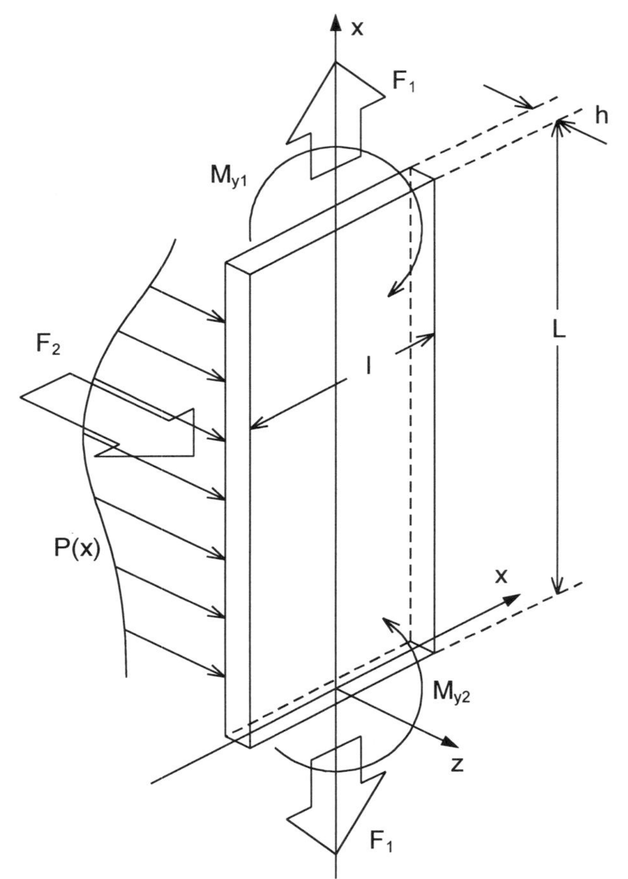

Figure 5.9: Elastic beam

E, G, and ν all are properties of the material the beam is fabricated of.

If the beam of Fig. 5.9 is deflected responding to the transverse point load F_z, the distributed area-load $P_z(x)$, longitudinal forces F_x and bending moments M_{y1}, M_{y2} about axes parallel to the y-axis, the longitudinal strain at the observable surface is [24]

$$\varepsilon_x = \frac{\partial d_{x0}(x)}{\partial x} - \frac{1}{2}h\left(\frac{\partial^2 d_z(x)}{\partial x^2}\right) \tag{5.36}$$

Here $d_{x0}(x)$ is the displacement in x-direction of the central plane at $z = 0$ of the beam due to the longitudinal forces F_x and h denotes the beam thickness. The longitudinal stress σ_{xx} at the surface of the beam then is

$$\sigma_{xx} = E\left[\frac{\partial d_{x0}(x)}{\partial x} - \frac{1}{2}h\left(\frac{\partial^2 d_z(x)}{\partial x^2}\right)\right] \tag{5.37}$$

As long as the deflections remain small the bending moment at any x is

$$M_y = -\frac{Eh^3}{12}\left(\frac{\partial^2 d_z(x)}{\partial x^2}\right) \tag{5.38}$$

A *plate* is a component where the width l is in the same range as the length L, $l \approx L$, but the constant thickness h is small compared to L and l: $h \ll l$, $h \ll L$, Fig. 5.10. The

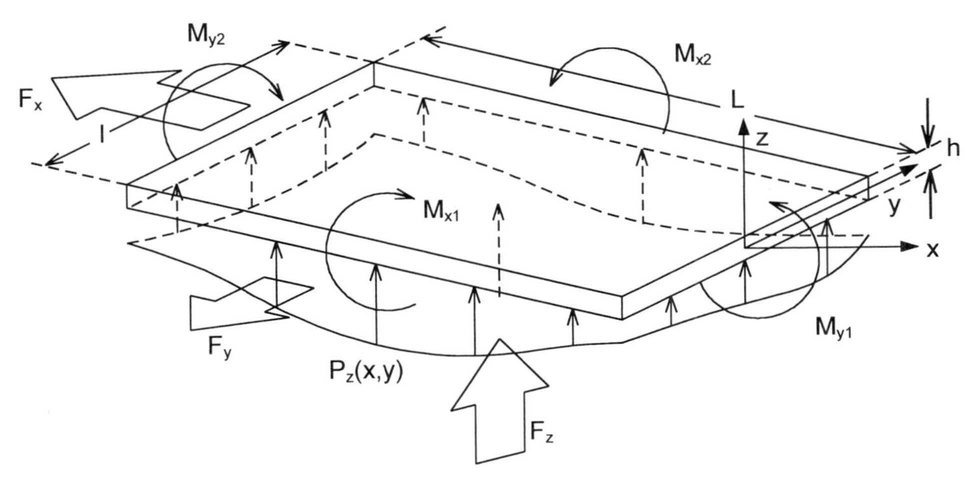

Figure 5.10: Elastic plate

plate is deformed e. g. by a distributed area-load $P_z(x, y)$, transverse forces F_z, axial forces F_x, F_y applied at its edges, as well as moments, M_x and M_y, acting at the edges. Then the strains at the observable surface are

$$\begin{aligned}
\varepsilon_x &= \frac{\partial d_{x0}(x,y)}{\partial x} - \frac{1}{2}h\left(\frac{\partial^2 d_z(x,y)}{\partial x^2}\right) \\
\varepsilon_y &= \frac{\partial d_{y0}(x,y)}{\partial y} - \frac{1}{2}h\left(\frac{\partial^2 d_z(x,y)}{\partial y^2}\right) \\
\gamma_{xy} &= \left(\frac{\partial d_{x0}(x,y)}{\partial x} + \frac{\partial d_{y0}(x,y)}{\partial y}\right) - h\left(\frac{\partial^2 d_z(x,y)}{\partial x \partial y}\right)
\end{aligned} \tag{5.39}$$

where $d_{x0}(x, y)$ and $d_{y0}(x, y)$ are the components of the in-plane translation of the central plane at $z = 0$ of the plate due to the axial forces F_x or F_y. The corresponding stresses are again $\sigma_{xx} = E\varepsilon_x$, $\sigma_{yy} = E\varepsilon_y$, $\tau_{xy} = G\gamma_{xy}$ and the bending moments per unit length are

$$\begin{aligned}
M_x &= -D\left(\frac{\partial^2 d_z(x,y)}{\partial x^2} + \nu\frac{\partial^2 d_z(x,y)}{\partial y^2}\right) \\
M_y &= -D\left(\frac{\partial^2 d_z(x,y)}{\partial y^2} + \nu\frac{\partial^2 d_z(x,y)}{\partial x^2}\right) \\
M_{xy} &= D(1-\nu)\frac{\partial^2 d_z(x,y)}{\partial x \partial y}
\end{aligned} \tag{5.40}$$

with $D = Eh^2/[12(1-\nu^2)]$ being the *flexural rigidity* of the plate.

In many applications the beams and plates are only affected by bending stresses, then $\partial d_{x0}/\partial x$ and $\partial d_{y0}/\partial y$ are negligible. For these cases the in-plane surface strains and stresses as well as the bending moments can be calculated from only the distribution of the normal displacement $d_z(x,y)$, which for the most holographic arrangements is the component to be measured with the highest accuracy [357, 358].

5.3.3 Numerical Differentiation

The determination of strains, stresses, and bending moments requires a *numerical differentiation* of the measured displacement data [119, 359, 360, 361, 362]. It is a well known fact that noise in the data is severely amplified by differentiation. Therefore we need highly accurate measured values at a large number of surface points for calculating the derivatives.

The direct approach to numerical differentiation is to compute *finite differences*: The differential quotient is approximated by a difference quotient, e. g.

$$\left(\frac{\partial d_z(x)}{\partial x}\right)_{x_i} \approx \frac{d_z(x_{i+1}) - d_z(x_i)}{x_{i+1} - x_i} \tag{5.41}$$

or

$$\left(\frac{\partial d_z(x)}{\partial x}\right)_{x_i} \approx \frac{d_z(x_{i-1}) - d_z(x_i)}{x_{i-1} - x_i} \tag{5.42}$$

or

$$\left(\frac{\partial d_z(x)}{\partial x}\right)_{x_i} \approx \frac{d_z(x_{i+1}) - d_z(x_{i-1})}{x_{i+1} - x_{i-1}} \tag{5.43}$$

For symmetry reasons the central difference approximation (5.43) is recommended, while at the edges of the evaluated data sets the forward (5.41) and the backward (5.42) differences have to be applied. At this stage the advantages of modern evaluation methods like phase stepping or Fourier transform evaluation compared to fringe counting or skeletonizing become obvious: displacement data now are determined directly at a rectangular dense grid of surface points giving constant denominators in (5.41) to (5.43) and thus constant accuracy. If on the other hand we have only access to the displacements at the fringe centers we get irregularly spaced points.

The *second derivative* at x_i we get by the finite difference of the forward and backward differences taken at the average positions between x_{i+1} and x_i and between x_i and x_{i-1}, resp. The approximation is

$$\left(\frac{\partial^2 d_z(x)}{\partial x^2}\right)_{x_i} \approx 2\frac{(x_i - x_{i-1})d_z(x_{i+1}) - (x_{i+1} - x_{i-1})d_z(x_i) + (x_{i+1} - x_i)d_z(x_{i-1})}{(x_i - x_{i-1})(x_{i+1} - x_{i-1})(x_{i+1} - x_i)} \tag{5.44}$$

For evenly spaced points having the constant distance $h = x_{i+1} - x_i$ this takes on the simple form

$$\left(\frac{\partial^2 d_z(x)}{\partial x^2}\right)_{x_i} \approx \frac{d_z(x_{i+1}) - 2d_z(x_i) + d_z(x_{i-1})}{h^2} \tag{5.45}$$

If in an actual experiment the data points are unevenly spaced or if they still contain some noise although being evenly distributed, it is good practice to fit a smooth curve to the displacement distribution before performing the numerical differentiation. If there are accurate displacement values but we have to interpolate between the points, cubic spline functions are an appropriate choice [24]. Cubic splines interpolate by third order polynomials in adjacent intervals with continuous second derivatives at the points where two intervals meet.

A smoothing of data distributions containing stochastic noise also may be accomplished by appropriate low-pass filtering or by fitting Bezier polynomials or other smooth curves by least squares methods.

5.3.4 Fringe Vector Theory

For the description of strain and rotation under more general circumstances we adopt the notation of the fringe locus function and fringe vectors of [88, 95, 96, 363, 364, 365, 366]. Suppose that an object undergoes a sufficiently small rigid body translation plus a *homogeneous deformation.* In homogeneous deformations the deformation of each object element is identical, e. g. a sphere deformed to an ellipsoid, or a cube deformed into a rectangular or trapezoidal parallelepiped. Prismatic bars subjected to simple tension or compression, or the expanding solids under uniform heating undergo homogeneous deformations. Generally the interference phase at each point P is given by (3.21)

$$\Delta\phi(P) = \boldsymbol{d}(P) \cdot \boldsymbol{e}(P) \tag{5.46}$$

This equation can be interpreted as a *fringe locus function* because constant values of $\Delta\phi(P)$ define fringe loci on the object surface. The observed fringes are interpreted as the intersections of fringe laminae in space with the object surface. If we investigate only small segments where the sensitivity vector $\boldsymbol{e}(P)$ can be regarded as constant, the fringes will be seen as equidistant straight lines on a flat object surface, indicating that the fringe laminae are equidistant planes. Let $\boldsymbol{r}$ be the space vector from an arbitrary origin of the coordinate system to the point P: $\boldsymbol{r} = (x_P, y_P, z_P)$, then we may write the fringe locus function $\Delta\phi(\boldsymbol{r})$ as a scalar product

$$\Delta\phi(\boldsymbol{r}) = \boldsymbol{K}_f \cdot \boldsymbol{r} \tag{5.47}$$

The vector $\boldsymbol{K}_f$ is the *fringe-vector* whose magnitude is inversely proportional to the spacing between the fringe laminae and whose direction is normal to them.

Let us assume a known value $\Delta\phi(\boldsymbol{r})$ of the fringe locus function at P having the space vector $\boldsymbol{r}$. Then the fringe locus function at a nearby point Q, described by the space vector $\boldsymbol{r} + \boldsymbol{\Delta r}_{PQ}$ is expressed by a Taylor series expansion as [158]

$$\Delta\phi(\boldsymbol{r} + \boldsymbol{\Delta r}_{PQ}) = \Delta\phi(\boldsymbol{r}) + \boldsymbol{\Delta r}_{PQ}^T \cdot \boldsymbol{K}_f + \frac{1}{2}\boldsymbol{\Delta r}_{PQ}^T \boldsymbol{T}_f \boldsymbol{\Delta r}_{PQ} \tag{5.48}$$

The second term on the right-hand side is the scalar product of the difference vector of positions with the fringe-vector which has to be the gradient of $\Delta\phi(\boldsymbol{r})$. $\boldsymbol{T}_f$ is the *fringe tensor* representing linear variations of $\boldsymbol{K}_f$, its consideration is only necessary when dealing also with inhomogeneous deformations and strains.

Calculating $\boldsymbol{K}_f$ by differentiating the fringe locus function

$$\Delta\phi = d_x e_x + d_y e_y + d_z e_z \tag{5.49}$$

which contains three products, we see that by applying the product rule of differentiation and this with respect to three coordinates we obtain eighteen terms which are written as the sum of two vector-matrix products

$$\boldsymbol{K}_f = \boldsymbol{F} \cdot \mathbf{e} + \boldsymbol{G} \cdot \mathbf{d} \tag{5.50}$$

While $\boldsymbol{e}$ and $\boldsymbol{d}$ are the sensitivity and displacement vectors, the matrix $\boldsymbol{G}$ accounts for the perspective variations of the sensitivity vector and $\boldsymbol{F}$ is the *deformation gradient matrix* which contains the derivatives of the displacement vector:

$$\boldsymbol{F} = \begin{pmatrix} \frac{\partial d_x}{\partial x} & \frac{\partial d_x}{\partial y} & \frac{\partial d_x}{\partial z} \\ \frac{\partial d_y}{\partial x} & \frac{\partial d_y}{\partial y} & \frac{\partial d_y}{\partial z} \\ \frac{\partial d_z}{\partial x} & \frac{\partial d_z}{\partial y} & \frac{\partial d_z}{\partial z} \end{pmatrix} \tag{5.51}$$

We recognize that $\boldsymbol{F}$ is composed of

$$\boldsymbol{F} = \boldsymbol{\mathcal{E}} + \boldsymbol{\Theta} \tag{5.52}$$

where $\boldsymbol{\mathcal{E}}$ is the symmetric matrix of strains and shears

$$\boldsymbol{\mathcal{E}} = \begin{pmatrix} \varepsilon_x & \frac{1}{2}\gamma_{xy} & \frac{1}{2}\gamma_{zx} \\ \frac{1}{2}\gamma_{xy} & \varepsilon_y & \frac{1}{2}\gamma_{yz} \\ \frac{1}{2}\gamma_{zx} & \frac{1}{2}\gamma_{yz} & \varepsilon_z \end{pmatrix} \tag{5.53}$$

and $\boldsymbol{\Theta}$ is the antisymmetric matrix of rotations

$$\boldsymbol{\Theta} = \begin{pmatrix} 0 & -\omega_z & \omega_y \\ \omega_z & 0 & -\omega_x \\ -\omega_y & \omega_x & 0 \end{pmatrix} \tag{5.54}$$

$\boldsymbol{F}$ can be decomposed into its symmetric part $\boldsymbol{\mathcal{E}}$ and the antisymmetric part $\boldsymbol{\Theta}$ by

$$\boldsymbol{\mathcal{E}} = \frac{1}{2}[\boldsymbol{F} + \boldsymbol{F}^T] \tag{5.55}$$

$$\boldsymbol{\Theta} = \frac{1}{2}[\boldsymbol{F} - \boldsymbol{F}^T] \tag{5.56}$$

With this we have a method to obtain the strains, shears, and rotations provided we have determined $\boldsymbol{F}$ by holographic measurement.

If the object's geometry is known, we can assign coordinates to the points of intersection of the object surface with the fringe laminae. In Fig. 5.11 let P, P_1, and P_2 be points on

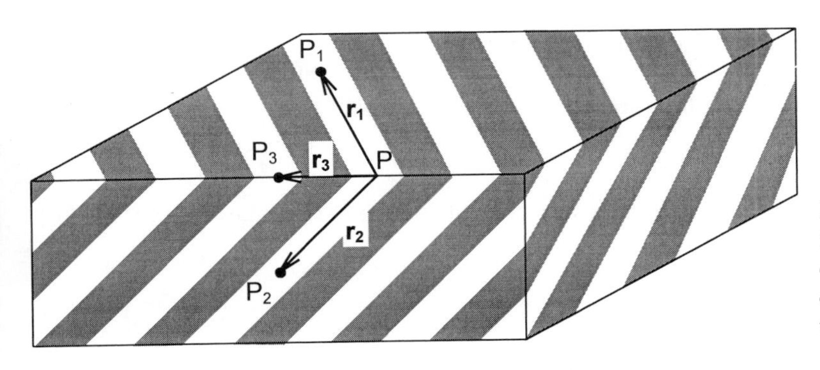

Figure 5.11: Straight fringes due to homogeneous deformation of the object

a common fringe, P_3 a point on an adjacent fringe. The vectors $\boldsymbol{r_1}$, $\boldsymbol{r_2}$, and $\boldsymbol{r_3}$ are the vectors from P to P_1, P_2, and P_3. The direction of the fringe vector $\boldsymbol{K}_f$ then is given by the unit vector $\boldsymbol{k}_f$

$$\boldsymbol{k}_f = \frac{\boldsymbol{r_1} \times \boldsymbol{r_2}}{|\boldsymbol{r_1} \times \boldsymbol{r_2}|} \tag{5.57}$$

and its magnitude $|\boldsymbol{K}_f|$ is

$$|\boldsymbol{K}_f| = \frac{\pi}{\boldsymbol{k}_f \cdot \boldsymbol{r_3}} \tag{5.58}$$

if the fringe at P_3 has higher interference order than that at P, otherwise we take $-\pi$ in the numerator.

To solve (5.50) for the deformation gradient matrix $\boldsymbol{F}$ we need three observations with different non-coplanar sensitivity vectors or even more observations and then solving by least squares. For each sensitivity vector we have to determine the corresponding fringe vector. As long as it is assumed that the sensitivity vectors do not vary, $\boldsymbol{G}$ may be neglected and $\boldsymbol{F}$ is obtained by inversion of (5.50).

If there is a substantial sensitivity vector variation, the displacements $\boldsymbol{d}$ have to be evaluated and $\boldsymbol{G}$ must be calculated from the known sensitivity vectors. Now we are able to correct $\boldsymbol{K}_f$ for perspective by

$$\boldsymbol{K}_{fc} = \boldsymbol{K}_f - \boldsymbol{G} \cdot \boldsymbol{d} \tag{5.59}$$

and $\boldsymbol{F}$ is determined by

$$\boldsymbol{F} = (\mathbf{e}^T \mathbf{e})^{-1} (\mathbf{e}^T \boldsymbol{K}_{fc}) \tag{5.60}$$

If we have not a three-dimensional object but a single plane surface in, say, the x-y-plane, only the x- and y-components of $\boldsymbol{K}_f$ contribute to the value of the fringe locus function. Therefore only the first two columns of $\boldsymbol{F}$ can be determined and in consequence only rotations about surface normals and the in-surface strains are processed. Three views, however, are still required in this case.

5.4 Hybrid Methods

The accurate measurement of displacements by holographic interferometry and computer-aided quantitative evaluation allows a subsequent numerical differentiation or other processing steps [108]. Thus these measurements can be combined with structural analysis methods like the finite element methods, boundary-element methods, or the techniques known under the heading of fracture mechanics. The refined analysis methods resulting from such combinations may be categorized as *hybrid methods*.

5.4.1 Finite Element Methods and Holographic Interferometry

The *finite element method* is a structural analysis method, where a technical structure is divided into a number of discrete elements which have a simple geometry. All these elements are connected via the nodes by which they are defined. The mechanical behaviour in each element can be calculated due to its simple geometry, then the program produces an overall solution that is compatible to all the nodes. Thus displacements, strains and stresses or the thermal behaviour of a structure under mechanical and/or thermal load can be approximated with good accuracy.

Holographic deformation analysis [367, 368, 369, 370, 371] or holographic vibration analysis [372] and finite element calculations can effectively be combined to reach a number of goals. One of the most important is the holographic verification of the finite element model [373, 374]. Especially for complex shaped or composite structures it is not an easy task to find a proper discretization and to choose the right material parameters. A comparison of the measured displacements with the calculated displacements can confirm the finite element model, which can then be used for, say, strain and stress calculations.

This strategy has been successfully followed in the investigation of vibration modes [375, 376] or stress distributions [323]. For the analysis of *adhesive bondings* it has been determined to what proportions the metal layers and the adhesive layers contribute to the deformation of the specimen [377]. A complete representation of the strain and stress conditions within the specimen has been obtained.

The temperature distribution and the deformation of a thermally loaded overlap adhesive bond with a local void in the adhesive layer was calculated by the finite element method and measured by holographic interferometry [378]. The combined evaluations have shown that the characteristic surface deformation above the defect in the internal adhesive layer is not caused by thermal expansion of the enclosed gas. The inhomogeneities in the surface deformation arise from the disturbed heat transfer in the defect area. So in this region we get locally higher temperature differences, which in consequence leads to locally different thermal deformations. These results explain the aptitude of *thermal load* for *defect detection* in holographic non-destructive testing.

The combination of holographic interferometry and finite element methods not only enables a defect detection but also a *defect validation*. In systematic calculations a catalogue of surface deformations in the region of internal cracks and voids in, say, steel with variations of defect type, length, orientation, volume or position, is compiled. For a given holographically measured displacement field above a defect, one starts with the best fitting displacement field from this catalogue. In an iterative process in a finite element model, the parameters of the simulated defect are varied. The displacement field is calculated over and over again with varied defects until the agreement with the measured displacement field is sufficiently good [323].

Of course, the calculation of a displacement field for a given discrete structure and loading by the finite element method is not a one-to-one mapping. Thus the inverse process of determining the discrete structure from the loading and the measured displacement field is not possible. Nevertheless, with the iterative method we get a defect which is representative of the equivalent class of all defects producing the same deformation under the specific applied load.

5.4.2 Boundary Element Methods and Holographic Interferometry

In the finite element method, the whole body to be tested is discretized into finite volumes connected at the nodes. Continuity of parameters which are not explicit variables is only warranted at the nodes and not at the borders between the elements. If we have other functions which fulfill exactly the differential equation in the whole region, we have no discretization errors in the interior of the body. Since only the boundary conditions have to be satisfied, the requirement that the boundary is discretized is in itself sufficient. The relating methods are called *boundary element methods*.

The boundary element methods can be combined advantageously with experimental methods like holography. Especially, if in practical applications the boundary conditions are too complicated to be described theoretically, they have to be measured. Having measured the displacements by, e.g., holographic interferometry, the boundary of two- or threedimensional regions is then divided into segments on which the displacements and strains are approximated by polynomials of the first degree. The stress components at prescribed internal points of the region are then calculated by means of the boundary element method.

In [379], this method is applied to transparent models manufactured from PMMA with roughened faces. The in-plane components of the displacement vectors are measured by double exposure double aperture speckle interferometry. The objects considered are a three-point loaded beam with an edge crack and a model of a large slab wall stiffened by a frame. Based on the measurements, exact values of stresses σ_x, σ_y, and σ_{xy} in the neighborhood of the crack tip are determined. In another application, the friction between the wall and its base is measured by applying the hybrid evaluation method of coherent optical measurement combined with the boundary element method.

5.4.3 Fracture Mechanics

In linear elastic *fracture mechanics* the influence of a *crack* or another defect on the damage of a technical structure is estimated. An important figure is the *stress intensity factor* K_I and its critical value K_{Ic}, the *fracture toughness*, which is a material property. The K_I-value can be determined holographically by first measuring the deformation field of the structure exhibiting a crack [380] followed by the determination of the boundary of the *plastic zone* arising above the crack during *tensile loading*. This is functionally related to the K_I-value [381, 382, 383]. Another method is based on the integration of the strain equations of Sneddon or Williams-Irwin. This method only requires a displacement measurement along a line perpendicular to the crack propagation direction [384]. In conjunction with these methods holographic interferometry has been used to determine the K_I-value for a CT 500 specimen with high accuracy and without any previous knowledge of the specific defect properties like its size and location.

Another criterion for *crack propagation* is the so-called *J-integral*, which is a figure independent of the path of integration. Crack propagation occurs if the J-integral exceeds a critical material parameter [383]. In [385], the determination of the J-integral is based on measurements of the displacement field by holographic interferometry. Power series estimations up to quadratic terms fulfilling the Lame-Navier equations are set up for the three displacement components. From the coefficients of the power series the integration can then be performed along a rectangular path to yield the J-integral.

5.5 Vibration Analysis

The measurement of vibrations is an important task in engineering, on the one hand to ascertain the operation of components, which should vibrate, like loudspeakers, ultrasonic transducers etc., on the other hand to check the behaviour of components, which have natural frequencies of response within the range of frequencies excited by the operation of an engine the component may be part of. The aims can be the prevention of fatigue failure or the detection of noise-generating parts or areas, to name just two. Of course non-contacting measurement methods which do not affect and bias the vibration are recommended. Since furthermore the amplitudes to be measured normally are in the range of the wavelengths of laser light, and since one is interested in the simultaneous measurement at a manifold of points, holographic interferometry is a suitable tool for analysing vibrations [386, 387]. It is worth to be noted that the first holographic interferograms were made of diffusely reflecting surfaces under vibration [10].

5.5.1 Surface Vibrations

The fundamental surface vibration is the *sinusoidal vibration* also called *harmonic vibration*. Here the displacement of each point of the vibrating surface is

$$\boldsymbol{d}(P,t) = \boldsymbol{d}(P)\sin\omega t \tag{5.61}$$

where $\boldsymbol{d}(P)$ is the vector amplitude and ω the circular frequency of the vibration. The vibrations of continua are described by partial differential equations. Generally the boundary conditions yield transcendental equations for the eigenvalues. Approximate solutions use the Rayleigh-quotients or the Ritz-method. For some special cases like vibrating cantilever beams, circular or rectangular shells or plates, closed solutions are known.

The solutions differ in the distribution of $\boldsymbol{d}(P)$ over the points P, which defines the *mode-shapes* of the specific vibration. This should be explained at the example of a vibrating rectangular plate being jointed at the edges. The differential equation is

$$\frac{\partial^2 d_z}{\partial t^2} = -\frac{N}{\rho h}\left(\frac{\partial^4 d_z}{\partial x^4} + 2\frac{\partial^4 d_z}{\partial x^2 \partial y^2} + \frac{\partial^4 d_z}{\partial y^4}\right) \tag{5.62}$$

with the stiffness of the plate $N = Eh^3/[12(1-\nu^2)]$, the thickness h, and the specific mass ρ. If the sidelengths are a and b, solutions are given by

$$d_z(x,y,t) = d_{max}\sin(\omega t + \phi_0)\sin(\frac{j\pi x}{a})\sin(\frac{k\pi y}{b}) \tag{5.63}$$

with the eigenvalues $\omega_{jk} = (j^2/a^2 + k^2/b^2)\pi^2\sqrt{N/(\rho h)}$ for $j,k = 1,2,\ldots$. The two extreme deformation states of a plate vibrating in the (j=2, k=2)-mode is shown in Fig. 5.12. The

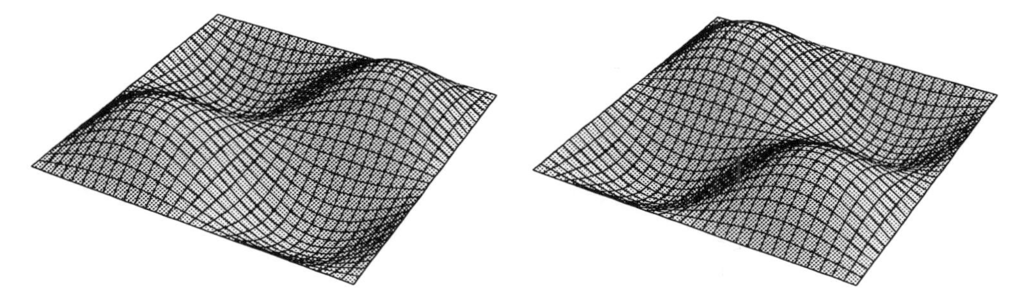

Figure 5.12: Vibrating rectangular plate

maximum amplitude at each point (x,y) is $|d_{max}\sin(j\pi x/a)\sin(k\pi y/b)|$. The loci where the maximum amplitude is zero all the time are the *nodes*, on the other hand the loci of local extrema of the amplitude are called *antinodes*. In Fig. 5.12 we have two nodal lines parallel to the edges and intersecting in the center of the plate. It is one task of holographic vibration analysis to identify the *mode-shape* (j,k) and the maximum amplitude of the amplitude distribution of a vibration.

In Fig. 5.12 all points vibrate in-phase, there is no phase shift between the oscillations of each point, except that when one antinode reaches its positive maximum, another takes its negative maximum, which may be interpreted as a phase shift by π. The phase relation between the vibrating points is another item to be analysed holographically. Fig. 5.13 shows some oscillatory states of a vibrating circular disc where the phase varies over 6π around the circumference. The phase is constant along each radius. There is only one nodal point in the center of the disc.

The sinusoidal vibration of (5.61) is an example for a separable object motion. *Separable object motions* are such motions which can be separated into a product of a displacement vector $\boldsymbol{d}(P)$ and a real temporal function $f(t)$. This $f(t)$ is $\sin(\omega t)$ in (5.61). Here the point P moves along a line defined by $\boldsymbol{d}(P)$. If the point moves along an arc, this motion can be described by the sum of two or three separable motions. Besides the sinusoidal vibration, $f(t)$ may represent damped harmonic vibrations [388] or other nonlinear vibrations [389, 390, 391].

In many applications the object surface executes a combination of N individual separable motions, e. g. a plate vibrating simultaneously in the $N = 2$ modes

$$d_z(P,t) = d_1(P)\sin(\omega_1 t + \phi_{01}) + d_2(P)\sin(\omega_2 t + \phi_{02}) \tag{5.64}$$

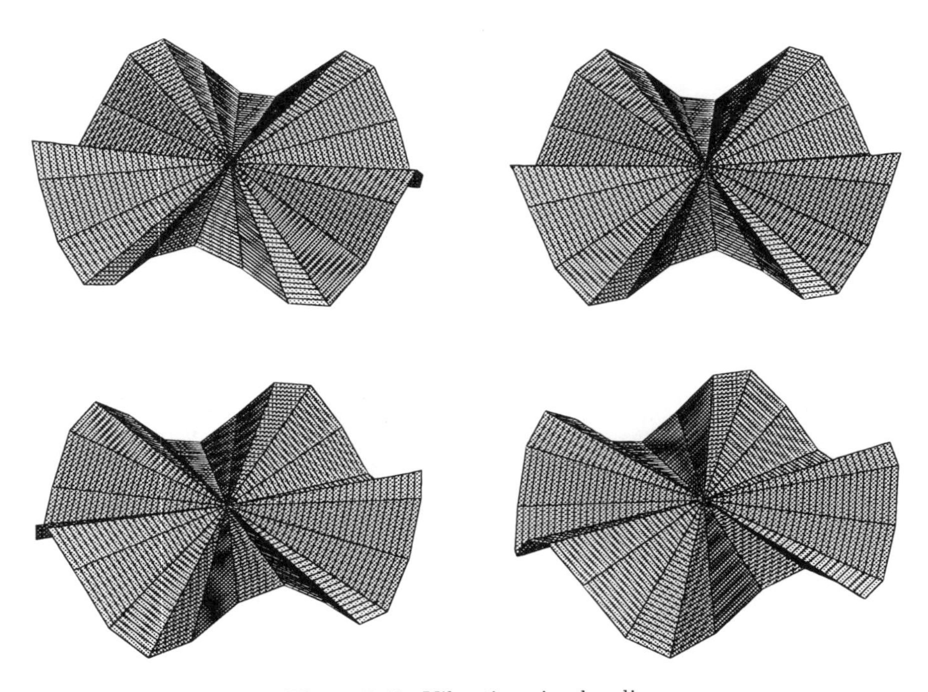

Figure 5.13: Vibrating circular disc

Even if the frequencies agree, $\omega_1 = \omega_2$, the combining modes may differ in mode shape, amplitude and phase. We speak of *dependent object motions* if ω_1/ω_2 is the ratio of two small integers; the motions are *independent* if there is no such ratio.

5.5.2 Stroboscopic and Real-Time Holographic Interferometry

Stroboscopic holographic interferometry consists of recording a hologram by using a sequence of short laser pulses which are synchronized with the vibrating object [392]. One can record a reference state of the object at rest and then illuminate with short pulses all the time the object is in the same position, preferably when the antinodes undergo the maximum displacement. But in most applications the pulses are fried when the object is in its maximum positive displacement and again when it is at its maximum negative displacement. If the motion is $\boldsymbol{d}(P)\sin\omega t$, pulses should be at $\omega t = \pi/2$ and $\omega t = 3\pi/2$, and the same after periods of 2π until the total required exposure is reached. If the pulses are short enough, the result is the same as with double exposure holographic interferometry, that means we get cos-type fringes (3.4) and (3.21)

$$I(P) = 2I_1(P)\{1 + \cos[\boldsymbol{d}(P) \cdot \boldsymbol{e}(P)]\} \tag{5.65}$$

with stationary reference state or

$$I(P) = 2I_1(P)\{1 + \cos[2\boldsymbol{d}(P) \cdot \boldsymbol{e}(P)]\} \tag{5.66}$$

if the extreme positions are compared. For a sensitivity vector perpendicular to the surface the fringes indicate loci of constant amplitude of the vibration modes. There are two main drawbacks of this approach: First the vibration has to be monitored for the synchronization of the laser pulses, e. g. by a Michelson interferometer or a Doppler velocimeter measuring in one point of the surface, thus increasing the experimental complexity, second if the pulses are short, many are needed for exposure, thus a high mechanical stability is required. On the other hand if the pulse duration is increased, the cosine-type fringes are modulated by a sinc-function with the result that the visibility decreases [393]. An advantage of the stroboscopic method are the cosine-shaped fringes with constant contrast over all fringe orders. Furthermore they allow the introduction of phase shifting for an automatic evaluation of the interference phase with subfringe accuracy [394]. A vibration analysis by stroboscopic *heterodyne holographic interferometry* employing a two reference beam arrangement is presented in [395, 396].

The investigation of the object as it responds to different exciting frequencies is best done by *real-time holographic interferometry*. Hereby a reference state of the object at rest is recorded holographically and the processed hologram plate is exactly repositioned. During object vibration optical wavefields of the reconstructed reference state and of the actual state interfere and for sufficiently high vibration frequency ω the resulting interference pattern is of the form (3.7) according to (C.2)

$$I(P) = 2I_1(P)\{1 - \mathrm{J}_0[\boldsymbol{d}(P) \cdot \boldsymbol{e}(P)]\} \tag{5.67}$$

where J_0 is the zero-order *Bessel function* of the first kind. The nodal points are the darkest in the pattern, local intensity minima and maxima occur in turn between the zeros of J_0, Fig. 3.3. Because of the nature of $\mathrm{J}_0(\boldsymbol{d} \cdot \boldsymbol{e})$ the fringes exhibit low contrast, an effect that is further increased if we have not the same irradiance or polarization state between the two interfering wave fields. But the method is suitable for the quick identification of resonant frequencies by monitoring the fringe pattern while sweeping the excitation frequency continuously through the frequency range of interest.

A modification of the real-time method is the holographic subtraction [397]. A first exposure is made of an object at rest, the second during vibration while an additional phase shift π is introduced. The resulting interference pattern is

$$I(P) = 2I_1(P)\{1 - \mathrm{J}_0[\boldsymbol{d}(P) \cdot \boldsymbol{e}(P)]\}^2 \tag{5.68}$$

which exhibits the same number and positions of the fringes as in the real-time method but due to the square with higher visibility.

5.5.3 Time Average Holographic Interferometry

The experimentally easiest and most often applied holographic method for vibration analysis is the *time average holographic interferometry* [10, 398, 399, 400, 401]. The object is recorded holographically with a single exposure, employing a CW-laser using an exposure time T, which is long compared with the period of the vibration, e. g. for harmonic vibrations $T \gg 2\pi/\omega$. The exact exposure time is only directed by the emulsion of the hologram plate. The resulting intensity in the reconstructed image is according to (3.10) and (C.2)

$$I(P) = I_0(P) \mathrm{J_0}^2[\boldsymbol{d}(P) \cdot \boldsymbol{e}(P)] \tag{5.69}$$

At the nodes of the vibration modes we have the maximal intensity, $\mathrm{J}_0(0) = 1$, and dark fringes we have where $\boldsymbol{d}(P) \cdot \boldsymbol{e}(P)$ equals the arguments of the zeros of the zero-order Bessel function of the first kind, Table C.1. The bright fringes between these zeros which do not correspond to nodal lines exhibit less intensity compared with the zero fringe. Thus the nodal lines can easily be identified. As an example consider a plane vibrating surface and a holographic arrangement with illumination and observation normal to the surface. If each point can be assumed to oscillate only in normal direction $\boldsymbol{d}(P) = (0, 0, d_z(P))$, then $\boldsymbol{e}(P) = 4\pi/\lambda$ according to (3.20) and the amplitudes at the centers of dark fringes are easily calculated by

$$d_z(P) = b_m \frac{\lambda}{4\pi} \tag{5.70}$$

where b_m is the m-th zero of J_0, Table C.1. The order of the fringe is determined by starting the counting with $m = 1$ for the dark fringe adjacent to the zero fringe, the nodal line.

For thin plates or shells the vibration vector can be assumed to be perpendicular to the surface, any in-plane components can be neglected. Then the time average fringes are interpreted as contour lines of the vibration modes. If for more general objects the direction of the vibration amplitude vector is not known a priori, at least three holographic observations with different sensitivity vectors have to be performed and a system of three linear equations or a least squares system has to be solved, see Sec. 5.1.

The interpretation of the brightest fringe as a nodal line nevertheless can be erroneous for more complex structures. Consider the pure torsional vibration of a cylinder. For illumination and observation directions parallel to the optical axis, which intersects the hologram, the surface line facing the hologram appears as the brightest fringe. But this is because at these points the vibration vector is orthogonal to the sensitivity vector [402].

Typical holographic interference patterns have been recorded using the time average method, Fig. 5.14. The object is a square plate rigidly clamped at all edges and excited acoustically by a loudspeaker, driven by various frequencies.

5.5.4 Temporally Modulated Reference Wave

In stroboscopic holographic interferometry, Sec. 5.5.2, the object and the reference wave fields have been modulated by temporal functions $f_{obj}(t)$ and $f_{ref}(t)$, which depend on the object vibration frequency and phase. This constituted a special case of the more general and powerful concept of temporal modulation of the wavefields. It is not necessary to have $f_{obj}(t) = f_{ref}(t)$ as in the stroboscopic method. In most applications it is only the reference wave which is modulated, this is considered in the sequel.

There are several ways to modulate the reference wave during the recording of the hologram, offering a number of benefits: increased sensitivity for small vibration amplitudes, detection of the relative phase between the oscillating points across the surface, compensation for extraneous object motion, etc.

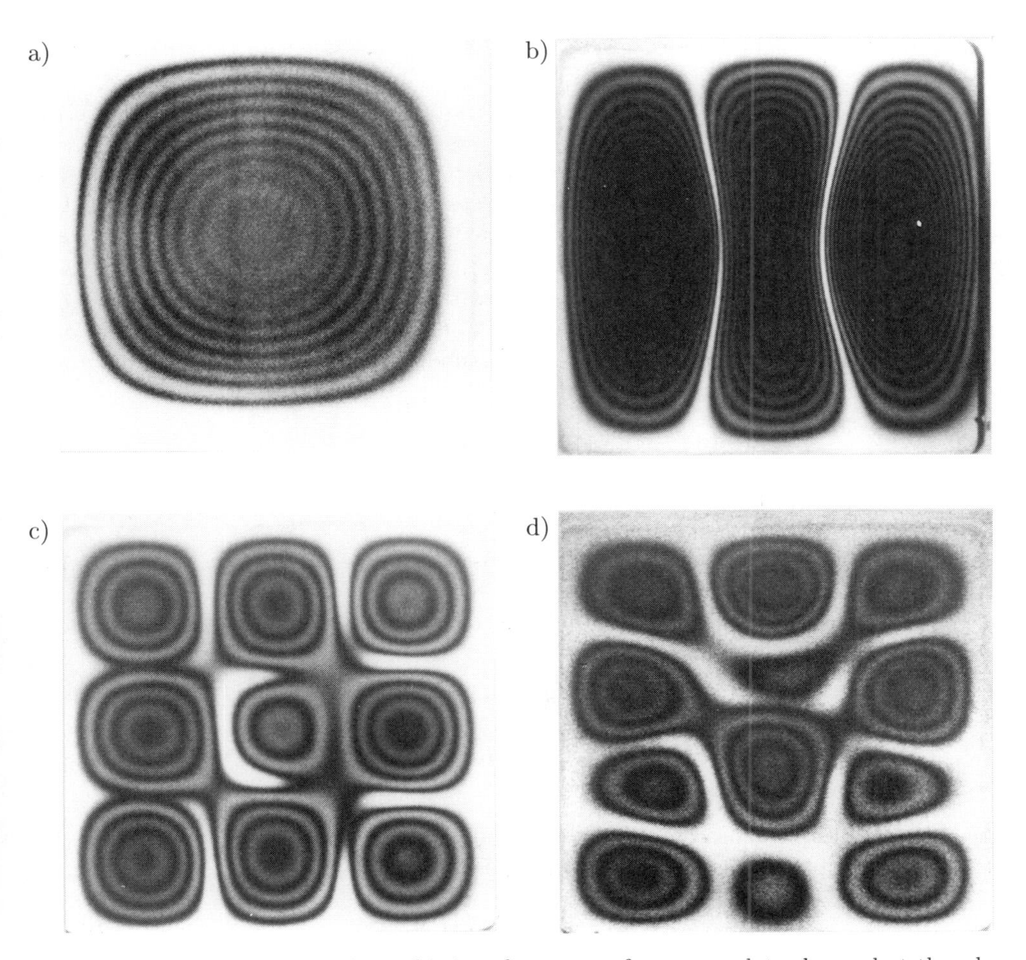

Figure 5.14: Time average holographic interferograms of a square plate clamped at the edges

If the object wave E_P and the reference wave E_R are modulated by $f_{obj}(t)$ and $f_{ref}(t)$, resp., the resulting intensity in the hologram plane at time t is according to (2.116) proportional to

$$I(t) = |f_{obj}(t)E_P + f_{ref}(t)E_R|^2 \tag{5.71}$$

Assuming linear recording, the amplitude transmittance of the hologram is, see (2.119)

$$T = \alpha - \beta \int_0^T |f_{obj}(t)E_P + f_{ref}(t)E_R|^2 dt \tag{5.72}$$

For reconstruction the hologram is illuminated with the continuous unmodulated reference wave E_R, see (2.125). The interesting third term of (2.125) describes the complex amplitude of a wave field proportional to

$$M_T = \frac{1}{T}\int_0^T f_{obj}(t) E_P f^*_{ref}(t) dt \tag{5.73}$$

The expression M_T often is called the *characteristic function* [24, 25, 158, 403], but the concept of characteristic functions will not be stressed extensively in this book. The resulting intensity is $I = |M_T|^2$. The M_T for some ways of modulating the reference wave now should be investigated in more detail:

Let the object surface vibrate harmonically with frequency ω, then the object wave is proportional to $\exp[\mathrm{i}\boldsymbol{d}(P) \cdot \boldsymbol{e}(P) \sin\omega t]$, while $f_{obj}(t) = 1$. In *frequency translated holography* the frequency of the reference wave is modulated by an integer multiple $n\omega$ of the object vibration frequency ω

$$f_{ref}(t) = \exp(\mathrm{i}n\omega t) \tag{5.74}$$

The resulting characteristic function is proportional to

$$M_T = \frac{1}{T}\int_0^T \exp[\mathrm{i}\boldsymbol{d}(P) \cdot \boldsymbol{e}(P) \sin\omega t] \exp(-\mathrm{i}n\omega t) dt \tag{5.75}$$

Using the identity (C.4) and reversing the order of integration and summation we get

$$\begin{aligned} M_T &= \frac{1}{T}\int_0^T \sum_{m=-\infty}^{\infty} \mathrm{J_m}[\boldsymbol{d}(P) \cdot \boldsymbol{e}(P)] \exp(\mathrm{i}m\omega t) \exp(-\mathrm{i}n\omega t) dt \\ &= \sum_{m=-\infty}^{\infty} \mathrm{J_m}[\boldsymbol{d}(P) \cdot \boldsymbol{e}(P)] \frac{1}{T}\int_0^T \exp[\mathrm{i}(m-n)\omega t] dt \end{aligned} \tag{5.76}$$

If the exposure time is long compared with the object vibration period, $T \gg 2\pi/\omega$, the integral vanishes for all m except for $m = n$, so that

$$M_T = \mathrm{J_n}[\boldsymbol{d}(P) \cdot \boldsymbol{e}(P)] \tag{5.77}$$

or

$$I(P) = \mathrm{J_n}^2[\boldsymbol{d}(P) \cdot \boldsymbol{e}(P)] \tag{5.78}$$

This result is consistent with the unmodulated time average case having $n = 0$, see (5.69). Proportionality factors influencing only the overall brightness have been omitted for convenience.

Frequency translated holography is used to increase the sensitivity for vibrations with small as well as with large amplitudes: Small amplitudes here are such that the interference phase $\Delta\phi(P) = \boldsymbol{d}(P) \cdot \boldsymbol{e}(P)$ remains small, $\Delta\phi(P) \ll 1$. In time average holography with no modulation $\mathrm{J_0}^2(0)$ is unity and has slope zero. So no significant intensity variation in the bright field will result. On the other hand $\mathrm{J_1}^2(0)$ has a positive slope in the dark field, see Fig. C.1, yielding visible intensity variations even for small amplitudes. The smallest detectable amplitude was estimated as $2.7 \times 10^{-4}\lambda$ [404]. In the case of large vibration amplitudes we take advantage of the fact that the locations of the zeros of the Bessel functions are spreaded apart for increasing order n, Fig. C.1. This results in a decreasing number of fringes for the same amplitudes with increasing n, meaning too high fringe densities can be effectively avoided [405].

A general temporally periodic object wave may be considered as composed of its Fourier series terms, having frequencies $\omega_0 + m\omega$, $m = 0, \pm 1, \pm 2, \ldots$ with ω_0 the frequency of the used laser light. The reference wave in frequency translated holography has the frequency $\omega_0 + n\omega$. So only the frequency component with $m = n$ will produce a time average hologram, because only this one is coherent to the reference wave. Thus we have a method for temporal filtering by selecting single frequency components from a periodic object motion [24, 405, 406].

In *amplitude modulation holography* the amplitude of the reference wave is modulated with the same frequency as the object vibrates, but with a controllable phase difference ψ with respect to the vibration of a selected object point P, $f_{ref}(t) = \cos(\omega t - \psi)$. In this case the n of (5.74) is 1. The resulting intensity is proportional to

$$I(P) = \mathrm{J_1}^2[\boldsymbol{d}(P) \cdot \boldsymbol{e}(P)] \cos^2 \psi \tag{5.79}$$

A sequence of holograms, recorded with amplitude modulation at varied phase ψ, will display contours of constant relative phase [407].

A further method to obtain phase information about the object vibration is *phase modulation holography.* The phase of the reference beam is modulated at the frequency ω of the vibrating object with a modulation depth of Ω_R. The modulation function is

$$f_{ref}(t) = \exp(\mathrm{i}\Omega_R \sin \omega t) \tag{5.80}$$

which gives the characteristic function [24]

$$\begin{aligned} M_T &= \frac{1}{T}\int_0^T \exp[\mathrm{i}\boldsymbol{d}(P) \cdot \boldsymbol{e}(P) \sin(\omega t - \phi_0)] \exp(-\mathrm{i}\Omega_R \sin \omega t) dt \\ &= \mathrm{J_0}\{[(\boldsymbol{d}(P) \cdot \boldsymbol{e}(P))^2 + \Omega_R^2 - 2\boldsymbol{d}(P) \cdot \boldsymbol{e}(P)\Omega_R \cos \phi_0]^{1/2}\} \end{aligned} \tag{5.81}$$

The implication of this characteristic function is, that the relative phase ϕ_0 of the vibration at each point is encoded in the fringe pattern. Relative phases [408] as well as amplitudes [409] are measured based on this principle. For simplicity of discussion assume $\phi_0 = 0$. Then we have an intensity

$$I(P) = |M_T(P)|^2 = \mathrm{J_0}^2(\boldsymbol{d}(P) \cdot \boldsymbol{e}(P) - \Omega_R) \tag{5.82}$$

Now the loci of bright zero fringes are controllable by the user, since they appear where $\boldsymbol{d}(P) \cdot \boldsymbol{e}(P) = \Omega_R$. Further insight is gained when the modulation (5.80) includes a phase term ϕ_R. The dark fringes in the interferogram then are characterized by

$$b_m^2 = \boldsymbol{d}(P) \cdot \boldsymbol{e}(P) + \Omega_R^2 - 2\boldsymbol{d}(P) \cdot \boldsymbol{e}(P)\Omega_R \cos(\phi_0 - \phi_R) \tag{5.83}$$

where b_m denotes the m-th zero of $\mathrm{J_0}$ [410]. If two holographic interferograms are produced in this way with phases ϕ_{R1} and ϕ_{R2}, they may be superimposed. At the intersections of fringes of equal order m we have

$$\phi_0 = \frac{\phi_{R1} + \phi_{R2}}{2} \pm m\pi \tag{5.84}$$

Using this approach one gets phase contours by simple visual inspection. This work has been extended [411] to bright fringes so that also bright-bright and bright-dark combinations can be used. Hereby phases intermediate to those of (5.84) can be inspected. Objects excited with swept sinusoidal vibration or randomly have been investigated by time average holography with a mechanically excited reference beam. Modes and amplitudes of randomly excited objects were analysed by a method called *spectroscopic holography* [412].

The holographic interferometry using phase modulation described by (5.82) is closely related to general motion compensation by reference waves modulated by the object motion [329, 331]. If the vibration amplitudes in the vicinity of a point P are large, a reference wave modulated by a mirror fixed at P and undergoing the same motions will yield a vibration measurement relative to this point.

5.5.5 Numerical Analysis of Time Average Holograms

Numerical analysis of cosine fringes by phase shifting or Fourier transform methods allows the determination of interference phases even between the fringe intensity maxima and minima with high accuracy. While a lot of research has been performed to automate the evaluation of cosine fringes, not a great deal has been done to automate Bessel-type fringe interpretation.

One approach is to convert J_0-fringes into sinusoidal fringes by stroboscopic techniques, and to apply heterodyning [395] or phase stepping [413]. The real-time method of vibration analysis was combined with heterodyne [396] and phase step evaluation [414]. In those methods vibration amplitudes must still be interpolated from integer fringe orders. The direct numerical extraction of vibration amplitudes from time average interference patterns as presented in [415] is given in the sequel.

The intensity of a time average hologram is (3.10)

$$I(x, y) = I_0(x, y)\mathrm{J}_0{}^2\left(\Delta\phi(x, y)\right) \tag{5.85}$$

where $I_0(x, y)$ is the irradiance of the object surface. The interference phase is $\Delta\phi(x, y) = \boldsymbol{d}(x, y) \cdot \boldsymbol{e}(x, y)$ with $\boldsymbol{d}(x, y)$ the vectorial displacement and $\boldsymbol{e}(x, y)$ the sensitivity vector in (x, y). As we have seen in Sec. 5.5.4, the J_0-fringes can be shifted by modulating the phase of either the object or the reference beam sinusoidally at the same frequency as the object vibration. This adds a phasor bias $\boldsymbol{\Omega}_R$ to the argument of the Bessel function. If the object is vibrating in only one vibration mode and if the phase of the sinusoidal beam modulation is adjusted to coincide with that of the object vibration the phasor bias becomes an additive term and the irradiance of a time average hologram reconstruction together with the unavoidable distortions is (5.82)

$$I(x, y) = a(x, y) + b(x, y)\mathrm{J}_0{}^2\left(\Delta\phi(x, y) - \Omega_R\right) \tag{5.86}$$

with $a(x, y)$ containing the background irradiance and $b(x, y)$ the multiplicative object surface irradiance. Eq. (5.86) has a form analogous to the phase sampling equation (4.16) for cosine fringes. Unfortunately the Bessel function of a sum cannot be expressed as a sum of terms as done in (4.20) for the cosine, so straightforward solutions are not possible. An iterative process for calculation of the phase $\Delta\phi$ of each (x, y) is outlined in [415], but this requires quite lengthy calculations.

A solution that makes use of the nearly periodic nature of the J_0-function employs the formula

$$\Delta\phi = \arctan \frac{(1 - \cos\Omega_R)(I_3 - I_1)}{\sin\Omega_R(I_1 - 2I_2 + I_3)} \tag{5.87}$$

see Table 4.1, where we use 3 reconstructions with the bias terms $-\Omega_R$, 0, $+\Omega_R$ having known values. The interference phase $\Delta\phi$ computed by (5.87) differs from the correct argument of the $J_0{}^2$-function. Due to the approximating function $J_0^*(x)$, see (C.6), this error approaches $\pi/4$ for large arguments. The error can be computed for any phase angle $\Delta\phi$ and any value of Ω_R to create a lookup table to convert the incorrect answers obtained from the use of (5.87) into the correct phase values. In practice, since the influence of $a(x, y)$ and $b(x, y)$ is eliminated inherently in the evaluation by (5.87), we can proceed with $a = 0$ and $b = 1$ when compiling the lookup table. For each specific Ω_R the intensities I_1, I_2, and I_3 corresponding to $-\Omega_R$, 0, and $+\Omega_R$ are computed by (5.86) for the desired values of $\Delta\phi$. Then (5.87) is solved for these I_1, I_2, I_3, resulting in a false interference phase $\Delta\phi^*$. These $\Delta\phi^*$ are tabulated with the corresponding correct values of $\Delta\phi$.

The phase data obtained by evaluation using (5.87) with subsequent correction employing the lookup table still has to be demodulated. The one pattern with no bias vibration now is used to identify the loci of the zero-order fringes. These are recognized by their high brightness relative to the rest of the fringes. So when working with time average hologram data we even have access to the abolute fringe order.

Difficulties still occur when two or more vibration modes lie sufficiently close in frequency, so that both are excited at the same time. It seems reasonable to suppose, that data recorded with the bias phase at a number of phase angles could be used to extract both the amplitude and the phase of the vibration [415]. In this direction work still has to be done.

5.6 Holographic Contouring

The holographic interference pattern arises from the superposition of at least two states of a reflected or refracted wavefield. In holographic deformation measurements the two states are given by the scalar product of the displacement vector field and the sensitivity vectors, (3.21). While the displacement vectors describe the variation of the surface point positions between the interferometrically measured states, the sensitivity vectors, which are given by the used laser wavelength, the directions of illumination and observation, and the geometry of the measured surface, remain constant.

This concept is inverted for *holographic contouring*. Contouring in general means the modulation of the image of a three-dimensional object by fringes corresponding to contours of constant elevation with respect to a reference plane [168, 416, 417]. For holographic contouring the object is not displaced, but the two states which by superposition form the fringe pattern are produced by variation of the sensitivity vector in (3.4) and (3.21).

5.6.1 Contouring by Wavelength Differences

The length of the sensitivity vector depends on the wavelength λ, see (3.20). Of course a change in the wavelength varies the optical path length. The wavelength can be changed in predetermined discrete steps with e. g. an argon-ion-laser or continuously with a dye-laser. For contouring by the *two-wavelength method* [418, 419, 420, 421, 422, 423, 424, 425] the optical arrangement shown in Fig. 5.15 is employed. It uses plane illumination and

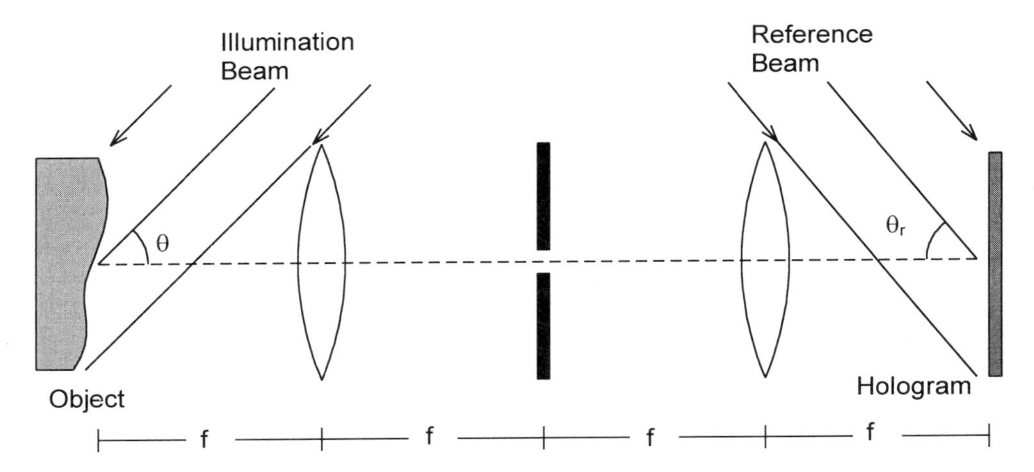

Figure 5.15: Arrangement for wavelength-difference contouring

reference waves, and a telecentric viewing system with an image plane hologram.

Let λ be the wavelength utilized in recording of the hologram and then apply the real-time method with wavelength λ'. The points P' of the reconstructed image are shifted relative to the really existing point P according to (2.145) - (2.147). The lateral displacements of the points can be eliminated by means of plane recording and reconstructing reference beams and shifting back the reconstructed image by tilting the reference beam by an appropriate amount such that the condition

$$\lambda \sin \Theta_r' = \lambda' \sin \Theta_r \tag{5.88}$$

is satisfied [85]. Θ_r and Θ_r' represent the initial and final angular positions of the reference beam. Another way of eliminating the lateral shift requires to bring the object very close to the hologram plate, meaning to record an image plane hologram of the object, Fig. 5.15. Under these assumptions we have the situation of (2.151) - (2.153), so that there is a fictive displacement vector

$$\boldsymbol{d} = P' - P = \begin{pmatrix} x_P' - x_P \\ y_P' - y_P \\ z_P' - z_P \end{pmatrix} = \begin{pmatrix} 0 \\ 0 \\ \frac{1}{\mu} z_P - z_P \end{pmatrix} \tag{5.89}$$

with $\mu = \lambda'/\lambda$. By the special choice of the configuration the observation unit vector is $\boldsymbol{b}(P) = (0, 0, 1)^T$ and the illumination unit vector is $\boldsymbol{s}(P) = (-\sin\Theta, 0, -\cos\Theta)^T$,

Fig. 5.15. Thus the sensitivity vector is

$$\boldsymbol{e}(P) = \frac{2\pi}{\lambda}(\sin\Theta, 0, 1 + \cos\Theta)^T \tag{5.90}$$

and the resulting phase difference, (3.21) is

$$\begin{aligned}\Delta\phi(P) &= \boldsymbol{d}(P)\cdot\boldsymbol{e}(P) \\ &= \frac{2\pi}{\lambda}\left(\frac{1}{\mu}z_P - z_P\right)(1+\cos\Theta) \\ &= 2\pi\frac{\lambda-\lambda'}{\lambda\lambda'}(1+\cos\Theta)z_P \end{aligned}\tag{5.91}$$

The two wavefronts interfere and produce fringes corresponding to *contours* of constant altitude. The fringe planes intersect the object in a direction parallel to the hologram plane. The contour sensitivity is given by the depth difference Δz

$$\Delta z = \frac{\lambda\lambda'}{(\lambda-\lambda')(1+\cos\Theta)} \tag{5.92}$$

which induces a change of 2π in the phase difference $\Delta\phi$.

As an example, Fig. 5.16 shows a statuette, that was contoured by the wavelength difference method using a dye-laser.

5.6.2 Contouring by Refractive Index Variation

A method closely related to the two-wavelength method is the *contouring by refractive index variation*, also called the *immersion method* [85, 426, 427]. Here the wavelength is not modified by changing the frequency of the laser, but by a change in the speed of light according to

$$\nu\lambda = nc \tag{5.93}$$

n is the refractive index of a transparent material the light is passing. (2.10) is the special case for vacuum, $n = 1$. The schematic of the immersion method is illustrated in Fig. 5.17. The object is placed in a glass tank filled with a transparent gas or liquid having refractive index n. While the first recording of the object is performed with the medium at refractive index n, the second exposure of the double exposure method is taken when the medium is replaced by another having refractive index n'. In the real-time method during reconstruction of the first holographically stored wavefield the object rests in the medium of refractive index n'. In both methods the produced interference fringes depend on the difference of the refractive indices and the distance $z(P)$ between the object surface and the tank wall.

The object is assumed to be illuminated by a plane wave in a direction perpendicular to the glass plane. Observation is carried out using a telecentric imaging system. Thus different deflections of the rays by the different refractive indices are avoided. The optical phase difference giving rise to fringe formation is

$$\Delta\phi(P) = \frac{4\pi}{\lambda}(n - n')z(P) \tag{5.94}$$

Figure 5.16: Wavelength difference contouring

The contour interval Δz producing a change in the phase difference $\Delta\Delta\phi = 2\pi$ is

$$\Delta z = \frac{\lambda}{2(n - n')} \tag{5.95}$$

If the liquid is a mixture of water and alcohol, and the refractive index is adjusted by the mixture ratio, the method is called *grog method* [48].

5.6.3 Contouring by Varied Illumination Direction

In the preceding subsections the sensitivity vector between the two interfering states was changed by a variation of the wavelength. Of course the sensitivity vector can be changed by altering the directions of illumination or observation.

If the illumination point S is changed to S' between the two exposures of the double exposure method, the resulting optical path length change δ and the interference phase $\Delta\phi$ can be calculated analogously to (3.13) - (3.21). For $\boldsymbol{d}_S(S) = S' - S$, we get

$$\Delta\phi(P) = -\frac{2\pi}{\lambda}\boldsymbol{d}_S(S) \cdot \boldsymbol{s}(P) \tag{5.96}$$

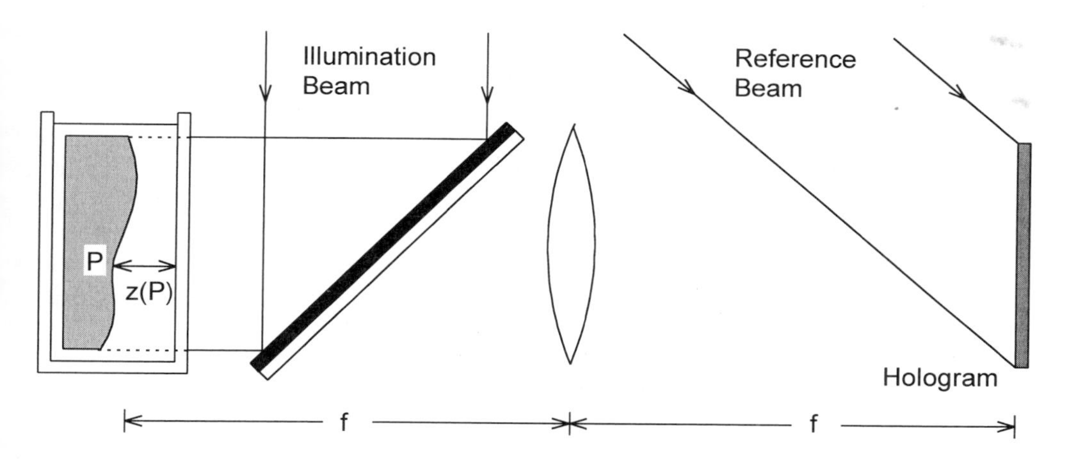

Figure 5.17: Arrangement for contouring by the immersion method

This means that the object is intersected by fringe surfaces which consist of a set of rotationally symmetric hyperboloids, their common foci being the two points of illumination. The fringe pattern is independent from the point of observation. The longer the distance between object and illumination source the more flat are the intersecting surfaces that are approximately parallel to the illuminating beams. Collimated beams produce equidistant parallel flat surfaces. The distance Δh of two such surfaces is

$$\Delta h = -\frac{\lambda}{2\sin\frac{\Theta}{2}} \tag{5.97}$$

where Θ is the angle between the two illumination directions. The analogy to the results of 2.2.1, (2.124) is obvious. Exactly the same result would be produced if holography was not used at all, but rather the object illuminated from the two points simultaneously, this procedure being called *projected fringe contouring*.

If the angle of illumination is changed by translating the object between the two exposures in the proper direction, contouring surfaces of nearly any orientation can be produced [428, 429]. This approach may be combined with a variation of the illumination direction [430, 431, 432].

Theoretically the double exposure method with two observation points might be performed by moving the hologram slightly between the two exposures. But since the intersecting surfaces now are parallel to the line of sight, this attempt is useless for contouring purposes. However, if the two states are recorded on different holograms in *sandwich hologram interferometry*, contour lines can be obtained by mutually shifting of the plates [433].

5.6.4 Contouring by Light-in-Flight recording

A conceptually different approach to contouring of three-dimensional objects is provided by the holographic *light-in-flight recording* and reconstruction [434, 435, 436, 437, 438,

439]. The idea behind holographic light-in-flight recording is the equivalence of a short temporal coherence and a fictitious extremely fast shutter or short light-pulse to produce a motion picture of a propagating optical wavefront. In each small region of a hologram only those parts of an object will be recorded for which the path length from the laser to this small region via the object does not differ from the path length of the reference beam by more than the temporal coherence length of the laser light used for the recording [434, 440].

If the coherence length is short, only those parts of a large object are recorded and appear brightly reconstructed for which a near-zero path difference holds. The object surface seems to be intersected by imaginary interference surfaces in the form of ellipsoids. One focus of these ellipsoids is the source point of the illuminating spherical wave, point A in Fig. 5.18, the other focus is the point in the holographic plate used for observation,

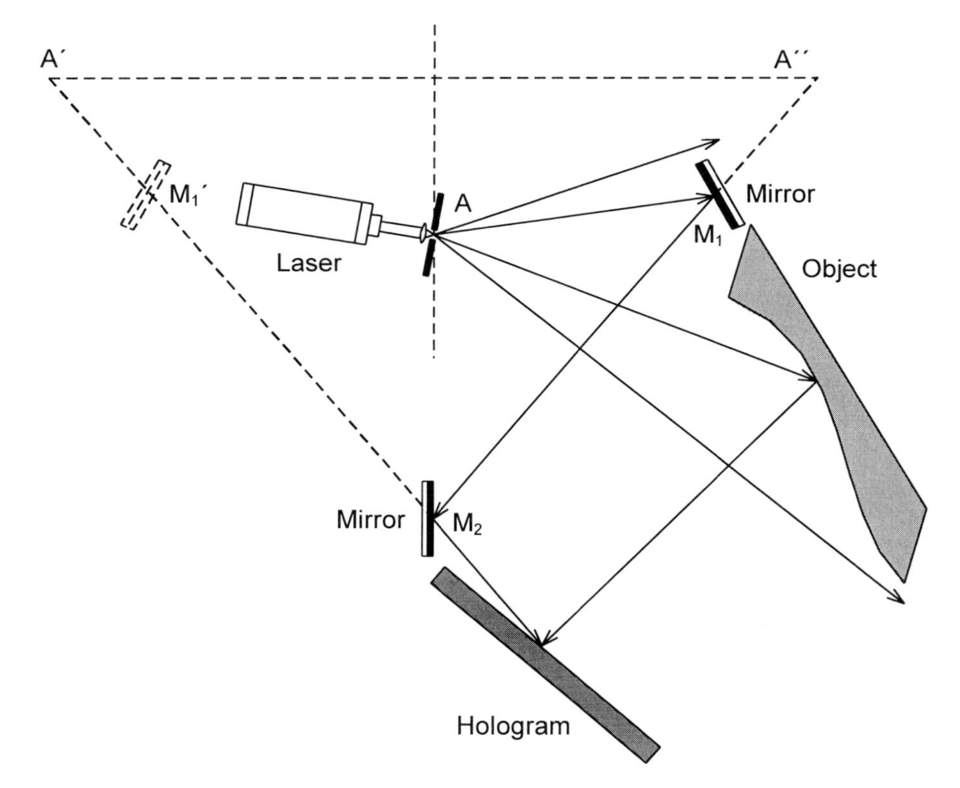

Figure 5.18: Arrangement for holographic light-in-flight recording

point H in Fig. 5.18. To each observation point in the hologram plate there corresponds an ellipsoid representing zero path length difference between object and reference beams, namely $\overline{AP} + \overline{PH} = \overline{AM_1} + \overline{M_1M_2} + \overline{M_2H}$ in Fig. 5.18.

When the point of observation H is moved along the holographic plate, the bright fringe of zero path length can move across the object. Only if H is varied in a way that does not change the zero path length for a specific object point P, this P remains stable. Thus to each object point P there exists a hyperboloid in space representing this zero path length. The two foci of these hyperboloids are the studied object point P and the virtual point source A′ of the reference beam. By moving the observation point along the hologram in a way that it crosses a number of hyperboloids, one sees the fringe intersecting different parts of the object surface as a continuous high-speed motion picture displayed at arbitrary low speeds.

For *contouring* by holographic light-in-flight recording the geometry of the holographic setup has to be optimized in such a way that the ellipsoids in the volume of the three-dimensional object can be approximated by flat surfaces, Fig. 5.19. The object is seen

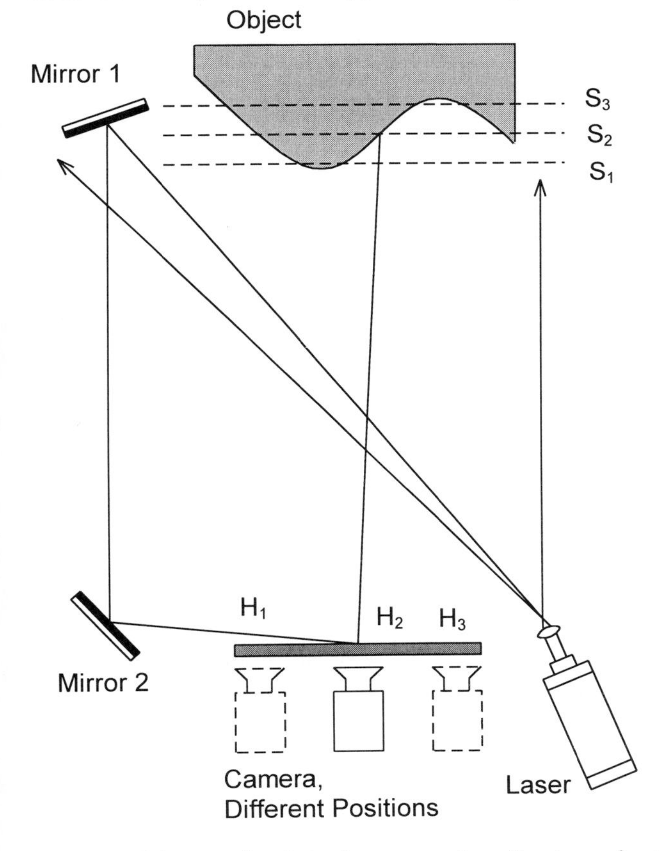

Figure 5.19: Contouring by holographic light-in-flight recording and reconstruction

intersected by one flat interference surface S, whose depth is varied during reconstruction as the observation point is shifted, Fig. 5.19. Here when looking from H_i, one only sees fringes corresponding to the intersection of surface S_i with the object.

The method seems especially suited for a computer-aided-evaluation. There is just a single fringe corresponding to one ellipsoid, no zero-orded-fringe problem arises. The depth resolution can be controlled by the shift of the observation point during reconstruction. The residual deviations from flatness of the ellipsoids can be taken into account numerically. An evaluation of the contouring lines of light-in-flight holography utilizing an image processing system is described in [441].

5.7 Measurement Range Extension

There are metrologic problems to be solved by holographic interferometry, where the methods introduced so far exhibit some drawbacks. The lack of the correct sign and the absolute phase, see Sec. 4.1, may be such a disadvantage. One would wish to measure with a wavelength much larger than the common laser-wavelengths to overcome these deficiencies. In other applications one is interested only in the in-plane-displacements, while the main sensitivity of the measurement is for out-of-plane displacements. The object may be loaded by centrifugal forces during rotation, for holographic interferometric measurements the interfering wavefronts must be steady. Since in holography only points of a rough surface interfere whose microstructures are identical in a small neighborhood, with conventional holographic interferometry different objects of the same macrostructure cannot be compared.

Solutions to these problems exist. The price one has to pay is generally a more complicated technical and optical arrangement and procedure to be used for the experimental measurements. But these efforts enable a substantial extension of the measuring range of holographic interferometry.

5.7.1 Two-Wavelength Holographic Interferometry

The holographic interferometric measurement of large deformations often leads to high fringe densities which cannot be resolved any more. A longer wavelength, by a factor 10 to 100, would be desirable. Another effect of a long wavelength would be a partial solution to the absolute phase problem, as long as the maximal displacement remains less than this wavelength. But a long wavelength, e. g. the $10.6\mu m$ of a CO_2-laser, would cause experimental difficulties since ordinary optical elements are no longer transparent, hologram plates have no sensitivity in this range and the observation requires additional experimental efforts. On the other hand these problems can be circumvented by using two-wavelength holography [442].

In *two-wavelength holographic interferometry* two double-exposure interferograms of the same object undergoing the same deformation are taken with different wavelengths, say λ_1 and λ_2. In a first approach [442], which is recommended if high fringe densities are expected, the fringe pattern belonging to λ_1 is photographed, the developed transparency is replaced into the holographic arrangement and is illuminated by the interference pattern resulting from λ_2. A moiré pattern will be obtained, which we get by multiplication of the two overlayed and filtered intensities.

Let the displacement lead to a phase difference $\Delta\phi(x, y)$, then the complex amplitudes are

$$E_j(x,y) = \mathrm{e}^{\mathrm{i}\frac{2\pi}{\lambda_j}\phi(x,y)} + \mathrm{e}^{\mathrm{i}\frac{2\pi}{\lambda_j}[\phi(x,y)+\Delta\phi(x,y)]} \qquad j = 1,2 \tag{5.98}$$

and the intensities are

$$I_j(x,y) = E_j(x,y)E_j^*(x,y) = 2 + \mathrm{e}^{\mathrm{i}\frac{2\pi}{\lambda_j}\Delta\phi(x,y)} + \mathrm{e}^{-\mathrm{i}\frac{2\pi}{\lambda_j}\Delta\phi(x,y)} \qquad j = 1,2 \tag{5.99}$$

where unit amplitudes were assumed. The moiré is

$$\begin{aligned} I_1(x,y)I_2(x,y) &= 4 + 2\mathrm{e}^{\mathrm{i}\frac{2\pi}{\lambda_1}\Delta\phi(x,y)} + 2\mathrm{e}^{-\mathrm{i}\frac{2\pi}{\lambda_1}\Delta\phi(x,y)} + 2\mathrm{e}^{\mathrm{i}\frac{2\pi}{\lambda_2}\Delta\phi(x,y)} \\ &+ 2\mathrm{e}^{-\mathrm{i}\frac{2\pi}{\lambda_2}\Delta\phi(x,y)} + \mathrm{e}^{\mathrm{i}2\pi(\frac{1}{\lambda_1}+\frac{1}{\lambda_2})\Delta\phi(x,y)} + \mathrm{e}^{-\mathrm{i}2\pi(\frac{1}{\lambda_1}+\frac{1}{\lambda_2})\Delta\phi(x,y)} \\ &+ \mathrm{e}^{\mathrm{i}2\pi(\frac{1}{\lambda_1}-\frac{1}{\lambda_2})\Delta\phi(x,y)} + \mathrm{e}^{-\mathrm{i}2\pi(\frac{1}{\lambda_1}-\frac{1}{\lambda_2})\Delta\phi(x,y)} \end{aligned} \tag{5.100}$$

Besides the d.c.- and the high-frequency terms the last two terms lead to an intensity proportional to

$$\cos[2\pi(\frac{1}{\lambda_1} - \frac{1}{\lambda_2})\Delta\phi(x,y)] \tag{5.101}$$

which would have been resulted if only the wavelength λ_{eq}

$$\lambda_{eq} = \frac{\lambda_1\lambda_2}{|\lambda_1 - \lambda_2|} \tag{5.102}$$

had been used. This wavelength is called *equivalent wavelength* [442, 443] or *synthetic wavelength* [444, 445, 446]. As an example, for the two lines of an Ar-laser of $\lambda_1 = 0.5145\mu m$ and $\lambda_2 = 0.4880\mu m$ the equivalent wavelength is $\lambda_{eq} = 9.4746\mu m$.

To avoid difficulties in the exact adjusting of the same deformation twice or to avoid spurious interference by air turbulence, the two interferograms can be recorded simultaneously on the same hologram plate. Two separated reference waves, one for each wavelength, are recommended to avoid the disturbing cross-reconstructions [447]. Then the interferograms are reconstructed and evaluated separately for the two wavelengths. According to (3.20) and (3.21) we have

$$\Delta\phi_j(P) = \frac{2\pi}{\lambda_j}\boldsymbol{d}(P)\cdot[\boldsymbol{b}(P) - \boldsymbol{s}(P)] \qquad j = 1,2 \tag{5.103}$$

The difference of the two evaluated interference phase distributions is

$$\begin{aligned} \Delta\phi_1(P) - \Delta\phi_2(P) &= 2\pi\left(\frac{1}{\lambda_1} - \frac{1}{\lambda_2}\right)\boldsymbol{d}(P)\cdot[\boldsymbol{b}(P) - \boldsymbol{s}(P)] \\ &= \frac{2\pi}{\lambda_{eq}}\boldsymbol{d}(P)\cdot[\boldsymbol{b}(P) - \boldsymbol{s}(P)] \end{aligned} \tag{5.104}$$

Thus we have a means to extend the range of unambiguity [448] or to reduce the sensitivity. Simultaneous recording avoids the influence of air turbulence, the remaining effect due to the different wavelengths can be neglected since air dispersion is small. Also chromatic aberration can be neglected, for only small wavelength differences are used. Theoretically the method can be expanded to multiple wavelengths yielding still larger equivalent wavelengths.

5.7.2 Holographic Moiré

The main sensitivity of holographic interferometry is in the direction of the bisector between the illumination and the observation direction, described by the sensitivity vector, 3.2, and thus for out-of-plane displacements. For measurement of the in-plane components without multiple observation from different directions and subsequent numerical analysis of the resulting interferograms one can employ the *holographic moiré* method [85, 363, 449]. Here the object is illuminated by collimated waves along two directions which are mutually coherent and symmetric to the surface normal, Fig. 5.20.

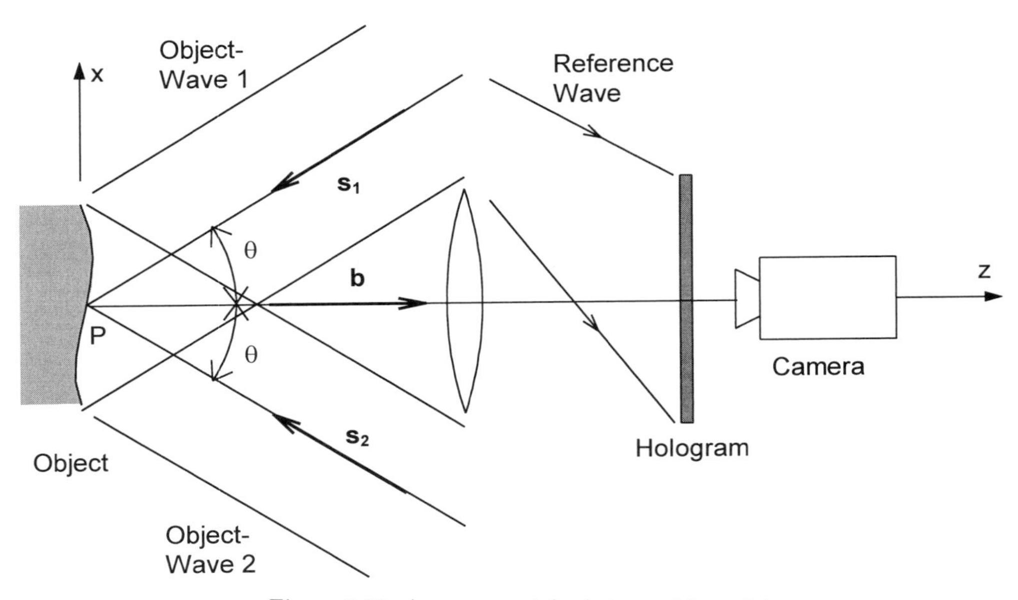

Figure 5.20: Arrangement for holographic moiré

Let the directions of the two object illumination waves in object point P be described by the unit vectors $\boldsymbol{s_1}(P) = (-\sin\theta, 0, \cos\theta)$ and $\boldsymbol{s_2}(P) = (\sin\theta, 0, \cos\theta)$. The observation direction is $\boldsymbol{b}(P) = (0, 0, 1)$. The complex amplitudes of the two illuminating object waves at P are A_1 and A_2 before the deformation and A_1' and A_2' after P has undergone the displacement $\boldsymbol{d}(P) = (d_x(P), d_y(P), d_z(P))$. This displacement causes the phase differences (3.20) and (3.20)

$$\begin{aligned} \Delta\phi_1(P) &= \frac{2\pi}{\lambda}\boldsymbol{d}(P)\cdot[\boldsymbol{b}(P) - \boldsymbol{s_1}(P)] \\ \Delta\phi_2(P) &= \frac{2\pi}{\lambda}\boldsymbol{d}(P)\cdot[\boldsymbol{b}(P) - \boldsymbol{s_2}(P)] \end{aligned} \tag{5.105}$$

so that $A_1'(P) = A_1(P)\mathrm{e}^{\mathrm{i}\Delta\phi_1(P)}$ and $A_2'(P) = A_2(P)\mathrm{e}^{\mathrm{i}\Delta\phi_2(P)}$. The intensity observed at point P when using the real-time method is [85]

$$\langle I(P)\rangle = \langle|\mathrm{e}^{\mathrm{i}\pi}A_1 + \mathrm{e}^{\mathrm{i}\pi}A_2 + A_1' + A_2'|\rangle \tag{5.106}$$

The reflected wave fields of the two illumination waves are uncorrelated, meaning $\langle A_i A_j^* \rangle = \langle A_i' A_j^{*\prime} \rangle = \langle A_i A_j^{*\prime} \rangle = \langle A_i' A_j^* \rangle = 0$ for $i \neq j$. Without restriction of generality we can assume equal real amplitudes $\langle |A_i|^2 \rangle = \langle |A_i'|^2 \rangle = \langle I_1 \rangle$ for $i = 1, 2$. Then (5.106) reduces to

$$\langle I(P) \rangle = 4\langle I_1(P) \rangle - 2\langle I_1(P) \rangle (\cos \Delta\phi_1(P) + \cos \Delta\phi_2(P)) \tag{5.107}$$

Introducing $\Phi(P) = (\Delta\phi_1(P) - \Delta\phi_2(P))/2$ and $\Psi(P) = (\Delta\phi_1(P) + \Delta\phi_2(P))/2$ we obtain

$$\langle I \rangle = 4\langle I_1 \rangle (1 - \cos \Phi \cos \Psi) \tag{5.108}$$

The argument (P) is omitted for clarity. From (5.105) it follows that

$$\begin{aligned} \Phi(P) &= \frac{2\pi}{\lambda} d_x(P) \sin\theta \\ \Psi(P) &= \frac{2\pi}{\lambda} d_z(P)(1 + \cos\theta) \end{aligned} \tag{5.109}$$

The loci of the moiré fringes resulting from the multiplication of the two patterns in (5.108) represent contour lines of the projection of the displacement vector $\boldsymbol{d}(P)$ onto the object plane in the direction containing the two beams [85]. The spacing of the fringes corresponds to an incremental displacement of $\lambda/(2\sin\theta)$.

To get clearly visible moiré patterns, generally the two multiplied patterns must have high density and have to be oriented in nearly the same direction. Their planes of maximum contrast must match. In practice these requirements can be fulfilled by the introduction of additional phase differences much larger than those generated by the object deformation to be measured. To obtain parallel equidistant fringes of high density localized on the object surface one may rotate the holographic plate around an axis parallel to its plane [450, 451], translate the hologram plate in its plane [452], rotate the reference beam accompanied by a translation of the hologram [453], or rotate the reference beam only [452, 454, 455].

Although the auxiliary fringes are necessary to generate the holographic moiré at all, they disturb the visual appearance of the moiré fringes. Therefore an optical or digital low-pass filter eliminating the high-frequency carrier fringes and passing only the low-frequency moiré fringes is applied in the preprocessing step of the evaluation. After this the fringe pattern can be evaluated by one of the diverse quantitative evaluation methods. An interesting combination of numerical preprocessing and quantitative evaluation by Fourier transform fringe pattern analysis which determines instantaneously both phase functions of the pattern is presented in [456].

5.7.3 Holographic Interferometry at Rotating Objects

The *vibration modes* of spinning components excited by *centrifugal forces* generally are different from those of stationary objects excited conventionally. Wheels, propellers, or turbine blades are just a few engineering components where the knowledge of the actual dynamic behaviour under real operating conditions helps to optimize the design and performance. Although the circumferential speed often reaches more than 100 m/s, holographic interferometry can be adapted to measure the deformation and vibration of such *rotating objects* [457, 458, 459, 460].

Consequently the holographic recording of moving objects, especially rotating objects, normally requires the use of a pulsed laser. To produce time-resolved holographic interferograms, [461] describes a system consisting of a multiple-pulsed Q-switched ruby laser and a rotating disk having radial slits with a constant angular separation. The disk is used to scan the reference beam along the holographic plate, thereby achieving spatial multiplexing. This system is a tool for full-field dynamic measurements.

If we intend to measure the displacements and deformations excited by the rotation, e. g. by centrifugal forces, or if we want to use a CW-laser for e. g. time average vibration analysis, we need a recording configuration which is insensitive to the rotational motion but sensitive to radial and normal displacements. Such an arrangement must have its illumination point as well as the observation point on the axis of rotation, Fig. 5.21, which here is the z-axis. Rotationally symmetric ellipsoids with illumination point S' and observation point B as focal points define the loci of constant optical path length. Since

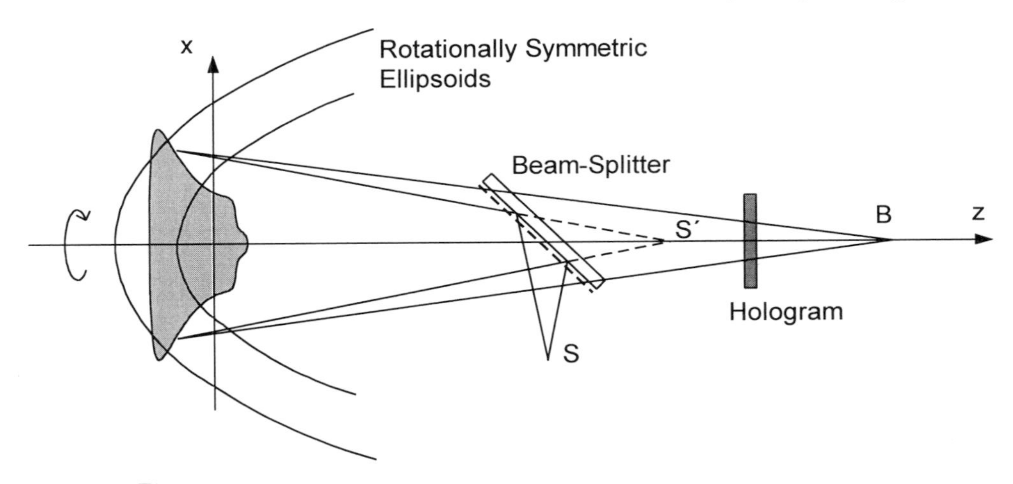

Figure 5.21: Holographic arrangement insensitive to in-plane rotations

object surface points undergoing only the in-plane rotation do not change the optical path lengths they do not contribute to the generation of the interference pattern. If furthermore the (virtual) illumination point S' and the observation point B coincide, the sensitivity vectors $\boldsymbol{e}(P)$ for all surface points P are parallel to the rotation axis, see (3.20), and in this case the arrangement is insensitive to all in-plane displacements, even to radial displacements. For an arrangement optimized in the described way the maximum rotation speed is only restricted by motion blur and resolution. That means the lateral movement in the image plane is limited to about half a speckle size, (2.111) and (2.112), to obtain an acceptable visibility.

A common method to keep the unwanted contributions of the rotational motions to the interferogram small, is the *object related triggering* of the double pulse laser [462]. By encoder discs or other non-contacting optical measurements the angular position of the object during the first of a double pulse is registered. The second pulse is triggered so that the object is in the same angular position, although during a different revolution, as with the first exposure. Pulse separation with this method must be rather high, i. e. some milliseconds, compared to conventional double pulse techniques. This object related triggering was successfully applied to measure forced vibrations of a rotating turbine blade model with maximum circumferential speed at the blade tip of 235 m/s. The measured deflection behaviour was compared to numerical results [463].

A holographic interferometer spinning synchronously with the object is shown in Fig. 5.22. In this *rotating interferometer* two holograms and two reference waves are

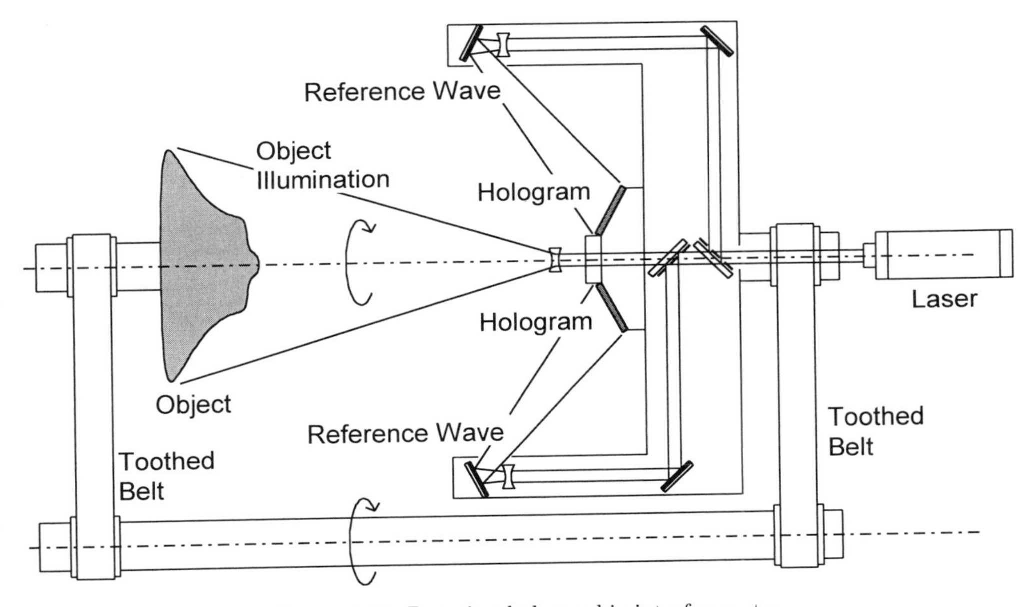

Figure 5.22: Rotating holographic interferometer

employed to achieve a better mechanical balance [462]. The object illumination comes from the laser beam passing through the hollow shaft of the interferometer. The axes of rotation of the object as well as of the holographic setup must be colinearly aligned. The synchronous rotation is achieved by mechanical coupling via toothed belts, Fig. 5.22, or electrooptical registration of the object rotation and electronic control of the interferometer rotation. A rotating interferometer as described allows to record double exposure holograms in any angular position contrary to the stroboscopic technique of the object related triggering method.

An auspicious approach to produce a stationary wavefield reflected from the rotating object is the compensation of the rotation by an *image derotator* [135, 464, 465]. Its principle is shown in Fig. 5.23. If one observes an image reflected from a roof edge prism, this image appears to rotate as soon as the prism is turned around its optical axis. The image rotates in reverse order to the prism but at twice the angular velocity of the prism

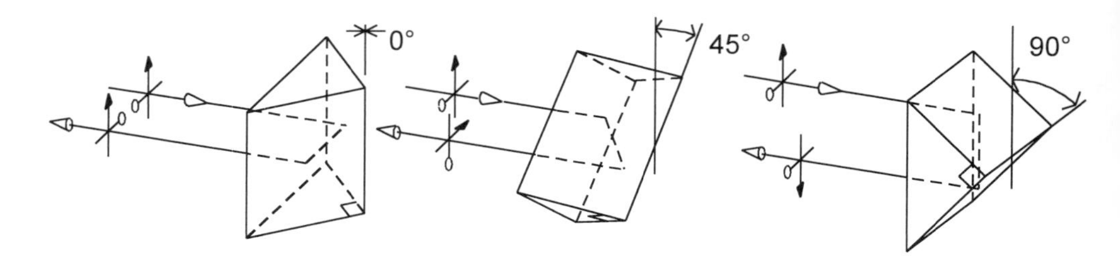

Figure 5.23: Image derotation by a roof edge prism

rotation. So if the prism is rotating with half the number of revolutions and in the same direction as the rotating object then the reflected image appears stationary to an observer [466]. Of course the axes of rotation must be colinearly aligned. A holographic

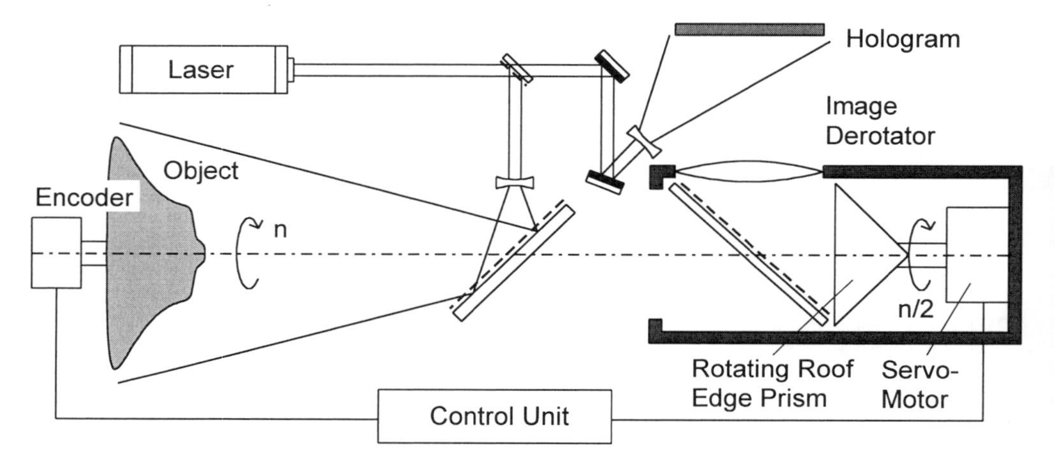

Figure 5.24: Holographic interferometric arrangement with image derotator

arrangement with an image derotator operating according to this principle is displayed in Fig. 5.24. The object is illuminated by a divergent wave field coming from a virtual point source located at the common axis of rotation. An encoder may be mounted on the shaft of the rotating component, or the object's rotational speed is recorded by a photocell detector or a tachogenerator to adapt the speeds of the object and the derotator prism. This prism is driven by a servo controlled motor, if the object rotates with n revolutions per minute, then the prism has to rotate with $n/2$ revolutions per minute. A fixed beamsplitter allows unhindered observation of the derotated image as well as the recording of holograms with double pulsed or continuous exposures. Using the image derotator, the rotating object can be viewed continuously in arbitrary angular positions: we have a non-stroboscopic system for freezing the wave field [467].

One application of holographic interferometry with rotation compensation by an image derotator was the investigation of the blade vibrations of the impeller of a radial compressor [468]. The radial impeller of 290 mm diameter consisted of 20 blades with every second blade cut back at the impeller inlet. Therefore 10 blades are observed along the rotational axis. The double exposure holographic interferogram of Fig. 5.25a was recorded

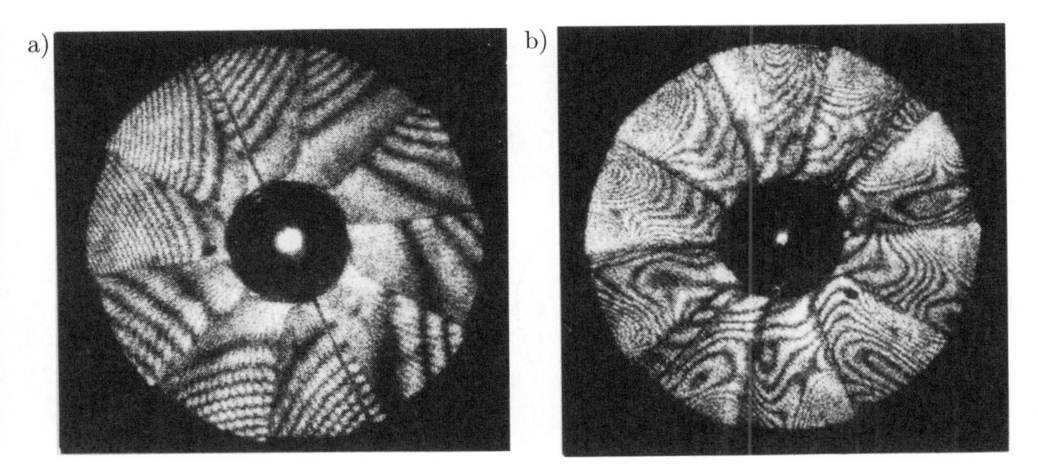

Figure 5.25: Holographic interferograms of a rotating radial impeller. (a) at 2,935 rpm, (b) at 13,450 rpm (Courtesy of J. Geldmacher, BIAS)

at a rotational speed of 2,935 rpm, and the one of Fig. 5.25b at 13,450 rpm. The resulting circumferential speed in this example reaches 204 m/s. The interferogram taken at 2,935 rpm was recorded with open compressor inlet. All blades vibrate in the first bending mode. At higher speeds, as in Fig. 5.25b a closed compressor inlet had to be used. So the derotator was placed outside the inlet tube, the impeller was viewed via a deflection mirror installed inside the tube. The resulting interferogram displays a superposition of different vibration modes.

5.7.4 Comparative Holographic Interferometry

One big advantage of holographic interferometry over conventional interferometry is its capability to compare the deformation states of rough diffusely reflecting surfaces. The technique is not restricted to specularly reflecting surfaces. But with the methods outlined so far it is only possible to compare states of one and the same surface: apart from the changes to be measured, the microstructure must remain identical. Different objects, although having the same macroscopic geometry, thus cannot be compared by these holographic techniques.

In non-destructive testing a common problem is to compare many produced components against a master piece which is guaranteed to have no defect. *Comparative holographic interferometry* now is a technique to compare the deformations of two specimens in respond to the same load or to compare the shapes of two specimens [320, 453, 469, 470]. These two specimens are macroscopically identical but have different microstructures.

The price one has to pay for this extra potential is the increased experimental effort and complexity.

In *difference holographic interferometry* two holograms are recorded of a master object, normally on separate hologram plates. The two holograms are taken one in the unloaded, one in the loaded state, Fig. 5.26a. Then a double-exposure holographic interferogram of

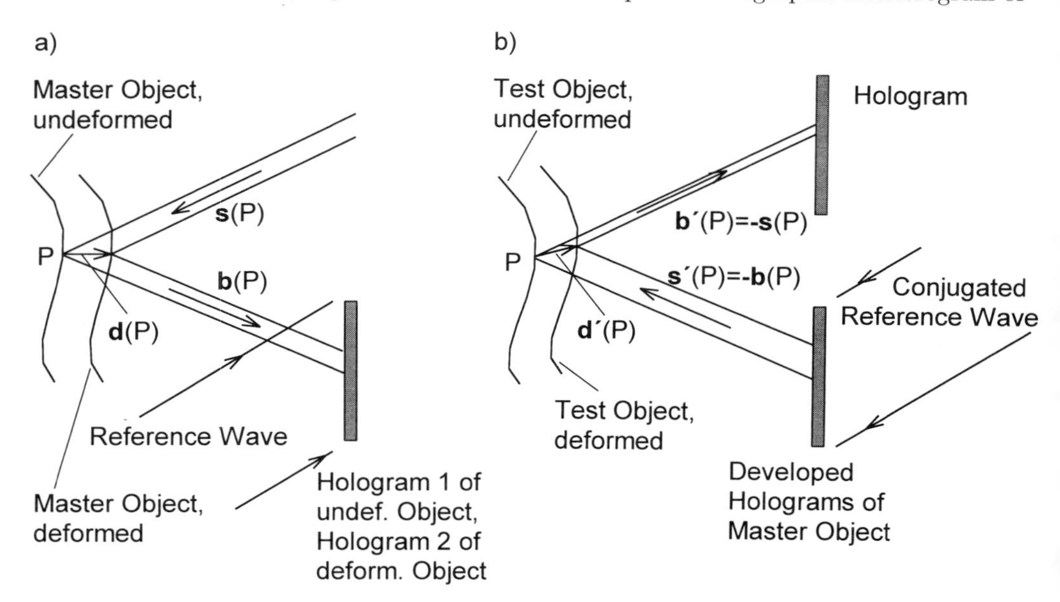

Figure 5.26: Difference holographic interferometry, (a) Recording the master holograms, (b) Recording the double-exposure test hologram

the object to be tested is produced. While illuminating its surface in the unloaded state by the real reconstruction of the unloaded master object and by illuminating the loaded test object by the reconstructed real image of the loaded master object. The real images of the master in both states are reconstructed by illuminating the repositioned developed holograms with the conjugate of the reference wave, Fig. 5.26b [320, 453, 471, 472, 473, 453].

For a quantitative description let $E_{1m}(P)$ be the wave field reflected from point P of the surface of the unloaded master piece, recorded on hologram 1

$$E_{1m}(P) = E_{01m}(P)\, \mathrm{e}^{\mathrm{i}\phi(P)} \tag{5.110}$$

The wave reflected from the master in the loaded state, stored in hologram 2, is

$$E_{2m}(P) = E_{02m}(P)\, \mathrm{e}^{\mathrm{i}(\phi(P) + \Delta\phi(P))} \tag{5.111}$$

Reconstruction with E_R^* instead of E_R yields $-\beta t_B |E_R|^2 E_P^*$ in (2.125), that means we illuminate the test object in its initial state with $E_{1m}^*(P) = E_{01m}(P)\, \mathrm{e}^{-\mathrm{i}\phi(P)}$ and in the loaded state with $E^*_{2m}(P) = E_{02m}(P)\, e^{-i(\phi(P)+\Delta\phi(P))}$. The reconstruction of the double exposed hologram of the test object gives the two waves

$$E'_1(P) = E'_{01}(P)\, e^{-i(\phi(P) - \phi'(P))} \tag{5.112}$$

and

$$E'_2(P) = E'_{02}(P)\, e^{-i(\phi(P) + \Delta\phi(P) - \phi'(P) - \Delta\phi'(P))} \tag{5.113}$$

which in analogy to (3.3) produce the interference pattern

$$I(P) = I_1(P) + I_2(P) + 2\sqrt{I_1(P)I_2(P)}\cos[\Delta\phi(P) - \Delta\phi'(P))] \tag{5.114}$$

Here $\Delta\phi(P)$ is the phase difference generated by loading the master object, $\Delta\phi'(P)$ that of loading the test object. According to (3.21) the phase difference $\Delta\phi(P)$ is given by $\Delta\phi(P) = \boldsymbol{d}(P) \cdot [\boldsymbol{b}(P) - \boldsymbol{s}(P)]$, Fig. 5.26a. The illumination direction $\boldsymbol{s}'(P)$ used to record the test object holographically is the reverse of the observation of the master object $\boldsymbol{s}'(P) = -\boldsymbol{b}(P)$ and if furthermore the interferogram is observed along the former illumination direction, $\boldsymbol{b}'(P) = -\boldsymbol{s}(P)$, Fig. 5.26b, then we have

$$\begin{aligned} \Delta\phi'(P) &= \boldsymbol{d}'(P) \cdot [\boldsymbol{b}'(P) - \boldsymbol{s}'(P)] \\ &= \boldsymbol{d}'(P) \cdot [-\boldsymbol{s}(P) + \boldsymbol{b}(P)] \end{aligned} \tag{5.115}$$

Altogether the resulting interferogram shows an intensity distribution belonging to the phase difference

$$\Delta\phi(P) - \Delta\phi'(P) = [\boldsymbol{d}(P) - \boldsymbol{d}'(P)] \cdot [\boldsymbol{b}(P) - \boldsymbol{s}(P)] \tag{5.116}$$

The holographic interferogram displays the difference of the displacement vectors between master and test object. As long as the load of the master can be reproduced exactly for the test object and the holograms can be repositioned precisely, possibly existing component defects lead to differing deformations which are reliably detected in the holographic difference interferogram.

So far the two exposures of the difference hologram have been taken on two separate holograms belonging to the two states of the master object. It is possible to record both states in a single hologram with separate reference waves, see Sec. 3.2.2. During reconstruction the related conjugate reference waves are employed. Also both master wavefronts may be recorded on a single plate with a single reference beam [320], but the resulting difference interference pattern becomes more complicated. Nevertheless the difference fringes can be resolved [474]. Besides measuring displacement differences this method also can be applied to detect small differences in the shapes of two objects [475, 476, 85], or in the context of measurements at phase objects [477].

Apart from the requirements of recording three separate holograms and repositioning the developed master holograms, difference holographic interferometry is principally a double-exposure method. The real-time comparison of the behaviour of two distinct surfaces is possible with the method of *comparative holographic moiré interferometry* [478, 479, 480]. To perform this method the images of master and test object are incoherently superimposed by an optical device. In the holographic setup sketched in Fig. 5.27

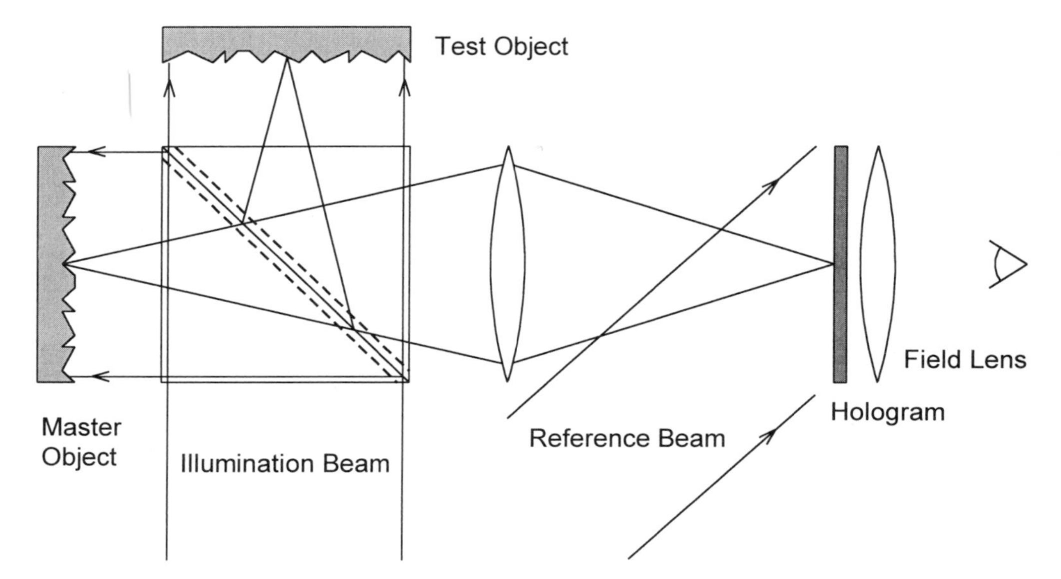

Figure 5.27: Arrangement for comparative holographic moiré interferometry

this is carried out by a Michelson type of arrangement using a beam splitter. The superposition of unloaded master and unloaded test object is recorded holographically. The reconstructed superposed wavefields then are compared in real-time with the superposition of the wavefields reflected from the loaded master and the loaded test object. The live character of the resulting fringes enables one to follow the evolution of the pattern with increasing load and to compensate for eventually unequal load or rigid body motions.

For a numerical description of comparative holographic moiré interferometry let E_{1m}, E_{2m}, E_{1t}, E_{2t} be the wavefields scattered from the unloaded master, the loaded master, the unloaded test, and the loaded test object, resp. Without restriction of generality we can assume equal reflectivities and thus equal real amplitudes, see Sec. 2.2.3.

$$\begin{aligned} E_{1m} &= E_{01}(P)\, \mathrm{e}^{\mathrm{i}\phi(P)} \\ E_{2m} &= E_{01}(P)\, \mathrm{e}^{\mathrm{i}(\phi(P) + \Delta\phi(P))} \\ E_{1t} &= E_{01}(P)\, \mathrm{e}^{\mathrm{i}\phi'(P)} \\ E_{2t} &= E_{01}(P)\, \mathrm{e}^{\mathrm{i}(\phi'(P) + \Delta\phi'(P))} \end{aligned} \tag{5.117}$$

Since master and test have not microscopically identical surfaces, their wavefields are uncorrelated, that means

$$\begin{aligned} 0 &= \langle E_{1m}E_{1t}^*\rangle = \langle E_{1m}E_{2t}^*\rangle = \langle E_{2m}E_{1t}^*\rangle = \langle E_{2m}E_{2t}^*\rangle \\ &= \langle E_{1t}E_{1m}^*\rangle = \langle E_{1t}E_{2m}^*\rangle = \langle E_{2t}E_{1m}^*\rangle = \langle E_{2t}E_{2m}^*\rangle \end{aligned} \tag{5.118}$$

Only E_{1m} and E_{2m} as well as E_{1t} and E_{2t} are correlated and can interfere.

The intensity $I(P)$ observed at any point in the image plane during observation of the reconstructed wavefields E_{1m} and E_{1t} together with the scattered fields E_{2m} and E_{2t} is

$$\begin{aligned} I(P) &= \langle |E_{1m}(P) + E_{2m}(P) + E_{1t}(P) + E_{2t}(P)|^2 \rangle \\ &= \langle |E_{1m}|^2 + |E_{2m}|^2 + |E_{1t}|^2 + |E_{2t}|^2 + E_{01}^2 \, \mathrm{e}^{-\mathrm{i}\Delta\phi(P)} + E_{01}^2 \, \mathrm{e}^{\mathrm{i}\Delta\phi(P)} \\ &\quad + E_{01}^2 \, \mathrm{e}^{-\mathrm{i}\Delta\phi'(P)} + E_{01}^2 \, \mathrm{e}^{\mathrm{i}\Delta\phi'(P)} \rangle \\ &= 4I_1(P) + 2I_1(P)\cos\Delta\phi(P) + 2I_1(P)\cos\Delta\phi'(P) \\ &= 4I_1(P)\left(1 + \cos\frac{\Delta\phi(P) + \Delta\phi'(P)}{2}\cos\frac{\Delta\phi(P) - \Delta\phi'(P)}{2}\right) \end{aligned} \tag{5.119}$$

The last line is the expression for an additive moiré. The high frequency fringes described by the cosine of the sum of the interference phases are modulated by the low frequency difference phase $\Delta\phi(P) - \Delta\phi'(P)$. For identical mechanical behaviour, $\Delta\phi(P) = \Delta\phi'(P)$, the resulting intensity is

$$I(P) = 4I_1(P)(1 + \cos\Delta\phi(P)) \tag{5.120}$$

No moiré fringes are recognized in this case.

Generally the master and test surfaces are illuminated from different directions, $\boldsymbol{s}$ and $\boldsymbol{s}'$, but observed in common direction $\boldsymbol{b}$. That means

$$\begin{aligned} \Delta\phi(P) &= \boldsymbol{d}(P) \cdot [\boldsymbol{b}(P) - \boldsymbol{s}(P)] \\ \Delta\phi'(P) &= \boldsymbol{d}'(P) \cdot [\boldsymbol{b}(P) - \boldsymbol{s}'(P)] \end{aligned} \tag{5.121}$$

and thus

$$\Delta\phi(P) - \Delta\phi'(P) = [\boldsymbol{d}(P) - \boldsymbol{d}'(P)] \cdot [\boldsymbol{b}(P) - \boldsymbol{s}(P)] - \boldsymbol{d}'(P) \cdot [\boldsymbol{s}(P) - \boldsymbol{s}'(P)] \tag{5.122}$$

In practice it is advantageous to have common illumination directions $\boldsymbol{s}$ and $\boldsymbol{s}'$, as in Fig. 5.27, then the second term in the right-hand side of (5.122) vanishes. The resulting moiré fringe pattern provides information about the difference of the displacement vectors along the direction of the sensitivity vector $\boldsymbol{b}(P) - \boldsymbol{s}(P)$ [478].

Comparative holographic interferometry and comparative holographic moiré interferometry enable the measurement of displacement differences or contour differences of distinct objects. This meets many requirements of holographic non-destructive testing. An improvement of the precision of quantitative evaluation and of the quality of the fringe patterns can be achieved by employing phase shifting techniques [481, 482] or Fourier transform evaluation [456] in these techniques.

5.7.5 Desensitized Holographic Interferometer

Flatness deviations normally are measured by Fizeau interferometers [48, 225]. For measurement of flatness deviations ranging from several to some 100 micrometers in [483] a *desensitized holographic interferometer* using a custom made holographic optical element is presented. The key optical component of this interferometer is the *diffractive optical element* which is recorded by two waves intersecting by the angle θ, Fig. 5.28a. At the

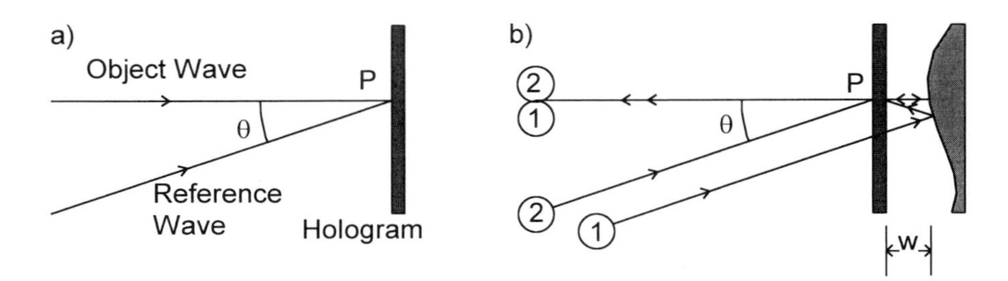

Figure 5.28: Desensitized holographic interferometer, (a) Recording of the DOE, (b) Testing of a surface

reconstruction stage, the mirrored surface to be tested is placed just behind this hologram. Two ray paths merit attention, Fig. 5.28b: Ray path 1 is transmitted in the zero diffraction order through the hologram, is reflected by the surface, and finally emerges back from the hologram through diffraction in order +1. Ray path 2 is first diffracted in the +1 order and after reflection by the object surface is transmitted through the hologram in the zero order. The optical path difference $\delta(P)$ between these two rays leaving the hologram in parallel is dependent on the width $w(P)$ of the air gap between hologram and surface

$$\delta(P) = 2w(P)(1 - \cos\theta) \tag{5.123}$$

The interference gives rise to fringes of equal thickness $w(P)$ between hologram and surface

$$w(P) = \frac{n\lambda}{2(1 - \cos\theta)} \tag{5.124}$$

For small angles θ the desensitization factor can approach values of about 100.

The analysis in [483] shows that as long as the mirrored surface lies close to the hologram and since the hologram itself performs a chromatic filtering, good contrast fringes are obtained even when the reconstruction is carried out in white light. Contrary to reconstruction with laser light, the final image is free from coherent noise arising from dust as well as from the roughness of the tested object. If the holographic plate is mounted on a holder that can be accurately translated normally to the object surface, the resuling fringes can be phase shifted, thus enabling the use of the phase shifting methods for quantitative evaluation.

5.8 Refractive Index Fields in Transparent Media

5.8.1 Refraction of Phase Objects

In a transparent nonpolar dielectric medium the speed of light primarily depends on the density of the medium and the wavelength of the light. The density is described by the *refractive index* n, which is the ratio between the *speed of light* in vacuum c_0, see (2.3), and the speed of light c in the medium

$$n = \frac{c_0}{c} \tag{5.125}$$

The refractive index may be spatially constant in the medium, then we speak of a *homogeneous medium*, or n varies from point to point, then we have a *nonhomogeneous medium*. In a nonhomogeneous medium there are regions with lower n, where light travels more rapidly than in regions of higher n, so wavefronts get distorted when propagating through this medium. The equation of a ray propagating through a medium with the refractive index distribution $n(x, y, z)$ is

$$\frac{d}{ds}\left(n(x, y, z)\frac{d\boldsymbol{r}}{ds}\right) = \nabla n(x, y, z) \tag{5.126}$$

where $\boldsymbol{r} = (x, y, z)$ is the vector of point (x, y, z) on the ray, s is the length coordinate along the ray and ∇ is the *gradient operator*, see (2.2)

$$\nabla = \left(\frac{\partial}{\partial x}, \frac{\partial}{\partial y}, \frac{\partial}{\partial z}\right) \tag{5.127}$$

For a homogeneous medium the *ray equation* (5.126) reduces to $d^2\boldsymbol{r}/ds^2 = 0$, whose solution is a straight line $\boldsymbol{r} = \mathbf{c_0}s + \mathbf{c_1}$ with constant vectors $\mathbf{c_0}$ and $\mathbf{c_1}$. Consequently in a nonhomogeneous medium the rays are curves, some of such curved rays are given in Fig. 5.29 together with some wavefronts.

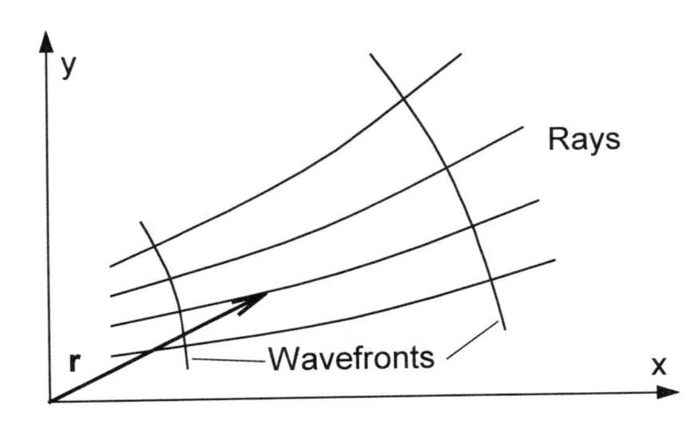

Figure 5.29: Curved rays and wavefronts in a nonhomogeneous medium

The path of a ray transmitted through a homogeneous transparent body embedded in another homogeneous medium - normally free space with the refractive index of air approximated by $n \approx 1$ - is governed by *Snell's law*

$$n_1 \sin\theta_1 = n_2 \sin\theta_2 \tag{5.128}$$

which describes the slopes of the path at the boundary between the media. n_1 and n_2 are the refractive indices in the two media, θ_1 and θ_2 are the angles between the normal to the interface and the rays in medium 1 and medium 2, resp.

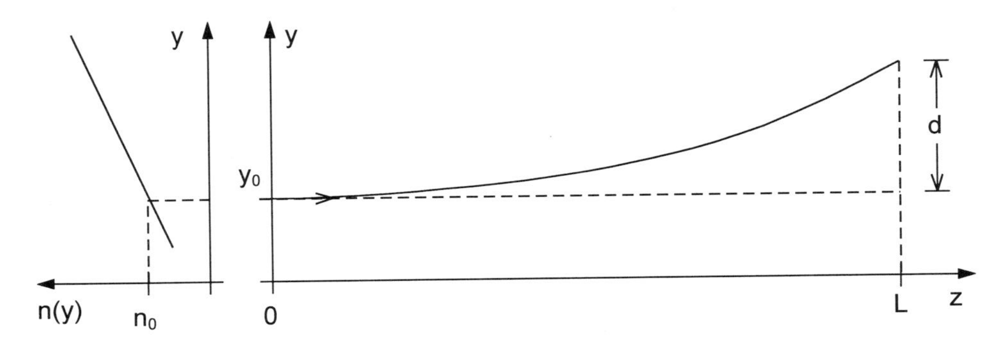

Figure 5.30: Ray propagating through a medium with $n = n(y)$

Things become more complicated when the refractive index field varies spatially. As a first example [24] assume a refractive index which varies only in one direction perpendicular to the incoming ray, say

$$n(x, y, z) = n(y) \tag{5.129}$$

Let the ray propagate in direction z and enter the volume of refractive index $n(y)$ at y_0, Fig. 5.30 The two vector components of (5.126) which are not identically zero are

$$\frac{d}{ds}\left(n(y)\frac{dy}{ds}\right) = \frac{d}{dy}n(y) \qquad \text{and} \qquad \frac{d}{ds}\left(n(y)\frac{dz}{ds}\right) = 0 \tag{5.130}$$

The differential line element is $ds = \sqrt{1 + (dy/dz)^2}dz$, so we combine (5.130) to

$$\frac{dn(y)}{dy} = n(y)\frac{y''}{1 + (y')^2} \tag{5.131}$$

with the prime denoting differentiation with respect to z. By assumption it is $y(0) = y_0$, $y'(0) = 0$, $n(0) = n_0$, so that the integration of (5.131) yields

$$1 + (y')^2 = \left(\frac{n}{n_0}\right)^2 \tag{5.132}$$

If the variation of n is linear, Fig. 5.30, $n(y) = n_0 + m(y - y_0)$, then we get the solution

$$y - y_0 = \frac{1}{2}\left(\frac{m}{n_0}\right)z^2 \tag{5.133}$$

where the first-order approximation $(y')^2 = 2(m/n_0)(y - y_0)$ is used. This means that the ray entering a linearly stratified medium will travel along a parabolic path. After leaving the test volume of length L the deflection d of the ray is

$$d = \frac{1}{2}\left(\frac{m}{n_0}\right)L^2 \tag{5.134}$$

and its slope is

$$y'(L) = \frac{m}{n_0}L \tag{5.135}$$

The optical pathlength δ of this ray in the test volume is

$$\delta = \int_0^L n(y)\sqrt{1 + (dy/dz)^2} dz \tag{5.136}$$

which is approximated by

$$\delta = n_0 L \left(1 + \frac{1}{3}\left(\frac{m}{n_0}\right)^2 L^2\right) \tag{5.137}$$

Even this simple example exhibits the bending of the rays when a spatial variation of the refractive index is present. This possibility has to be envisaged in the interpretation and evaluation of holographic interferograms to measure refractive index variations of transparent media.

As mentioned before, each interpretation or evaluation of an interference pattern requires an imaging system to form a real image, at least the lens of the observer's eye. The imaging system is represented by a single thin lens in Fig. 5.31. The refractive index field

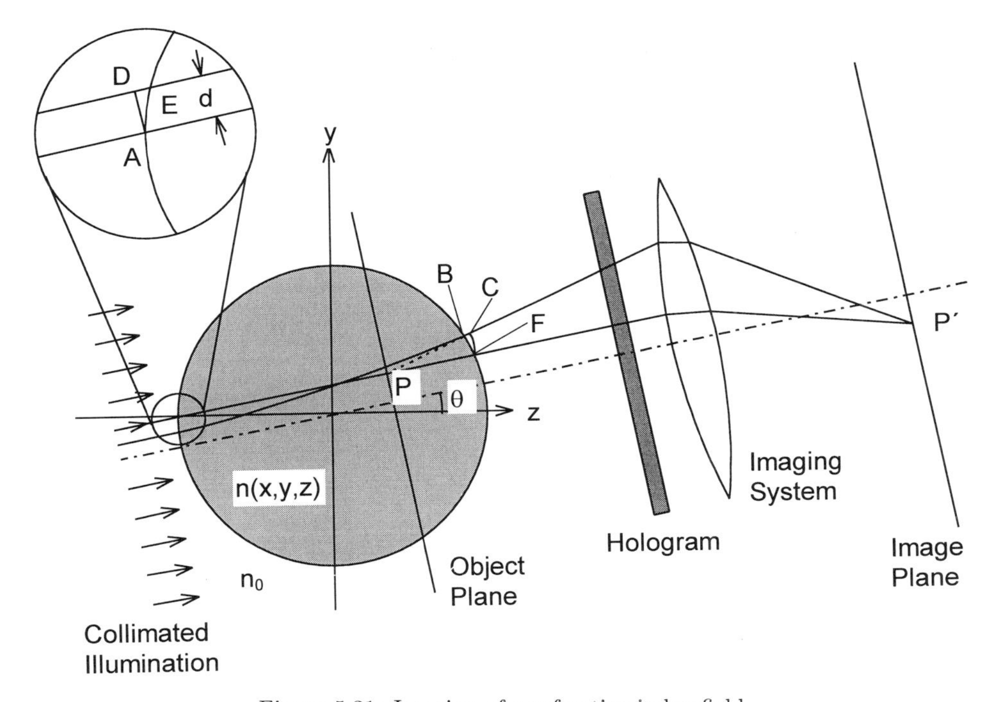

Figure 5.31: Imaging of a refractive index field

acting as the object to be measured is assumed to be contained in a circular region as proposed in App. B. A ray parallel to the optical axis enters the refracting region in point A, is bended and leaves the region at B, it passes through the surrounding homogeneous medium of refractive index n_0, hits the thin lens and is imaged onto point P' in the image plane. In double exposure holographic interferometry of course this role is played by the holographic reconstruction of the ray belonging to the exposure during the presence of the refracting object. At P' the ray interferes with another holographically reconstructed ray, this second one stemming from the exposure when no refracting object was present. To find the corresponding point P in the object plane, we can trace back with the common thin lens techniques. Another way is to extend the first ray from the refracting medium to the lens further along a straight line to the object plane [484], Fig. 5.31. This straight ray enters the refracting region at E and leaves it at F. These two rays of the collimated illumination are separated by the distance d.

To find the optical path length changes δ, or equivalently the optical phase difference $\Delta\phi$ between the rays interfering at P', we notice that no relative changes occur left of points A and D or to the right of points C and F. D is the intersection of the straight ray with the perpendicular to the first ray through A. C has the same distance from P as F has, Fig. 5.31. The optical pathlength difference δ therefore is

$$\delta = \int_A^B n(x,y,z)ds + n_0(\overline{BC} - \overline{DE} - \overline{EF}) \tag{5.138}$$

While the first term of (5.138) is the integral of the refractive index along the curved path $\widetilde{AB}$, the second term accounts for the different path length in the surrounding with refractive index n_0. For each viewing direction, defined by the angle θ to the arbitrary reference direction z, $\delta(\rho,\theta)$ or $\Delta\phi(\rho,\theta) = \delta(\rho,\theta)\cdot 2\pi/\lambda$ can be measured holographically. The ray equation (5.126) and the integral (5.138) together define the so called *path length transform*, which is a nonlinear integral transform of $n(x,y,z)$ and n_0. If the refractive index variation is sufficiently small the curved ray $\widetilde{AB}$ coincides with the straight line $\overline{EF}$, so that (5.138) reduces to

$$\delta = \int_E^F (n(x,y,z) - n_0)dl \tag{5.139}$$

with dl denoting the differential distance along line $\overline{EF}$. The *line integral transform* defined by (5.126) and (5.139) is mathematically equivalent to the *Radon transform* (B.2), see App. B.

5.8.2 Physical Quantities Affecting the Refractive Index Field

In most applications of holographic interferometry for measurements at transparent media it is not the refractive index distribution, which is of main concern, but another physical quantity. The value of this physical quantity is determined by the effect it has on the refractive index field. So in the following some relations between such physical quantities and the refractive index are given [24].

In *aerodynamics* and *flow visualization* the flow of compressible gases is studied, e. g. in wind tunnels or shock tubes. The quantity of interest is the density ρ in a gas, the mass per unit volume. Its relation to refractive index n is given by the *Gladstone-Dale equation*

$$n - 1 = K\rho \tag{5.140}$$

with the *Gladstone-Dale constant* K, which is a property of the gas. The Gladstone-Dale constant is nearly independent of temperature and pressure under moderate physical conditions and it is a weak function of wavelength [24]. Some values are given in Table 5.1 [24]. The Gladstone-Dale constant of a mixture of gases can be calculated as the mass-

Table 5.1: Gladstone-Dale constants of gases

Gas	K (m^3/kg) at $\lambda = .5145\ \mu m$	K (m^3/kg) at $\lambda = .6328\ \mu m$
Ar	0.175×10^{-3}	0.175×10^{-3}
O_2	0.191×10^{-3}	0.189×10^{-3}
He	0.196×10^{-3}	0.195×10^{-3}
CO_2	0.229×10^{-3}	0.227×10^{-3}
N_2	0.240×10^{-3}	0.238×10^{-3}

weighted average of the values for the component gases

$$K = \sum_i a_i K_i \tag{5.141}$$

with the mass fraction a_i and Gladstone-Dale constant K_i of the i-th component.

The density ρ of a gas in most cases of interest can be calculated from the pressure P, the molecular weight M and the absolute temperature T via the *ideal gas equation*

$$\rho = \frac{MP}{RT} \tag{5.142}$$

with the universal gas constant $R = 8.3143\ J/(mol\ K)$. The combination with (5.140) yields

$$n - 1 = \frac{KMP}{RT} \tag{5.143}$$

This can be used for holographic *temperature measurements.* If the temperature changes remain small, a linear relation between the change of the refractive index and the change of the temperature can be adopted. As an example for air at $288^o\ K$ and $0.1013\ MPa$ the Gladstone-Dale constant at $\lambda = 0.6328\ \mu m$ is $0.226 \times 10^{-3}\ m^3/kg$ and the molecular weight is 28.97. This results in

$$\frac{dn}{dT} = -0.9617 \times 10^{-6}\ K^{-1} \tag{5.144}$$

With even more precision the dependence of the refractive index of air from temperature at $\lambda = 0.6328\ \mu m$ is [24]

$$n = 1 + \frac{0.292015 \times 10^{-3}}{1 + 0.368184 \times 10^{-2} T} \tag{5.145}$$

and at $\lambda = 0.5145\ \mu m$

$$n = 1 + \frac{0.294036 \times 10^{-3}}{1 + 0.369203 \times 10^{-2} T} \tag{5.146}$$

with T in degree Celsius.

In *liquids* the refractive index is related to density ρ by the *Lorentz-Lorenz equation*

$$\frac{n^2 - 1}{\rho(n^2 + 2)} = \overline{r}(\lambda) \tag{5.147}$$

where $\overline{r}(\lambda)$ is the *specific refractivity*, which depends on the substance and the wavelength of light. There is no direct analogue to the ideal gas equation (5.142) in the case of liquids, instead empirical relations between refractive index and temperature must be used. Some are given in Table 5.2 [485, 24]. Quite accurate equations for water are

Table 5.2: Dependence of refractive index from temperature in liquids

Liquid	$-dn/dT(^o\ K^{-1})$ at $\lambda = .5461\ \mu m$	$-dn/dT(^o\ K^{-1})$ at $\lambda = .6328\ \mu m$
Water	1.00×10^{-4}	0.985×10^{-4}
Methyl alcohol	4.05×10^{-4}	4.0×10^{-4}
Ethyl alcohol	4.05×10^{-4}	4.0×10^{-4}
Isopropyl alcohol	4.15×10^{-4}	4.15×10^{-4}
Benzene	6.42×10^{-4}	6.40×10^{-4}
Toluene	5.55×10^{-4}	5.55×10^{-4}
Nitrobenzene	4.68×10^{-4}	4.68×10^{-4}
c-Hexane	5.46×10^{-4}	5.43×10^{-4}
Acetone	5.31×10^{-4}	5.31×10^{-4}
Chloroform	5.98×10^{-4}	5.98×10^{-4}
Carbon tetrachloride	5.99×10^{-4}	5.98×10^{-4}
Carbon disulfide	7.96×10^{-4}	7.96×10^{-4}

$$n = 1.3331733 - (1.936\, T + 0.1699\, T^2) \times 10^{-5} \tag{5.148}$$

for $\lambda = 0.6328\ \mu m$ and

$$n = 1.337253 - (2.8767\, T + 0.14825\, T^2) \times 10^{-5} \tag{5.149}$$

for $\lambda = 0.5145\ \mu m$ where T again is measured in degree Celsius.

In *transparent solids* the refractive index depends on the state of stress, which is termed the *stress-optical effect*. Furthermore the thickness of a component is changed when it is stressed. Both effects together allow a determination of strains and stresses of transparent solid objects or models, as it is done in the field of photoelasticity.

Let a thin plane specimen of thickness h be subjected to a tensile force F. Since all stresses lie in the x-y-plane, the specimen is in the state of plane stress. Due to the transverse contraction the material undergoes a strain ε_z in the z-direction

$$\varepsilon_z = \frac{\Delta h}{h} \tag{5.150}$$

where Δh is the change in thickness. This transverse strain in an elastic material is related to the stress field by the *Poisson ratio* ν and the *modulus of elasticity* E

$$\varepsilon_z = \frac{-\nu}{E}(\sigma_1 + \sigma_2) \tag{5.151}$$

where σ_1 and σ_2 are the principal stresses. The principal stresses are mutually orthogonal and lie in the x-y-plane. If the state before stressing the specimen and the one during stressing are holographically recorded and reconstructed, the optical pathlength difference δ yielding the interference is

$$\delta = n(h + \Delta h) - n_0 h - \Delta h \tag{5.152}$$

Here the term $n(h + \Delta h)$ is due to the refractive index n of the stressed material, in the second term n_0 is the refractive index of the material in its unstressed state, and the third term is the pathlength Δh during the first exposure multiplied by 1, the refractive index of the surrounding air.

The refractive index n is related to the state of stress by the two-dimensional *Maxwell-Neumann stress-optical law*

$$\begin{aligned} n_1 - n_0 &= A\sigma_1 + B\sigma_2 \\ n_2 - n_0 &= B\sigma_1 + A\sigma_2 \end{aligned} \tag{5.153}$$

Here A and B are the *stress-optical coefficients* of the material, the n_i are the refractive indices for light polarized in the direction of σ_i, $i = 1, 2$. For materials of low stress-optical sensitivity we have $A \approx B$ and $n_1 \approx n_2 = n$, so that for optically isotropic material (5.153) reduces to

$$n - n_0 = A(\sigma_1 + \sigma_2) \tag{5.154}$$

With this and (5.151) the optical pathlength difference is

$$\begin{aligned} \delta &= [\varepsilon_z(n_0 - 1 + A(\sigma_1 + \sigma_2)) + A(\sigma_1 + \sigma_2)]h \\ &= [\varepsilon_z(n_0 - 1 - \frac{AE}{\nu}\varepsilon_z) - \frac{AE}{\nu}\varepsilon_z]h \end{aligned} \tag{5.155}$$

Neglecting the small term $(AE/\nu)\varepsilon_z^2$, we obtain

$$\delta = (n_0 - AE/\nu - 1)h\varepsilon_z \tag{5.156}$$

The quantity $n_0 - AE/\nu$ can be considered to be the *effective refractive index* of the material.

In *plasma diagnostics* one deals with a plasma, a collection of atoms, ions, and electrons. In plasmas a number of electrons is separated from the nucleus. The refractive index of a plasma is the sum of the refractive indices of the atoms, ions, and electrons weighted by their number densities. The refractive indices of atoms and ions are described by the Gladstone-Dale equation and are of the same order. They are only weakly dependent on the wavelength of the probing light. On the other hand the refractive index n_e of the electron gas is

$$n_e = \sqrt{1 - \frac{N_e e^2 \lambda^2}{2\pi m_e c^2}} \tag{5.157}$$

Here N_e is the number density of electrons, namely the number of electrons per unit volume, e is the electron charge, m_e is the mass of an electron and c is the speed of light. Evaluation of the constants gives

$$n_e = \sqrt{1 - 8.92 \times 10^{-14} \lambda^2 N_e} \tag{5.158}$$

or approximated to the first order

$$n_e - 1 = -4.46 \times 10^{-14} \lambda^2 N_e \tag{5.159}$$

with λ in centimeters and N_e in cm^{-3} [24].

These equations show that the electron gas is very dispersive. Furthermore the contribution to the refractive index of a plasma per electron is an order of magnitude greater than that per atom and is of opposite sign. Therefore the electron gas dominates the refractive index in moderately as well as in highly ionized plasmas.

5.8.3 Two-Dimensional Refractive Index Fields

In holographic interferometric measurements at transparent media the evaluation consists first in extracting the interference phase distribution $\Delta\phi(x, y)$ as given in (3.22) by any of the methods presented in Chap. 4 [486]. For the discussion here we assume a propagation of light in the z-direction, Fig. 3.6.

The easiest case is an object with a refractive index varying in only one, say the y-direction, which is orthogonal to the z-direction, see Sec. 5.8.1. A typical application is the measurement of thermal boundary layers [171]. Let the object have length l in the z-direction and by assumption a constant refractive index along l. Then the pathlength difference is simply

$$\begin{aligned} \delta(x, y) &= \int (n(x, y, z) - n_0) dz \\ &= (n(y) - n_0) l \end{aligned} \tag{5.160}$$

or

$$\Delta\phi(y) = \frac{2\pi}{\lambda}[n(y) - n_0] l \tag{5.161}$$

The fringe spacing is determined by the gradient of n. If the refractive index varies linearly

$$n(y) = n_0 + my \tag{5.162}$$

we get equally spaced parallel fringes, Fig. 5.32a, at

$$y = \frac{N\lambda}{ml} \qquad N = 0, 1, 2, \ldots \tag{5.163}$$

with N denoting the fringe order.

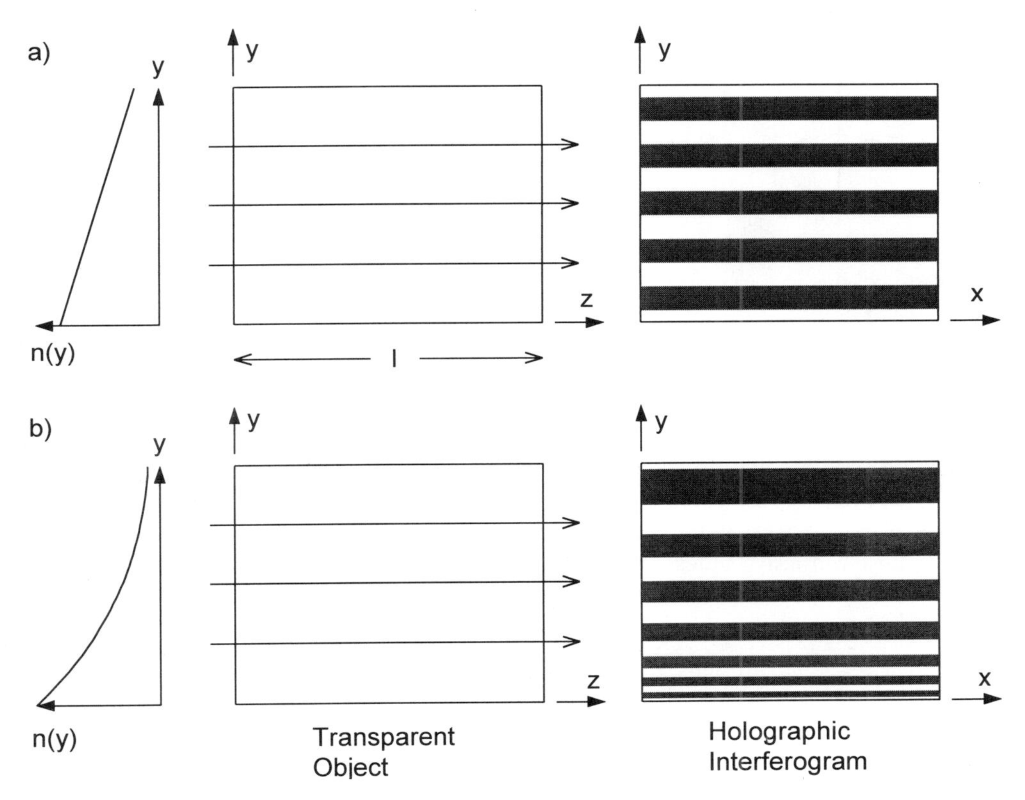

Figure 5.32: Linearly (a) and nonlinearly (b) varying refractive index fields (in y-direction) and resulting holographic interferograms

For a nonlinear refractive index distribution, e. g. the exponential one

$$n(y) = n_0 + m\mathrm{e}^{-ay} \tag{5.164}$$

we get parallel straight fringes with large spacing in regions of a small gradient and small spacing in regions of a high gradient of the refractive index, Fig. 5.32b.

5.8.4 Holographic Interferometry of Circular Symmetric Refractive Index Fields

In *circular symmetric phase objects* the refractive index is a function of the radius r only [487]. Typical objects are flows around cones, jets, thermal plumes, flames, or plasmas.

Their form may be cylindrical or spherical. According to Fig. 5.33 it is

$$dz = \frac{r\,dr}{\sqrt{r^2 - x^2}} \tag{5.165}$$

so the optical pathlength difference δ is

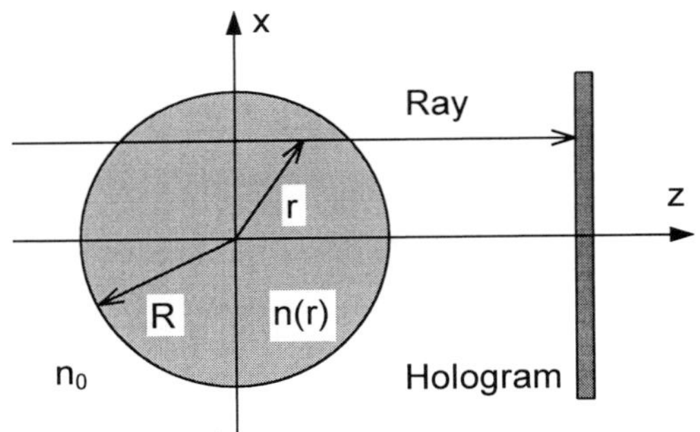

Figure 5.33: Geometry of a circular symmetric phase object

$$\delta(x) = 2\int_x^R \frac{[n(r) - n_0]r}{\sqrt{r^2 - x^2}} dr \tag{5.166}$$

with $n(r)$ the refractive index inside the object of radius R, and n_0 outside. For phase objects whose refractive index decays smoothly to the ambient n_0 the object radius can be taken as $R \to \infty$. So for the interference phase then we have with $f(r) = n(r) - n_0$

$$\frac{\Delta\phi(x)}{2\pi}\lambda = 2\int_x^\infty \frac{f(r)r}{\sqrt{r^2 - x^2}} dr \tag{5.167}$$

The right-hand side of this equation is the *Abel transform* of $f(r)$. Its inversion formula is

$$f(r) = \frac{\lambda}{2\pi^2}\int_r^\infty \frac{d\Delta\phi(x)}{dx} \frac{1}{\sqrt{r^2 - x^2}} dx \tag{5.168}$$

The problem now is to find an effective way to perform this inversion on the interference phase values measured at discrete locations.

A first simple approach divides the object into discrete annular elements of constant width Δr and assumes a uniform refractive index in each element [24]. This leads to a set of simultaneous linear algebraic equations which are solved by the common methods of matrix inversion [485]. More refined methods use refractive indices which vary linearly with r in each annular element [488] or employ representations of $f(r)$ by sampling series [489, 490]. Methods based on the inversion of formula (5.168) use interpolated phase data and numerical differentiation.

A fast and efficient method utilizes a Fourier decomposition of the interference phase and calculates the Abel inversion of each spatial frequency component [491]. The Fourier coefficients are obtained from an FFT-routine, App. A.6.

5.8.5 Multidirectional Recording of Asymmetric Refractive Index Fields

The determination of *asymmetric refractive index fields* requires the analysis of a large number of holographic interferograms by methods of *computer aided tomography*, see App. B. Each of the reconstructed holographic interference patterns has to correspond to a different viewing direction [50]. Therefore holographic interferometry is ideally suited for rendering tomographic data, since just a single hologram allows the realization of a number of viewing directions. Thus although many interferograms are required, the number of necessary holograms remains limited. Some arrangements to record multiple holograms for subsequent tomographic reconstruction are proposed in Fig. 5.34 [492].

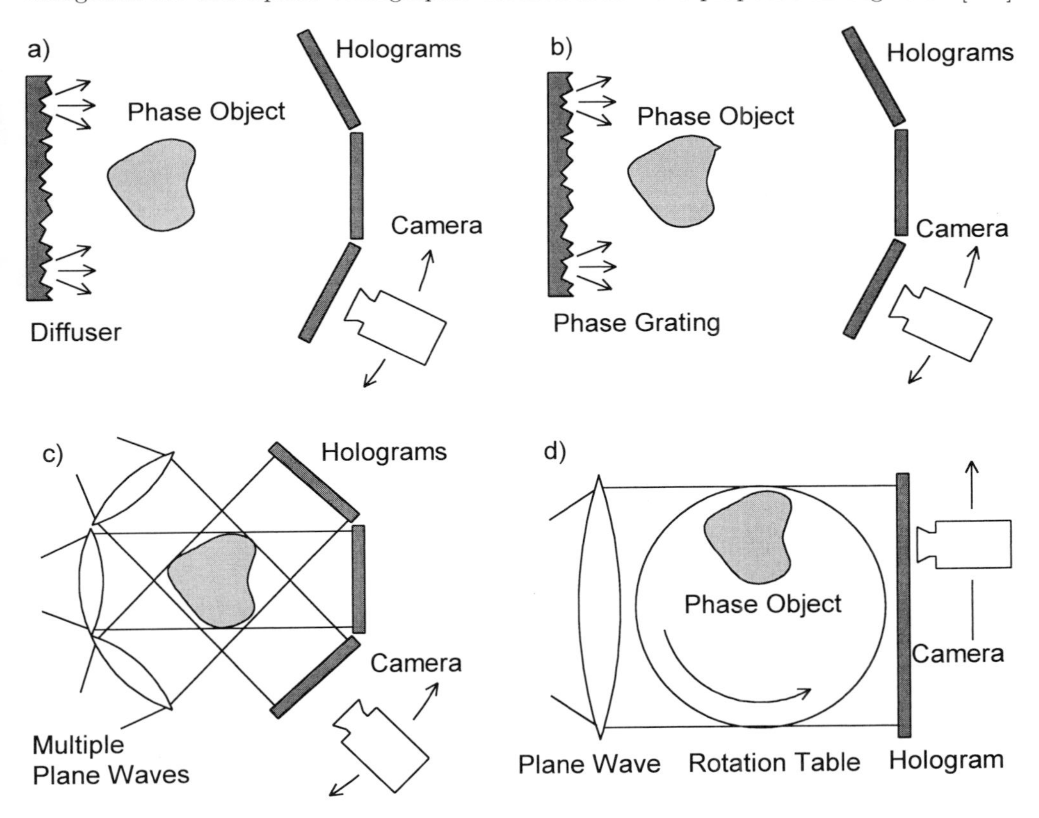

Figure 5.34: Arrangements for recording multiple views of phase objects (Reference waves omitted for clarity)

Fig. 5.34a shows the diffuse illumination via a diffuser of e. g. ground glass. The range of viewing directions is only limited by the size of the diffuser and the aperture of the holograms. But speckles may become a problem as soon as due to complicated localization of the fringes the interferograms have to be observed with a small aperture, see Sec. 3.3.

This possible disadvantage is circumvented by using a phase grating, Fig. 5.34b, which diffracts several plane waves out of the impinging plane wave. In Fig. 5.34c the individual plane waves are produced separately. A fixed plane object wave and a rotating object field are used in the arrangement of Fig. 5.34d. In all cases one has to consider whether a transient or a steady or at least repeatable refractive index distribution is present. The arrangements of Figs. 5.34a, b, and c allow a simultaneous recording of all holograms, the holograms in Fig. 5.34d are recorded sequentially. The system of Fig. 5.34c also enables a sequential registration.

Quite another way of registration of the refractive index variation may be taken by using the holographic *light-in-flight recording* and reconstruction of optical wavefronts, see Sec. 5.6.4. With this method wavefronts can be visualized as they pass a transparent object, [434]. The spatial distortion of a plane or spherical wavefront induced by the refractive index field can be measured this way, Fig. 5.35.

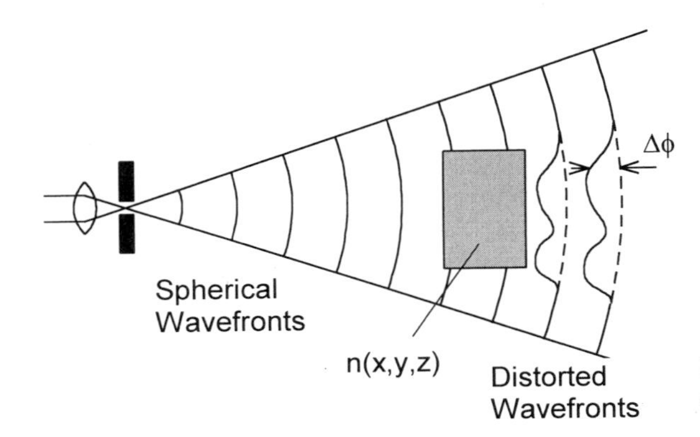

Figure 5.35: Registration of wavefront distortion by light-in-flight recording

In the treatment of reconstruction of asymmetric refractive index fields we have to consider two cases: The first is when ray bending due to refraction is minor and can be neglected. Then the integral defining the pathlength difference

$$\delta(x, y) = \int_s f(x, y, z) ds \tag{5.169}$$

with $f(x, y, z) = n(x, y, z) - n_0$ can be evaluated along straight lines s. We speak of the *refractionless limit*. The second case is when we have to take into account ray bending by strongly refracting fields.

5.8.6 Tomographic Reconstruction in the Refractionless Limit

Reconstructions for both aforementioned cases are based on the tomographic methods introduced in App. B [493]. While most of the methods work well with multidirectional data collected along views subtending the whole angle of 180°, in holographic interferometry the angular range is often less than 180°. The angle of view is restricted by the aperture of the hologram, the angular aperture of the illuminated diffuser, and the extent of the transparent object, Fig. 5.34.

Several reconstruction procedures have been introduced and compared with regard to their capability to work with restricted angles of view in [494, 495]. The first approach in [494] is the *Fourier synthesis*, which fills the spatial frequency plane with values along lines through the origin, as they are given by the *Fourier slice theorem*, (B.7). But this method suffers from the necessity of interpolation when in practice data are collected only for a finite number of views and positions, as pointed out in App. B.2.

The problem of the values not uniformly distributed in the frequency plane are avoided by the *direct inversion* of (5.169) as introduced by [496, 497] and adopted by [494]. This approach is equivalent to the one introduced in App. B.3. Its practical implementation, when discrete data are at hand, corresponds to the methods of App. B.4.

These two methods require optical pathlength data collected over the whole 180° angle of view. The next methods can be applied even when data are available only for viewing angles less than 180°. A number of such methods represent the $f(x,y)$ of (5.169) in each plane $z = const.$ by a series expansion

$$f_e(x,y) = \sum_{m=0}^{M-1} \sum_{n=0}^{N-1} a_{mn} H_{mn}(x,y) \tag{5.170}$$

Together with (5.169) we get after interchanging the order of integration and summation

$$\delta(x,y) = \sum_{m=0}^{M-1} \sum_{n=0}^{N-1} a_{mn} \int_s H_{mn}(x,y) ds \tag{5.171}$$

This shows that it is advantageous to choose generating functions $H_{mn}(x,y)$ which are easily integrated along arbitrary straight lines.

A convenient series expansion is directed by the Whittaker-Shannon *sampling theorem*, which states that a properly sampled band-limited function can be exactly represented by a linear combination of sinc-functions, see App. A.5

$$f_e(x,y) = \sum_{l=-\infty}^{\infty} \sum_{k=-\infty}^{\infty} f\left(\frac{l}{2B_x}, \frac{k}{2B_y}\right) \operatorname{sinc}\left[2B_x\left(x - \frac{l}{2B_x}\right)\right] \operatorname{sinc}\left[2B_y\left(y - \frac{k}{2B_y}\right)\right] \tag{5.172}$$

where $\operatorname{sinc}(x) = [\sin(\pi x)]/(\pi x)$ and B_x, B_y are the bandwidths in x- and y-direction. The sinc-function approach leads to a set of algebraic equations [494]

$$\delta(\rho_i, \theta_j) = \sum_{m=0}^{M-1} \sum_{n=0}^{N-1} W_{mn}(\rho_i, \theta_j) f_e(\Delta x\, m, \Delta y\, n) \tag{5.173}$$

where the $W_{mn}(\rho_i, \theta_j)$ are defined by

$$W_{mn}(\rho_i, \theta_j) = \begin{cases} \sqrt{1+\tan^2\theta_j}\,\Delta x \operatorname{sinc}[(\rho_i \sec\theta_j + \Delta x m \tan\theta_j - \Delta y n)/\Delta y] \\ \qquad\qquad \text{for } 0 \le |\tan\theta_j| \le \Delta y/\Delta x \\ \sqrt{1+\tan^2\theta_j}\,\frac{\Delta y}{\tan\theta_j} \operatorname{sinc}[(\rho_i \sec\theta_j + \Delta x m \tan\theta_j - \Delta y n)/(\Delta y \tan\theta_j)] \\ \qquad\qquad \text{for } \Delta_y/\Delta x < |\tan\theta_j| < \infty \\ \Delta y \operatorname{sinc}[(\rho_i + \Delta x m)/\Delta x] \qquad \text{for } |\tan\theta_j| = \infty \end{cases} \tag{5.174}$$

The θ_j represent the different angular views, the ρ_i the pathlength values in each projection. Δx and Δy denote the spacing of the pathlength values in each direction. The system of equations (5.173) can be solved if at least $M \times N$ pathlength data are measured. Tests have shown that reliable results are obtained when redundant data are used, meaning much more than $M \times N$ values of $\delta(\rho_i, \theta_j)$ are present. This is particularly true, if only restricted angles of view are possible. The overdetermined system of linear equations then is solved by the method of Gaussian least squares.

The next method investigated in [494], the so called *grid method* uses a set of rectangular elements, with constant refractive index in each element. A similar approach is described in Sec. B.5. As with the sinc-method, we get a set of algebraic equations whose solution represents an approximation to the actual $f(x,y)$.

A Fourier transform approach to reconstruction from data over a restricted angular view is given by the *frequency plane restoration*. If multidirectional data are given only for directions between $-\gamma$ and $+\gamma$, Fig. 5.36a, the resulting frequency samples also are only

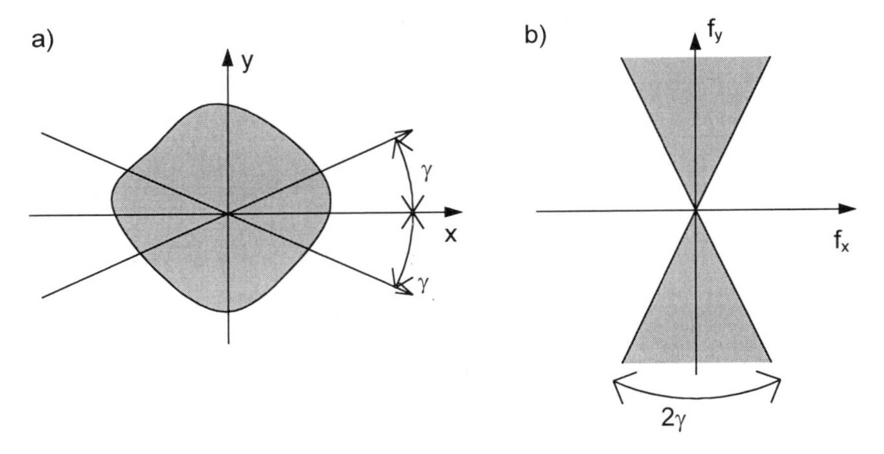

Figure 5.36: Restricted angular views, (a) spatial domain, (b) frequency domain

within an angular range of 2γ, the shaded region in Fig. 5.36b. The determination of the Fourier transform values outside of this known region is based on the fact that the Fourier transform of a continuous, bounded, and spatially limited function is analytic. If the Fourier transform can be uniquely determined over any finite domain, the entire transform can be determined using analytic continuation [494]. Therefore a truncated summation over sampling functions in the frequency domain is defined with the frequency values of the whole frequency plane as coefficients. This leads to a set of algebraic equations, which can be solved for the unknown coefficients.

The last reconstruction method addressed in [494] is an iterative method for solution of an underdetermined system of linear equations. Of the manifold of possible solutions the one with minimum variance is chosen. Nevertheless the method is also feasible for overdetermined systems of equations.

The six reconstruction techniques mentioned above have been compared with simulated data as they may arise in fluid temperature measurements [494]. These comparisons have given the following results: If an 180° angle of view is available, all techniques produce good reconstructions. Especially the sinc-method, the direct inversion and the frequency restoration achieve a good accuracy, while direct inversion requires the least computer time. If the angle of view is less than 180°, e. g. 45°, the frequency plane restoration provides the most accurate reconstruction. It has been observed that if the angle of view is decreased, the degree of redundancy necessary for a reliable reconstruction increases. On the other hand if in an experiment the amount of data is fixed, this relation limits the number of degrees of freedom, and therefore the achievable resolution.

Independent from the specific technique used for the reconstruction one has to reflect upon the data sampling rates. Theoretically only band-limited functions can be reconstructed exactly, although in practice we have spatially limited refractive index fields which therefore are not band-limited. But if the Fourier components of this field are sufficiently small outside some finite region in the frequency plane, an *effective bandwidth* B can be defined. The sampling theorem now guarantees a reliable reconstruction, if the spacing between consecutive ρ_i is less than $1/(2B)$ in the corresponding θ_j direction. In defining the angular separation $\Delta\theta$ between the views one has to consider that the evaluated points in the frequency plane are lying along radial lines, so they have maximum separation at the effective band limits. If we consider two adjacent radial lines oriented near $\pi/2$ and data should be sampled at the Nyquist rate corresponding to the effective band limit B_y, we meet the condition

$$B_y \tan(\Delta\theta) \leq 1/L_x \tag{5.175}$$

where L_x is the object's extent in x-direction. The angular spacing between views near $\pi/2$ must obey

$$\Delta\theta \leq \arctan\left(\frac{1}{B_y L_x}\right) \tag{5.176}$$

and similar for the other direction. In the vicinity of $\pi/2$ the tan-function is steepest, so this is a conservative estimate.

Samples are more close near the origin of the frequency plane than near the band limits. So if the views are chosen to satisfy (5.176) the data will be oversampled. Altogether approximately $8B_xB_yL_xL_y$ samples will be used. Compared to the space-bandwidth product of the representation of the refractive index distribution, which is $4B_xB_yL_xL_y$, we find out a data redundancy of about 2 [494]. The space-bandwidth product is a measure of the total number of degrees of freedom, and is invariant under Fourier transformation.

5.8.7 Tomographic Reconstruction of Strongly Refracting Fields

In tomography codes have been designed to perform an inversion in the case of the refractionless limit. If these codes are applied to pathlength transforms of *strongly refracting refractive index fields* appreciable errors may result. Although an analytical solution for radially symmetric strongly refracting fields exists, Sec. 5.8.4, there is no analytical solution for the general asymmetric case. Therefore in the sequel an iterative algorithm for reconstruction in this instance is presented [484].

Let $\widetilde{\delta}(\rho,\theta)$ be the pathlength difference along the ray bended by refraction as given in (5.138). Correspondingly let $\overline{\delta}(\rho,\theta)$ be the pathlength along the straight ray of (5.139). θ is the angle of the projection and ρ the coordinate along the projection, Fig. 5.31. Let the operator $\overline{P}$ express the straight line integral transform and $\widetilde{P}$ the bended line integral transform which maps $n(r,\phi) - n_0$ onto $\widetilde{\delta}(\rho,\theta)$ and $\overline{\delta}(\rho,\theta)$, resp. The iterative method is based on successive estimation of the deviation $D(\rho,\theta)$ between the straight ray and bended ray pathlength transforms

$$D(\rho,\theta) = \widetilde{\delta}(\rho,\theta) - \overline{\delta}(\rho,\theta) \tag{5.177}$$

It is assumed that $\widetilde{\delta}(\rho,\theta)$ and $\overline{\delta}(\rho,\theta)$ are defined in the same domain [484]. The iterative algorithm begins with (1) setting an initial estimate of the deviation $D_0(\rho,\theta)$. Then (2) the estimate of the refractionless path length transform $\overline{\delta}(\rho,\theta)$ is calculated from the measured pathlength differences $\widetilde{\delta}(\rho,\theta)$ and the assumed deviation

$$\overline{\delta_i}(\rho,\theta) = \widetilde{\delta}(\rho,\theta) - D_i(\rho,\theta) \tag{5.178}$$

The corresponding refractive index field $n_i(r,\phi) - n_0$ is (3) reconstructed by a numerical inverse line integral transformation

$$n_i(r,\phi) - n_0 = \overline{P}^{-1}[\overline{\delta_i}(\rho,\theta)] \tag{5.179}$$

Using computational ray tracing (4), according to (5.126) the pathlength transform of the estimated field is calculated

$$\widetilde{\delta_i}(\rho,\theta) = \widetilde{P}[n_i(r,\phi) - n_0] \tag{5.180}$$

A new estimate of the deviation (5) is calculated by

$$D_i(\rho,\theta) = \widetilde{\delta_i}(\rho,\theta) - \overline{\delta_i}(\rho,\theta) \tag{5.181}$$

and the algorithm proceeds at step (2). This iterative procedure continues until the change in $D_i(\rho,\theta)$ or the difference between two successive reconstructed fields is smaller than a predetermined value.

The line integral transform to be inverted in step (3) is expressed as a Fourier series within a circular domain of radius R, $\rho/R \leq 1$

$$\overline{\delta_i}(\rho,\theta) = \sum_{m=0}^{M-1}\sum_{n=0}^{N-1} A_{mn} g_{mn}(\rho) \mathrm{e}^{\mathrm{i}m\theta} \tag{5.182}$$

A finite number of coefficients A_{mn} are found by inverting an overdetermined system of algebraic equations of the form (5.182) with discrete ρ_j and θ_j by a least squares method. Once the coefficients A_{mn} have been calculated, the reconstructed $f(r,\phi) = n(r,\phi) - n_0$ can be presented as a Fourier series within the circular domain $r/R \leq 1$

$$f(r,\phi) = \sum_{m=0}^{M-1}\sum_{n=0}^{N-1} A_{mn} f_{mn}(r) \mathrm{e}^{\mathrm{i}m\theta} \tag{5.183}$$

Appropriate functions $g_{mn}(\rho)$ and $f_{mn}(r)$, which are line-integral transform pairs are discussed in [484].

The iterative algorithm was applied to numerically simulated data as well as in experiments measuring strongly refracting boundary layers. Refractive index fields with multiple maxima and quite steep gradients have been successfully reconstructed, even when refraction was strong enough to bend some rays by as much as 27°. Nevertheless operator interaction was needed to detect computational ray crossing [24]. Path length data contaminated by this effect were eliminated [498].

The iterative algorithm for correction of errors caused by ray bending also is employed in [499], where optical tomography for flow visualization of the density field around a revolving helicopter rotor blade is investigated. The tomographic reconstruction there is based on the convolution backprojection employing a Shepp-Logan-kernel for filtering, see App. B.3.

Further approaches to reconstruction of asymmetric refractive index fields which bend rays are the application of perturbation analysis [500], which is feasible for mildly refracting objects or the approach via inverse scattering: If diffraction is mild, the Rytov or Born approximations to the wave equation can be used [501, 502].

More problems encountered in tomographic reconstruction of refractive index fields are caused by limited interferometric data. Physical constraints such as test section enclosures may restrict the angular views to less than the desired 180°, and/or a portion of the probing rays may be blocked [499]. Limited data reconstructions are sensitive to noise and produce geometric distortions with various artifacts. An approach to reconstruct under such circumstances is published in [503]: The so called complementary field method is an iterative one and incorporates a priori information effectively. Another problem occurs when an opaque object is present within the field under study, e. g. the test model about which flow is being studied in aerodynamic testing. A solution by illuminating the embedded objects and holographic recording of the wave field scattered by the object's surface is presented in [504]. By this technique furthermore the fringe localization surface is compressed as well as displaced and the number of fringes is doubled. The compression of the localization surface enables the observer to use a larger numerical aperture than would be possible otherwise, see Sec. 3.3.5.

To summarize, it can be stated that the approaches to reconstruction of refractive index distributions with ray bending, published up to now, seem to be designed for specialized cases with no guarantee to work under general conditions. Much more needs to be done in the future in this important, challenging, and interesting field [486, 498].

5.8.8 Resonance Holographic Interferometry

The *refractive index* n as well as the *absorption coefficient* κ of gases show a characteristic dependence on the wavelength near the *resonances* of the molecules, Fig. 5.37, an effect often termed *abnormal dispersion.* This can be used for holographic interferometric measurement of *species concentrations*, e. g. in combustion research [505]. The experiments are performed with two dye-lasers, driven by the same pump-laser. They generate two

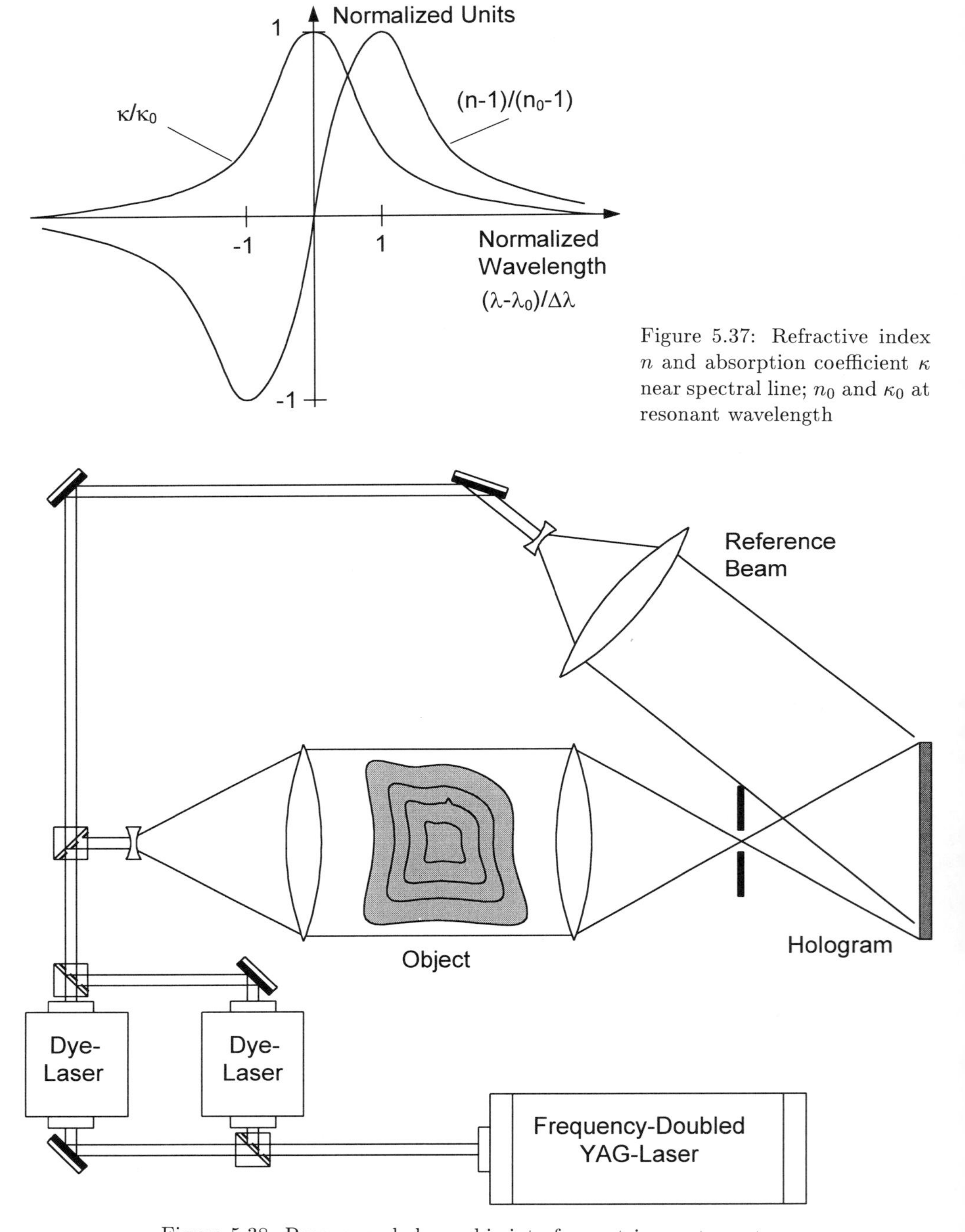

Figure 5.37: Refractive index n and absorption coefficient κ near spectral line; n_0 and κ_0 at resonant wavelength

Figure 5.38: Resonance holographic interferometric spectrometer

beams, one with wavelength λ_0 on resonance and the other, λ, slightly off resonance, Fig. 5.38. Optimal tuning is one beam to the minimum and the other to the maximum of n. The beams are combined to produce a single beam of both wavelengths. This beam is divided into object and reference beam with the expanded object beam passing the transparent object. The refractive indices of all species except the resonant species are essentially identical at the two wavelengths. So only the species showing resonance at the used wavelength is characterized by this method. The conversion from refractive index to species concentration N_i is given by [34, 505]

$$N_i = \frac{(n-1)[(\lambda - \lambda_0)^2 + \Delta\lambda^2]}{K f \lambda_0^3 (\lambda - \lambda_0)} \tag{5.184}$$

where $K = e^2/4\pi mc^2 = 2.24 \times 10^{-14}$ is a constant, f is the oscillator strength for the line, and $\Delta\lambda$ is the half-width of the absorption line at half maximum. A further advantage of resonance holographic interferometry is that the effects of turbulence or gradients in a flow are effectively eliminated.

5.9 Defect Detection by Holographic Non-Destructive Testing

In *holographic nondestructive testing* (HNDT) the inspected sample is loaded and the resulting deformations are made visible as fringe patterns. Local faults, like voids, cracks or other defects, lead to typical local deformations which deviate from the global deformation, schematically shown in Fig. 5.39. Thus the faults are detected by their characteristic local interference patterns [506].

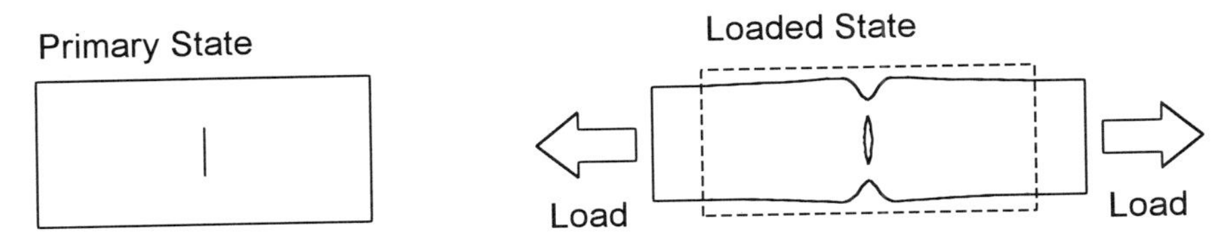

Figure 5.39: Deformation above internal defect (schematically)

While a lot has been done in computerized quantitative evaluation of holographic interferograms, little is reported on automatic computer aided detection of local fringe patterns characterizing material defects. The main reasons are the manifold possible global fringe patterns, the variety of local patterns typical of the defects, and the difficulty to translate into computer software the knowledge of the experienced personnel judging the interferograms.

The few known approaches for automatic qualitative evaluation are based on dividing the whole pattern into a number of sections and comparing the fringe densities in these sections. Actual approaches to computerized fault detection in HNDT use modern software concepts, like knowledge based expert systems [507] and artificial neural networks [71]. The basics of all these attempts are outlined in the following.

5.9.1 Classification of Defects

The approaches to automatic HNDT try to detect faults and flaws by their characteristic partial patterns directly in the holographic interferogram, but not by performing a quantitative evaluation and then investigating the deformation field. Therefore it is primarily of interest how the special interference patterns indicating defects look like. Some typical interference patterns caused by defects are shown schematically in Fig. 5.40 [507].

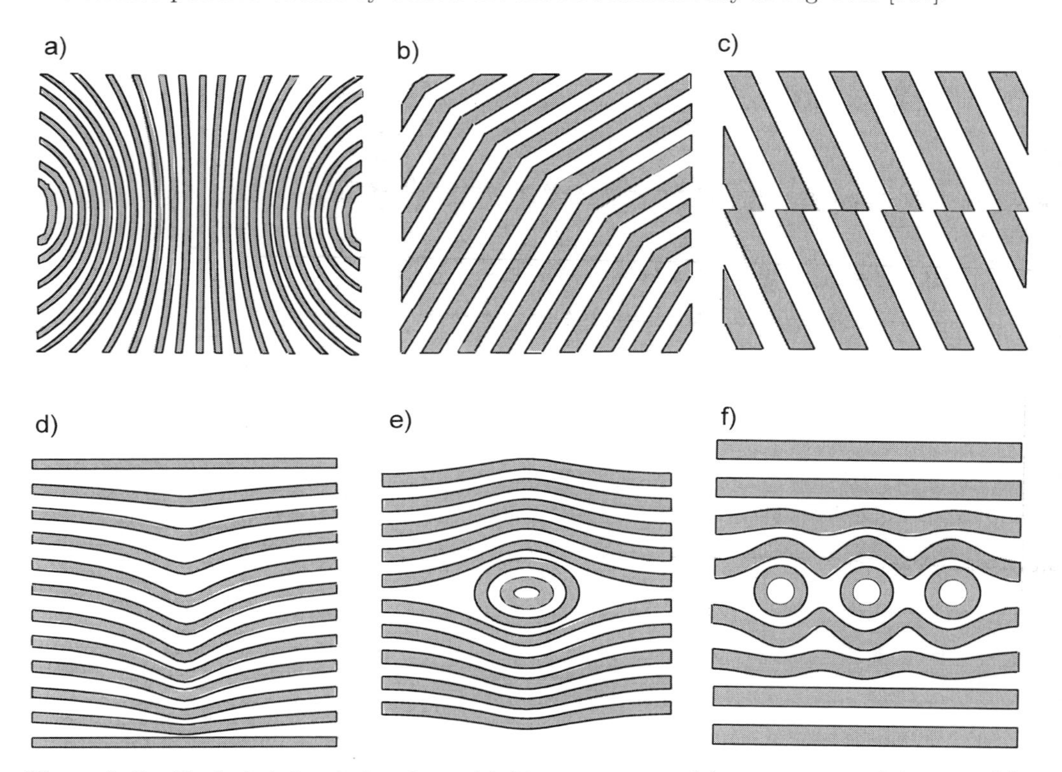

Figure 5.40: Typical defect induced partial fringe patterns: (a) compression, (b) bend, (c) displacement, (d) groove, (e) bull's eye, (f) eye chain

The compression of fringes shown in Fig. 5.40a manifests in a locally higher fringe density. It may be caused by weak points or areas, subsurface voids or separation of layers in compound materials. The bending of the fringes in Fig. 5.40b exhibits a noncontinuous change in the fringe direction although the fringes themselves are continuous. The reason may be a bending or buckling of the structure, local debonds, or the abrupt change of material properties along a line. Even the interference fringes have no continuous paths in Fig. 5.40c, they are displaced or broken. This normally indicates a crack reaching up to the inspected surface. But for this interpretation one has to guarantee a smooth surface. The same pattern can be produced at the edge of two plane surfaces differing in height. The groove of Fig. 5.40d has a locally varying direction and curvature, it may stem from a subsurface crack. The typical bull's eye pattern, Fig. 5.40e, which is often formed by circular or elliptical closed fringes, is caused by local debonds, voids, or inclusions of various forms and types. The eye chain of Fig. 5.40f with the regular occurrence of comparable partial patterns often specifies systematic errors in the production of the tested material, e. g. in fiber reinforced plastic components.

Of course this list is far from being exhaustive, also the underlying defects cited here are only the most obvious. The actual form not only depends on the type of the defect, but also on its orientation and amplitude, as well as on the loading parameters and the vectorial sensitivity of the holographic arrangement. By preparing specimens with artificial defects, real holographic interference patterns from all the shown classes of partial patterns have been generated, Fig. 5.41 [509].

5.9.2 Data Reduction for Automatic Qualitative Evaluation

Central step in each solution of a pattern recognition problem is the reduction of the registered input data to a tractable number of parameters, which resemble the significant features of the input data. In our context we have to reduce from the hundredthousands of pixels of the recorded holographic interference pattern to some parameters which still can discriminate interferograms with characteristic partial patterns indicating a material defect from those without. A first stage of data reduction may be the skeletonizing introduced in Sec. 4.3.

A data reduction strategy, that often is successful in digital image processing, is to take the 2D Fourier transform. The real amplitude spectrum already is translation invariant. If we describe the amplitude spectrum in polar coordinates and integrate all spectral values along the angles from 0° to 360° (0° to 180° also suffice) at each radius, we get a rotation- and translation-invariant 1D spectrum. If needed, a further Fourier transform may produce even a scale-invariant set of features.

Since we are interested only in the existence of a defect induced partial pattern, but not primarily in its location, orientation, or size; translation-, rotation-, and scale-invariant features are exactly what we need. But the outlined Fourier transform based strategy here gives no significant results. The reason is the averaging effect of the global 2D Fourier transform. Small partial patterns may cover only a few pixels of all the pixels of the interferogram. During calculation of each spectral value a weighted sum of the intensities of all pixels is formed, the contribution of a small partial pattern is suppressed by the averaging effect. Especially if some broadband noise is present in the interferogram, a defect induced local pattern will not contribute significantly to the spectrum.

The few published approaches for automatic qualitative evaluation are based on dividing the whole pattern into a number of sections and comparing the fringe densities in these sections. In [510], the interference patterns of pressure vessels are partitioned into squares and the number of fringes in each square is counted. Based on holographic interferograms of proven intact specimens, each square gets a minimum and maximum acceptable fringe count. The fringes are counted automatically in each square, regardless of their orientation. If all fringe counts fall between the predetermined thresholds, the pressure vessel is accepted, otherwise it is rejected as a defective one.

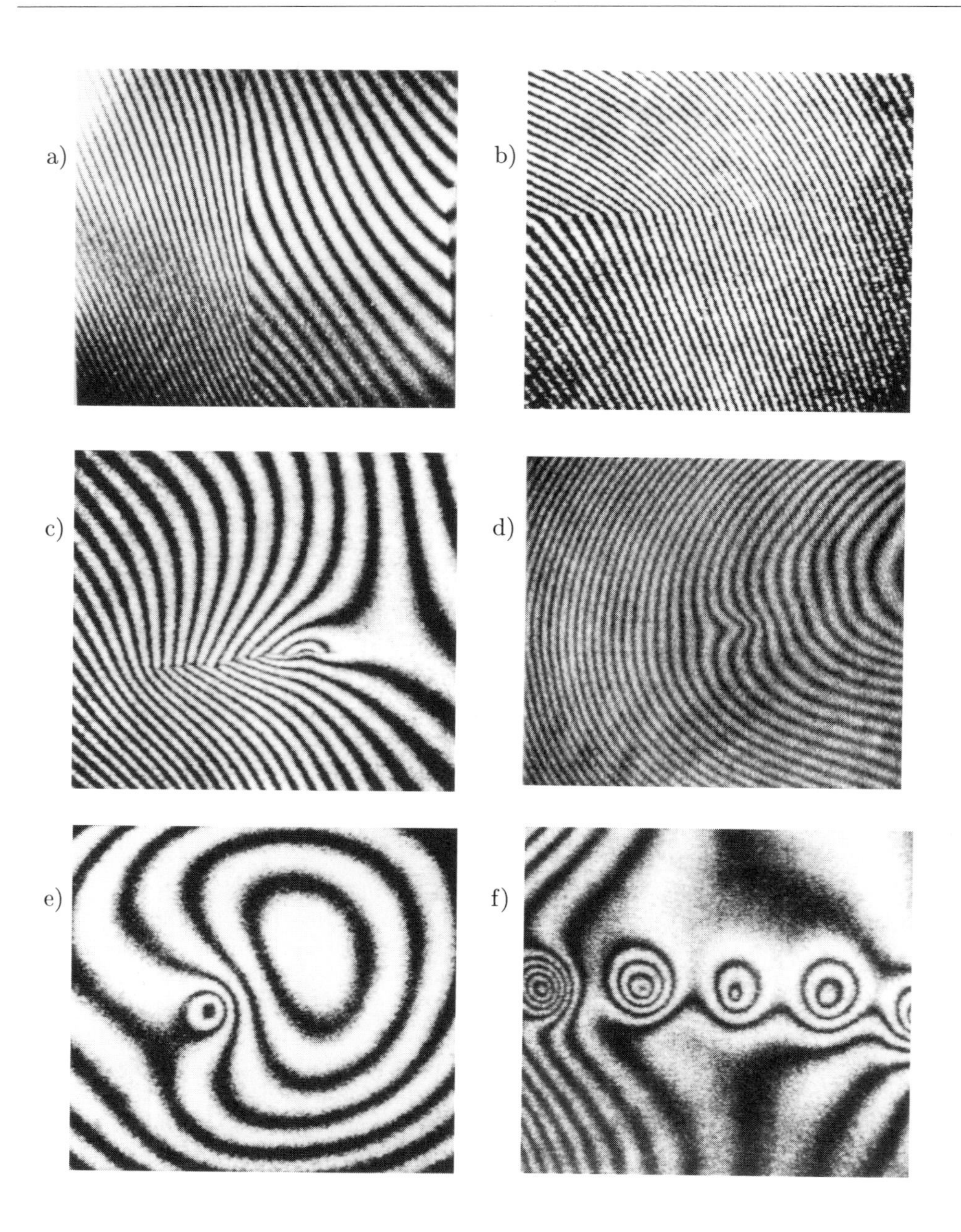

Figure 5.41: Experimentally generated defect induced fringe patterns (Courtesy of W. Osten, BIAS): (a) compression, (b) bend, (c) displacement, (d) groove, (e) bull's eye, (f) eye chain

The sections, the patterns are divided in, can even be degenerated rectangles, like rows or columns [511, 512]. In [512], fringe peaks are counted along horizontal lines. If the number of peaks exceeds a threshold in one line, fringes are counted along short vertical columns centering at this horizontal line until again a threshold is reached. To ensure that a closed ring pattern is detected, the fringes are counted along inclined vectors as well. A fault is considered to be detected if the fringe count in each direction is above the threshold value. The process can be repeated to detect multiple defects. The procedure is applied to the detection of debrazes in brazed cooling panels.

In [511], Fourier amplitude spectra are calculated one-dimensionally along lines and columns or two-dimensionally in small rectangles. From these spectra an average amplitude spectrum together with an acceptance band broader than its band-width, is determined. If the spectrum along one line, column, or rectangle differs significantly from the average spectrum, then a locally higher or lower fringe density must have occurred, indicating a defect.

A feature selection scheme that associates parameters to sections of the pattern, which was successfully applied in an artificial neural network approach to qualitative evaluation, is presented in the following. It is based on the fact that the intensity distribution of the defect induced partial patterns, which are searched, is varying more rapidly than the interference pattern in its defect free neighborhood. But the local variation must not be compared to an averaged global variation, since the fringe density changes continuously even in the non-defect case due to the loading and the varying sensitivity of the holographic arrangement. This change may lead to a higher fringe density in a non-defect area than the density of a defect in a low fringe density area [71]. So the comparison of intensity variation in a small area is restricted to its immediated neighborhood.

For the determination of the intensity variation at each pixel (x, y) the slopes a and b of the two-dimensional plane

$$I(x,y) = ax + by + c \tag{5.185}$$

tangential to the intensity distribution $I(x, y)$ are calculated by Gaussian least squares based on its eight neighbors. The slopes are

$$\begin{aligned} a = \tfrac{1}{6}[\; & I(x+1,y-1) - I(x-1,y-1) + I(x+1,y) - I(x-1,y) + I(x+1,y+1) \\ & -I(x-1,y+1)] \\ b = \tfrac{1}{6}[\; & I(x-1,y+1) - I(x-1,y-1) + I(x,y+1) - I(x,y-1) + I(x+1,y+1) \\ & -I(x+1,y-1)] \end{aligned} \tag{5.186}$$

a and b may have different or equal signs, Fig. 5.42, so the maximum slope at (a, b) is approximated by

$$s(x,y) = \max\{|a+b|, |a-b|\} \tag{5.187}$$

To remain in the range of 8 bits in image processing, we truncate this expression

$$s(x,y) = \min\{\max\{|a+b|, |a-b|\}, 255\} \tag{5.188}$$

The whole pattern now is partitioned into non-overlapping areas, e. g. the pattern of 512×512 pixels is divided into 8×8 areas of 64×64 pixels each. Each area E gets the

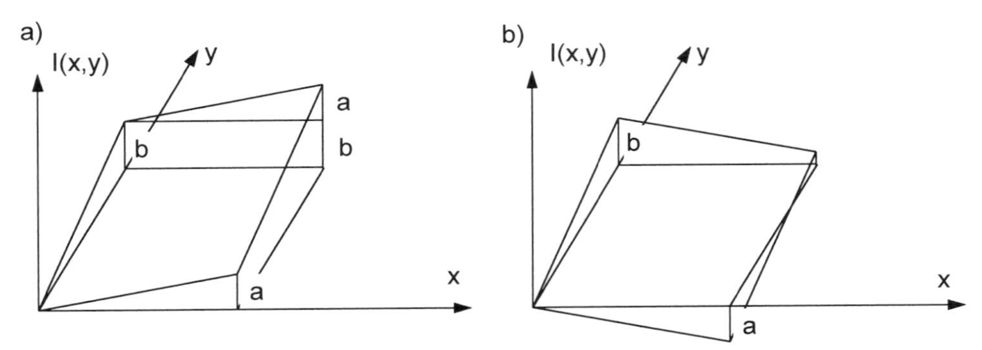

Figure 5.42: Local slope for equal signs of a and b (a), and different signs (b)

parameter $k(E)$, which is the maximal slope of all its pixels

$$k(E) = \max\{s(x,y) : (x,y) \in E\} \tag{5.189}$$

Fig. 5.43a shows an interference pattern with a defect induced local variation in an 8-

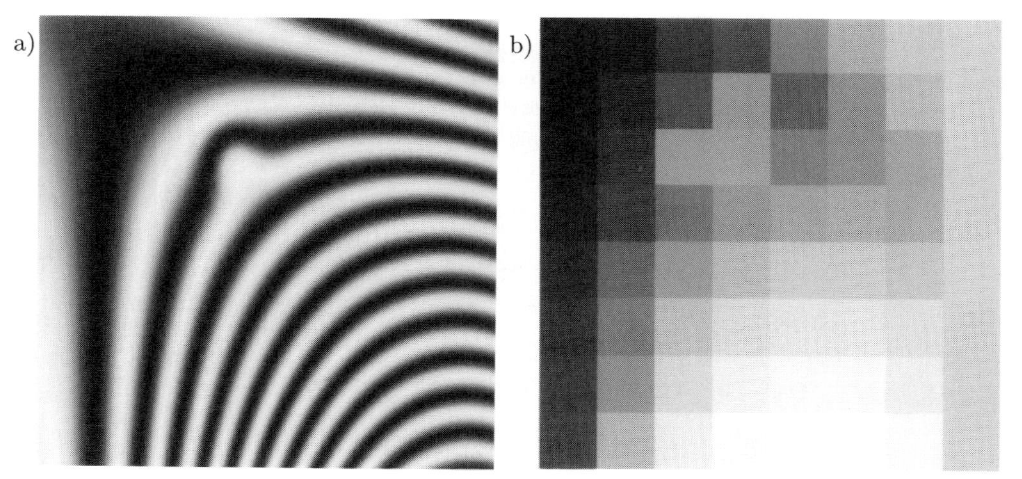

Figure 5.43: Feature selection: (a) interference pattern, (b) maximal slopes in 8×8 areas as gray-values,

bit gray-scale display. The parameters $k(E)$ of a partition into 8×8 areas E are given as numbers in Fig. 5.43c and as gray-values in Fig. 5.43b. One can notice the effect of higher fringe density and thus higher slope at the defect-free lower margin of the pattern compared to the defect area in the center of the pattern.

A defect induced local variation of the fringe density is present in area E, if the parameter $k(E)$ differs significantly from the $k(E')$ of the neighboring areas E'. To detect this, we apply a Laplace-filter to the image of area parameters. If the neighboring areas of E are denoted by A to I according to

c)

9	12	24	37	52	65	73	78
10	19	42	61	42	57	69	77
13	25	63	63	53	57	64	77
17	29	48	60	64	66	64	77
20	44	59	69	75	76	71	77
24	48	70	82	87	87	81	77
28	59	81	92	96	98	88	76
32	67	91	104	108	107	96	77

d)

	-46	32	112	-109	-19	4	
	-41	157	71	-46	-23	-32	
	-57	-28	-14	-7	4	-53	
	37	22	7	9	13	-37	
	-1	26	27	21	24	-2	
	31	35	17	3	34	5	

Figure 5.43: Feature selection: (c) maximal slopes in 8×8 areas as numbers, (d) Laplace-values

A	B	C
D	E	F
G	H	I

two possible realizations of the Laplace-filter are

$$f_1(E) = 4k(E) - k(B) - k(D) - k(F) - k(H) \tag{5.190}$$

or

$$f_2(E) = 8k(E) - k(A) - k(B) - k(C) - k(D) - k(F) - k(G) - k(H) - k(I) \tag{5.191}$$

Fig. 5.43d shows the Laplace-values for all areas having 8 neighboring areas. A defect now is indicated by a high Laplace-value. The highest Laplace-values of the pattern, regardless where they occur, are translation-invariant as well as rotation-invariant features.

Defects which cause high density fringes even of low contrast are detected by this procedure as long as they are confined to one area or extend over only a few areas.

In a first attempt an automatic defect detection by a threshold comparison of the highest Laplace-value over all areas of the tested pattern was carried out. To define the optimal threshold, a sample set of 1000 holographic interferograms was simulated on computer, see Sec. 3.1.6, 500 with and 500 without defects. The threshold was chosen at the valley between the two modes of the bimodal histogram of the highest Laplace value of each pattern of the sample set. This approach was feasible as long as the partial patterns did not vary too much. All defect induced partial patterns had to fit into a single area.

Another approach to automatic defect detection is presented in [258]. There the variety of possible interference patterns is limited by experimental modifications, that produce only linear fringes. This holographic fringe linearization is obtained by swinging the object beam between the two exposures. Proper selection of the fringe frequency by adjusting the object beam swing and of the loading force creates a reconstructed image laced with linear fringes that have highly visible fringe shifts at the defect locations. These fringe shifts furthermore have characteristic Fourier signatures different from those of the linear fringe acting as a carrier.

5.9.3 Neural Network Approach to Qualitative Evaluation

The concept of *artificial neural networks* promises a reliable defect detection based on a number of typical examples because (1) a neural network has the intrinsic ability to learn from the input data and to generalize. (2) It is nonparametric and makes weaker assumptions about the input data distributions than traditional statistical (Bayesian) methods. (3) A neural network is capable of forming highly nonlinear decision bounds in the feature space.

Each neural network consists of a number of neurons which are the fundamental information processing units, Fig. 5.44. Every *neuron* has a number n of input paths numbered by $i = 1, \ldots, n$. The inputs x_i, $i = 1, \ldots, n$ are multiplied by *synaptic weights* w_{ji}, where

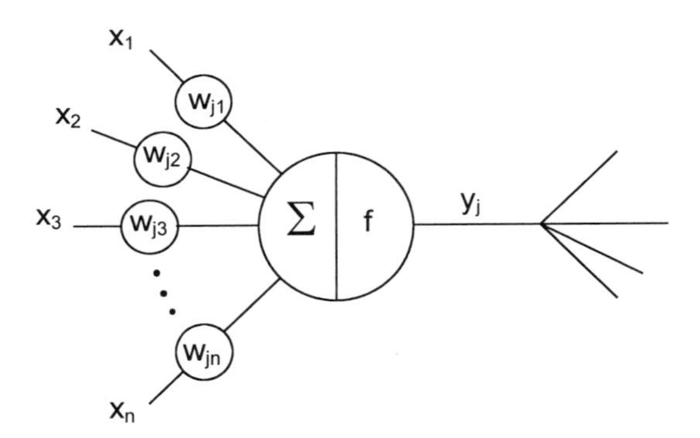

Figure 5.44: Processing element of an artificial neural network

j counts the neurons in the network. The weighted inputs are summed, the result is the internal activity level I_j

$$I_j = \sum_i w_{ji} x_i \tag{5.192}$$

This activity is modified by a transfer function f which can be a binary (0 and 1) or bipolar (-1 and +1) hard limiter, a threshold function with a linear range, the *sigmoid function* $f(I) = [1 + \exp(-I)]^{-1}$, or any nondecreasing bounded function else. The result

$$y_j = f\left(\sum_i w_{ji} x_i\right) \tag{5.193}$$

is given to the output path which may be branched to be connected to the input paths of other neurons or it may present the results of the processing of the whole network to the outside.

Usually the neurons of a neural network are organized into groups called layers. A typical network consists of an input layer of source elements, followed by one or more so-called hidden layers, and an output layer, Fig. 5.45. While the basic operation of a neuron

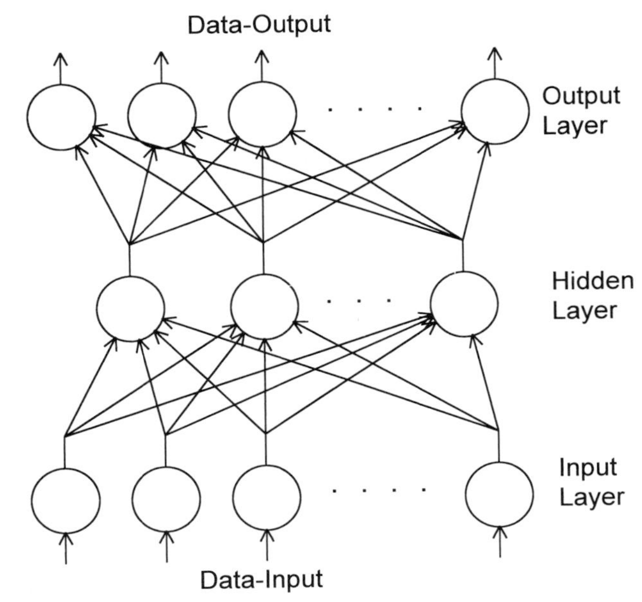

Figure 5.45: Artificial neural network with one hidden layer

is always the same, the different concepts of neural networks differ in their architecture, that is: the number of inputs and outputs, the number of layers, the number of neurons in each layer, the number of weights in each neuron, the way the neurons and corresponding weights are linked together within a layer or between the layers, or which neurons receive correction signals. Closely related to the architecture is the way information is fed to the network, its training. Although a single neuron is far from achieving relevant processing of information, it is the network of many interconnected neurons and the information stored in the synaptic weights that exhibits the far reaching problem solving capability. In brain information is processed in parallel by many neurons, this is actually performed sequentially in the computer software realizations of artificial neural networks.

To achieve an automatic detection of the characteristic local patterns of HNDT, a multilayer network trained by backpropagation learning at sample patterns has been implemented and tested [309, 71]. Since it is not possible to generate the whole training set experimentally, - one needs several thousands of holographic interferograms, - the training samples are calculated by computer simulation, see Sec. 3.1.6. For a given application the material and shape of the object, the type of loading, and the typical defects to be detected have first to be examined practically and theoretically. Based on these examinations a restricted number of experiments are performed to check the validity of the simulation program and help to optimize it.

The features of each interference pattern to be presented to the input neurons are the

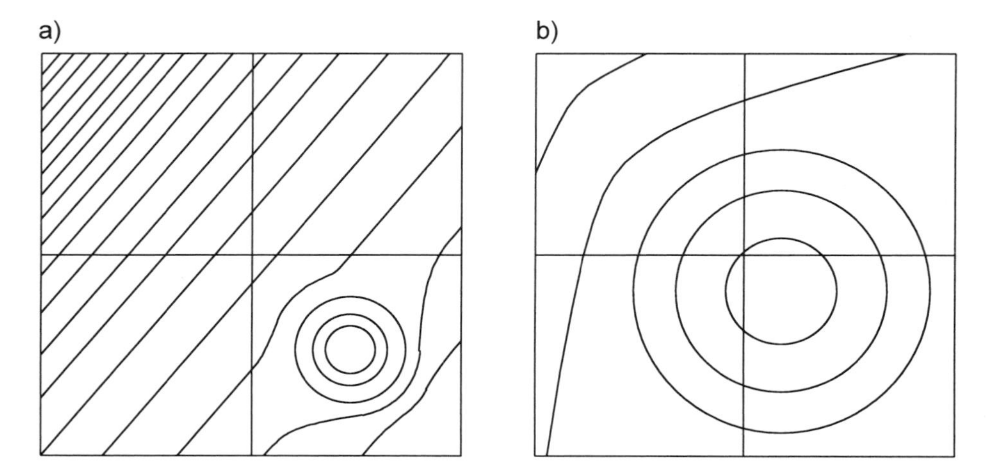

Figure 5.46: Area partitioning and defect size: (a) small defect and large areas, (b) large defect and small areas

four highest Laplace values introduced in Sec. 5.9.2, calculated for each partition of the 512×512 pixel pattern into 8×8, 16×16, 32×32, and 64×64 areas. Thus a total of 16 features were to be fed to a neural net with 16 input neurons. The reasons for the advantages of the multiple area partitions are explained with the help of Fig. 5.46. In Fig. 5.46a the defect defines the maximum fringe density of its area, but neighboring areas, especially the upper left, also contain high fringe densities, leading to a small Laplace value. Contrary to this, the defect of Fig. 5.46b defines the maximum slopes of several neighboring areas, so the Laplace values representing the difference of slopes between adjacent areas remains small.

The neural network structure that showed optimal performance in this special application, is shown in Fig. 5.47. Training of this network was performed by the backpropagation method for supervised learning. This technique, basically a gradient-descent method, propagates the input through the layers, determining the output according to the actual weights. The output is compared with the output expected for this specific input, which quantifies an error. This error then is propagated back through the network from the output layer to the input layer while modifying the weights with the objective of minimizing the global error. This process is repeated with all input samples of the training set, which are taken each one several times in a stochastic order.

The evaluation of actually measured interference patterns, after the training phase was successfully finished, is shown in Fig. 5.48. Three interference patterns were generated by the method of digital holography, see Sec. 4.7. In a preprocessing step the speckle noise is reduced by lowpass filtering. The three patterns are given in Figs. 5.48a-c, the maximal slopes in gray scale display for 8×8 pixel areas in Fig. 5.48d-f, for 16×16 pixel areas in Fig. 5.48g-i, for 32×32 pixel areas in Fig. 5.48j-l, and for 64×64 pixel areas in Fig. 5.48m-o. The corresponding Laplace-values in gray scale coding are shown in Figs. 5.48p-A.

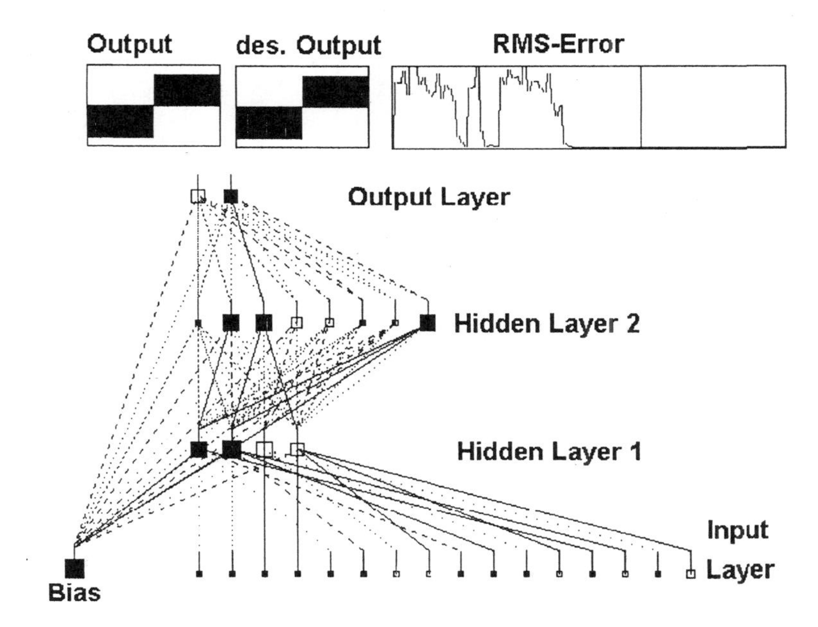

Figure 5.47: Neural network structure

The neural network detected a defect in the patterns of Figs. 5.48a and 5.48b, no defect in the pattern of Fig. 5.48c. The same result would have been achieved by a skilled human investigator.

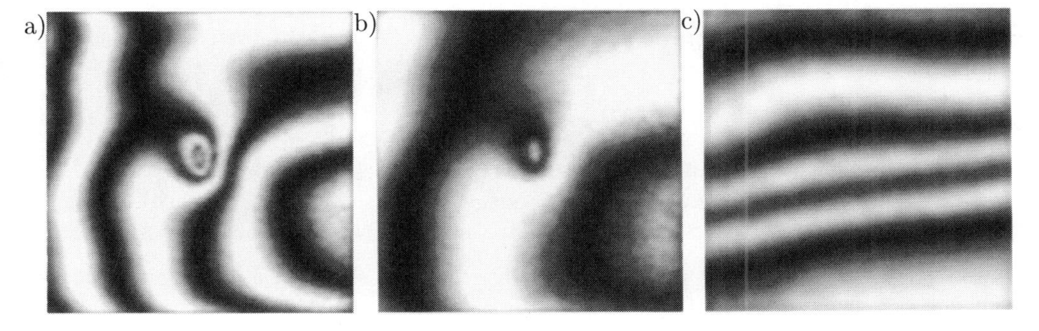

Figure 5.48: Evaluation by neural network, (a-c): interference patterns

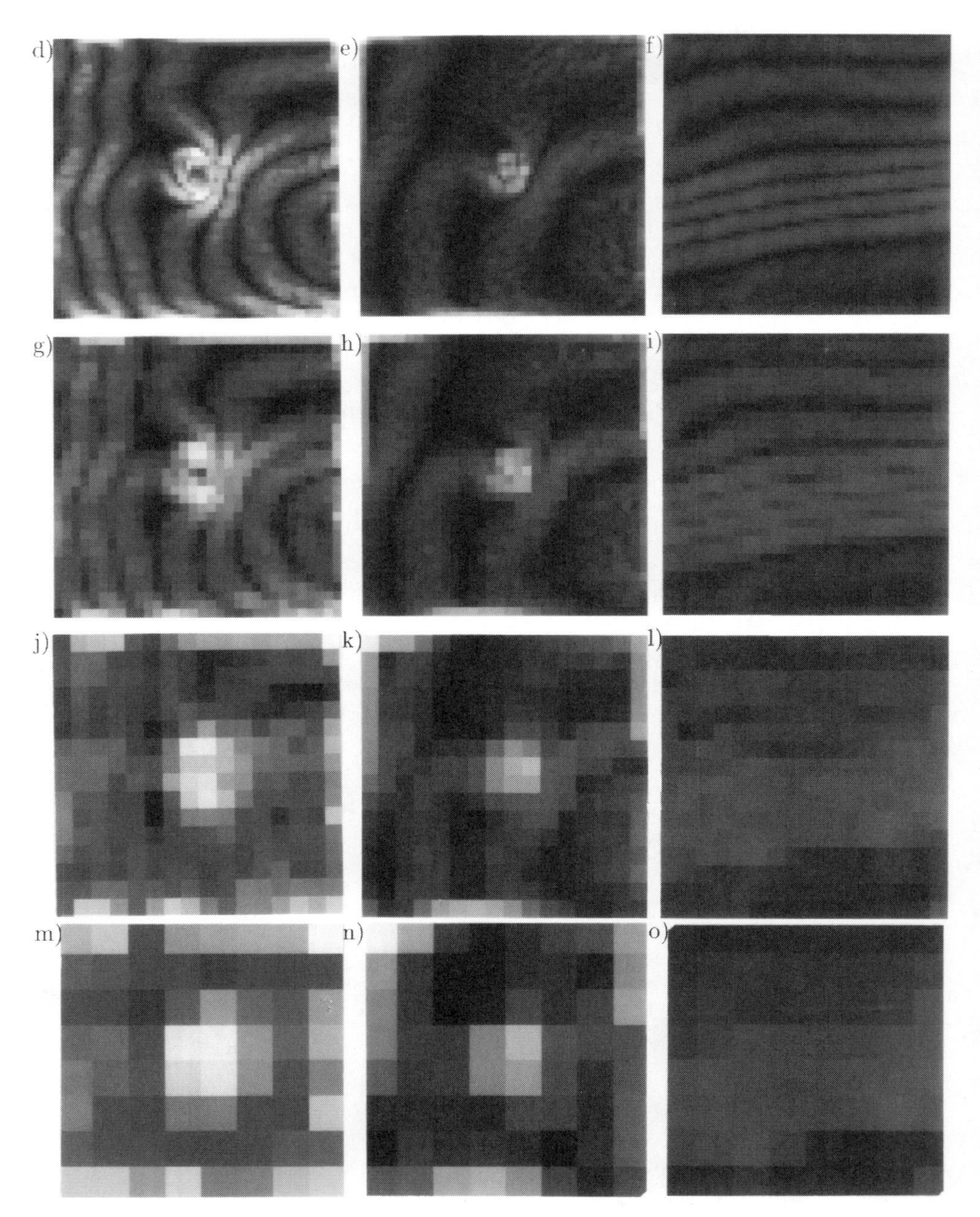

Figure 5.48: Evaluation by neural network, (d-f): maximal slopes in areas of 8×8 pixels, (g-i): maximal slopes in areas of 16×16 pixels, (j-l): maximal slopes in areas of 32×32 pixels, (m-o): maximal slopes in areas of 64×64 pixels

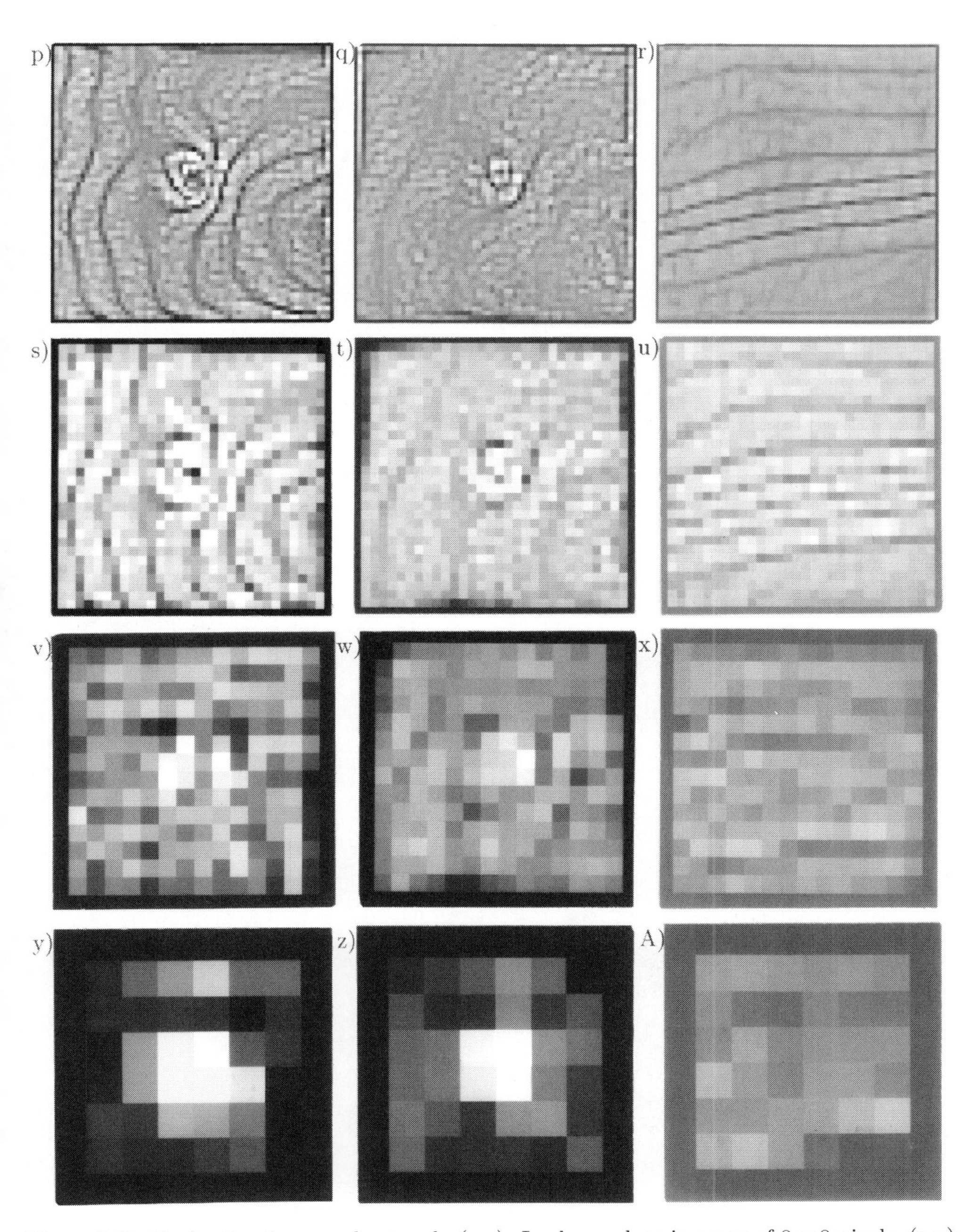

Figure 5.48: Evaluation by neural network, (p-r): Laplace-values in areas of 8×8 pixels, (s-u): Laplace-values in areas of 16×16 pixels, (v-x): Laplace-values in areas of 32×32 pixels, (y-A): Laplace-values in areas of 64×64 pixels

6

Speckle Metrology

In the preceeding chapters the speckles which always appear when coherent light is diffusely scattered or reflected, were viewed as a disturbance to be suppressed or eliminated. On the other hand the speckles can be treated as the fundamental carriers of information, and thus can be used for specific measurement techniques. A number of methods of speckle metrology are closely related to methods of holographic interferometry. Therefore holographic interferometry and speckle metrology are often presented together in a closed form. Here only a brief introduction to the main techniques should be given, a detailed presentation of speckle metrology is planned for a separate book in the series devoted exclusively to this topic.

Two principal approaches have to be distinguished in speckle metrology applied to e. g. deformation analysis of opaque diffusely reflecting surfaces: In speckle photography two reflected speckle fields are incoherently superposed to give information about an in-plane displacement, in speckle interferometry two interference fields are compared, each one generated by coherent superposition of the reflected wave field and a reference wave. The two fields to be compared correspond to the object states before and after the deformation.

The nature of the speckles and their statistics, which are of general interest also for holographic interferometry are described in detail in Sec. 2.5.

6.1 Speckle Photography

In *speckle photography* an opaque diffusely reflecting surface is illuminated by coherent light. The resulting speckle pattern is imaged by the lens of a photo-camera onto photographic film. The exposure results in a pointwise blackening of the film. When the surface point motion has a component in a direction normal to the optical axis, the speckle pattern follows this displacement component. A developed double exposure negative with the two exposures before and after the deformation, will consist of a manifold of speckle pairs. The distance between the points of each pair is proportional to the displacement component normal to the optical axis of the corresponding object point.

For reconstruction the double exposure negative, often called *specklegram*, is illuminated pointwisely by an unexpanded laser beam, Fig. 6.1. The point pairs in the small region

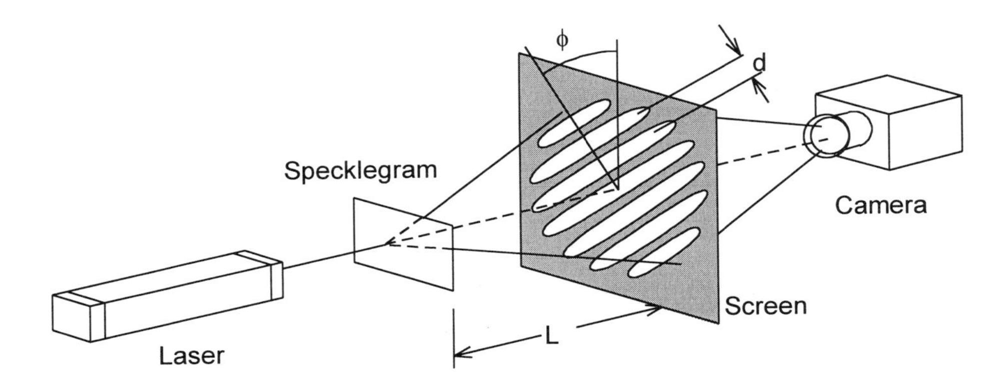

Figure 6.1: Reconstruction of Young's fringes in speckle photography

where the beam hits, act like the two apertures in *Young's double aperture interferometer*, see Sec. 2.3.2. We get parallel equispaced fringes with a spacing of $\lambda L/d$ at a screen, which is placed in a distance L from the specklegram. d is the distance in the pointpair in the specklegram, λ the used wavelength. The orientation of the fringes is orthogonal to the direction of the measured displacement component. By scanning the laser spot over the specklegram a two-dimensional field of displacement components can be measured. It should be stressed, that from a single specklegram one can get only two informations at each point: The modulus of the displacement projection onto a plane orthogonal to the optical axis, and the direction of this displacement component. The first is given by the distance of the Young's fringes, the second by the orientation of the fringes. There are a number of approaches to determine exactly the spacing and orientation of the fringes [513, 514, 515]. Good results have been obtained by locating the primary side lobes in the numerical 2D-Fourier spectrum of the fringe pattern [516, 517].

Contrary to the pointwise evaluation and scanning there is a full field measurement method [18], where the specklegram is placed in an optical Fourier-processor. A screen with a small hole is placed in the diffraction plane and passes only a small part of the spectrum, ideally a single spatial frequency. The resulting pattern in the image plane after such a filtering shows contours of equal displacement components in the direction given by the position of the filter. The filter position in the spatial frequency domain can now be varied by shifting the hole. It is obvious that in this way the measuring sensitivity can be varied, even after the speckle patterns have been originally recorded.

The main sensitivity of speckle photography is in the direction normal to the optical axis, that means for in-plane displacements of the surface points. The displacements generally have to be larger than the speckle size, which can be controlled by the aperture of the photo-camera.

6.2 Electronic and Digital Speckle Interferometry

The most important technique of speckle interferometry is the *digital speckle pattern interferometry (DSPI)*, originally called *electronic speckle pattern interferometry (ESPI)*, also known by the name *electronic holography*. It was invented independently by several groups [518, 519, 520]. The original aim was to overcome the time consuming wet-chemical processing of the silver halide holograms and to use electronic camera tubes instead. To adapt the micro-interference between object- and reference wave to the resolution of the cameras, colinear reference and object waves have to be employed and an imaging system has to be used, Fig. 6.2. The object surface is focused onto the camera target, what in

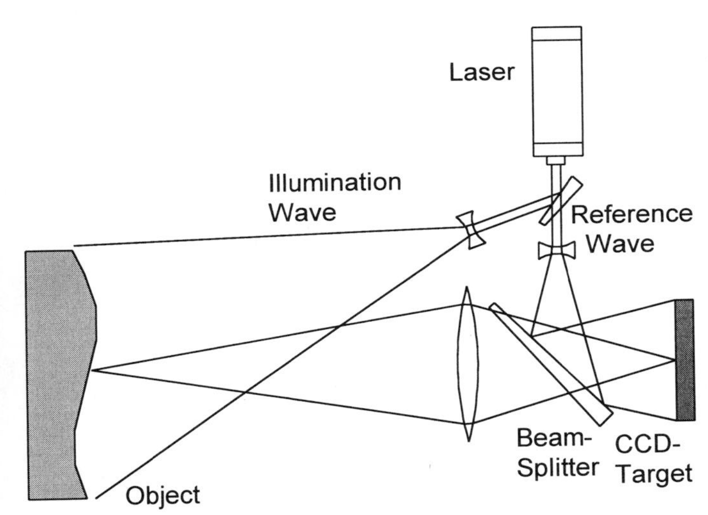

Figure 6.2: Arrangement for digital speckle interferometry

conjunction with the colinear reference wave results in large speckles which now can be resolved by the camera but degrade the resulting interference pattern. This disadvantage is accepted due to the nearly real-time recording and reconstruction, and will be less severe, if in future CCD-targets with more and smaller pixels are available.

The object wave field in the image plane (x, y), the plane of the camera-target, Fig. 6.2, is described by

$$E^{(ob)}(x,y) = E_0^{(ob)}(x,y)\, \mathrm{e}^{\mathrm{i}\phi^{(ob)}(x,y)} \tag{6.1}$$

where $E_0^{(ob)}(x,y)$ is the real amplitude and $\phi^{(ob)}(x,y)$ is the random phase due to the surface roughness, Fig. 6.3a. The colinear reference wavefield

$$E^{(ref)}(x,y) = E_0^{(ref)}(x,y)\, \mathrm{e}^{\mathrm{i}\phi^{(ref)}(x,y)} \tag{6.2}$$

is superposed. This reference wave may be a plane wave, a spherical wave, or an arbitrary reflected one, Fig. 6.3b. Only intensities are recorded by the TV-target, Fig. 6.3d

$$\begin{aligned} I_A(x,y) &= |E^{(ob)}(x,y) + E^{(ref)}(x,y)|^2 \\ &= I^{(ob)}(x,y) + I^{(ref)}(x,y) + 2\sqrt{I^{(ob)}(x,y)\; I^{(ref)}(x,y)}\cos\psi(x,y) \end{aligned} \tag{6.3}$$

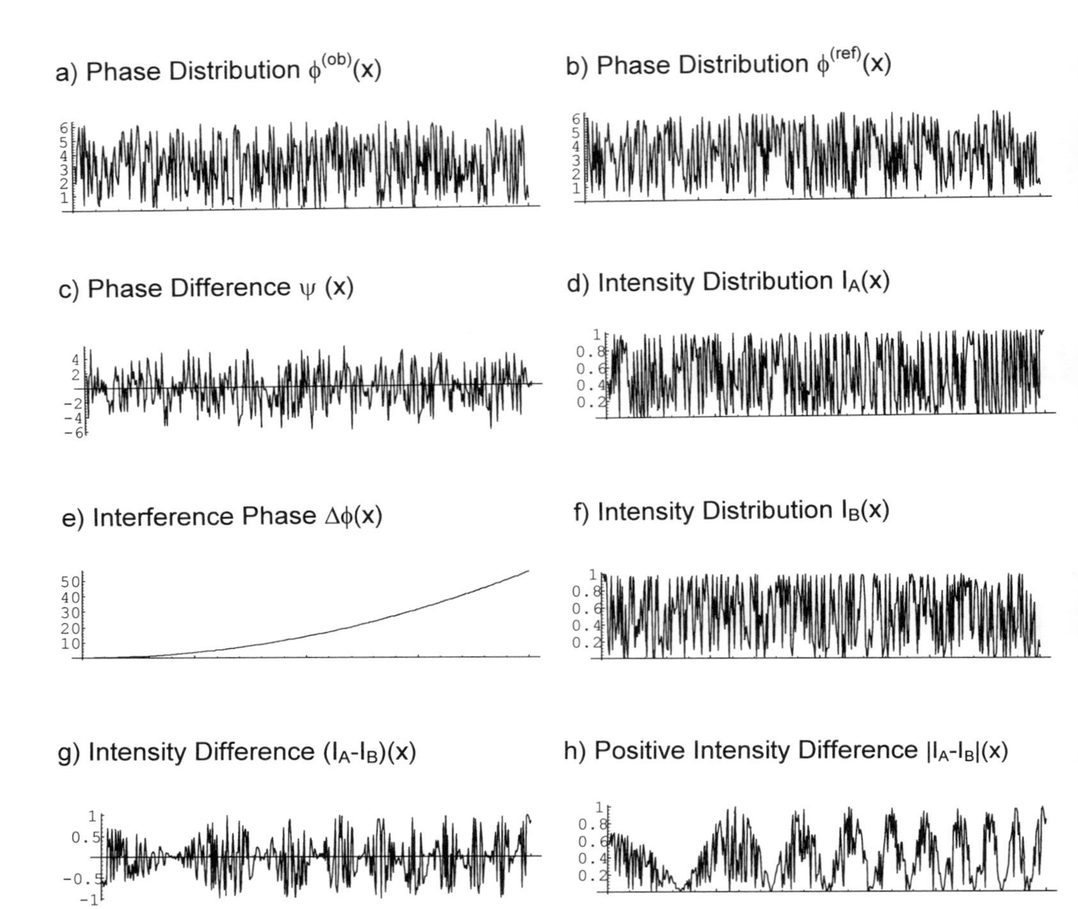

Figure 6.3: Digital speckle interferometry

This is the recorded, digitized, and stored speckle pattern with the stochastic phase difference $\psi(x, y) = \phi^{(ob)}(x, y) - \phi^{(ref)}(x, y)$, Fig. 6.3c. A deformation changes the phase $\phi^{(ob)}(x, y)$ of each point by $\Delta\phi(x, y)$, Fig. 6.3e, so that the wave field after deformation is

$$E^{(ob)\prime}(x, y) = E_0^{(ob)}(x, y)\, \mathrm{e}^{\mathrm{i}[\phi^{(ob)}(x, y) + \Delta\phi(x, y)]} \tag{6.4}$$

Superposition with the colinear reference wave leads to $I_B(x, y)$, Fig. 6.3f

$$I_B(x, y) = I^{(ob)\prime}(x, y) + I^{(ref)}(x, y) + 2\sqrt{I^{(ob)\prime}(x, y)\; I^{(ref)}(x, y)}\cos[\psi(x, y) + \Delta\phi(x, y)] \tag{6.5}$$

In the digital image processing system this second speckle pattern $I_B(x, y)$ is subtracted pointwisely in real time from the stored $I_A(x, y)$, where it is assumed that the deformation changes the phase but not the amplitude, meaning $I^{(ob)\prime}(x, y) = I^{(ob)}(x, y)$. The resulting difference is, Fig. 6.3g

$$\begin{aligned}(I_A - I_B)(x,y) &= 2\sqrt{I^{(ob)}(x,y)\; I^{(ref)}(x,y)}[\cos\psi - \cos\psi\cos\Delta\phi + \sin\psi\sin\Delta\phi](x,y)\\ &= 4\sqrt{I^{(ob)}(x,y)\; I^{(ref)}(x,y)}\sin\left[\psi(x,y) + \frac{\Delta\phi(x,y)}{2}\right]\sin\frac{\Delta\phi(x,y)}{2} \quad (6.6)\end{aligned}$$

To display this result in real-time on a monitor, positive intensities are obtained by taking the modulus $|I_A - I_B|$ or the square $(I_A - I_B)^2$, Fig. 6.3h. The square-root in (6.6) describes the background illumination. The first sine-term gives the stochastic speckle noise which varies randomly from pixel to pixel. This noise is modulated by the sine of the half phase difference induced by the deformation. This low frequency modulation of the high frequency speckle noise is recognized as an interference pattern. The relation between the displacement vector $\boldsymbol{d}(x,y)$ and the phase difference $\Delta\phi(x,y)$ is as in holographic interferometry, see (3.20) and (3.21)

$$\Delta\phi(x,y) = \frac{2\pi}{\lambda}\boldsymbol{d}(x,y)\cdot[\boldsymbol{b}(x,y) - \boldsymbol{s}(x,y)] \quad (6.7)$$

We see that the DSPI-patterns essentially contain the same information as the corresponding holographic interferograms. Thus their production requires the same precautions concerning vibration isolation and stability during the recording process. The results can be observed in real-time, due to the electronic recording there is no problem with the exact repositioning of a hologram plate.

6.3 Speckle Shearography

The requirement of vibration isolation can be dropped to a good extent when using the *speckle-shearing methods.* Here only the spatial variations of the displacement in a predetermined direction are measured, so the methods are rather insensitive to rigid body motions [521, 522]. In the following the method should be explained at the example of *digital shearography,* which attained some economic importance.

Again two wavefields, object and reference, as in DSPI interfere, but now two slightly spatially shifted speckle fields of the rough surface are superposed. The role of the reference wave is taken by one of the two mutually shifted object wave fields, one speaks of *self-reference.* The shifting of the speckle fields is performed by a shearing element, e. g. a wedge in front of one half of the imaging lens, two tilted glass plates, or a Michelson-interferometer-like arrangement with one mirror slightly tilted, Fig. 6.4.

Let the two wavefields be separated laterally by the mutual shearing Δx, then we get the two nearly colinear wavefields

$$\begin{aligned}E_1(x,y) &= E_{01}(x,y)\,\mathrm{e}^{\mathrm{i}\phi(x,y)}\\ E_2(x,y) &= E_{02}(x,y)\,\mathrm{e}^{\mathrm{i}\phi(x+\Delta x,y)} \quad (6.8)\end{aligned}$$

Their interference produces the speckle pattern

$$\begin{aligned}I_A(x,y) &= |E_1(x,y) + E_2(x,y)|^2 \quad (6.9)\\ &= I_1(x,y) + I_2(x,y) + 2\sqrt{I_1(x,y)\; I_2(x,y)}\cos\psi(x,y)\end{aligned}$$

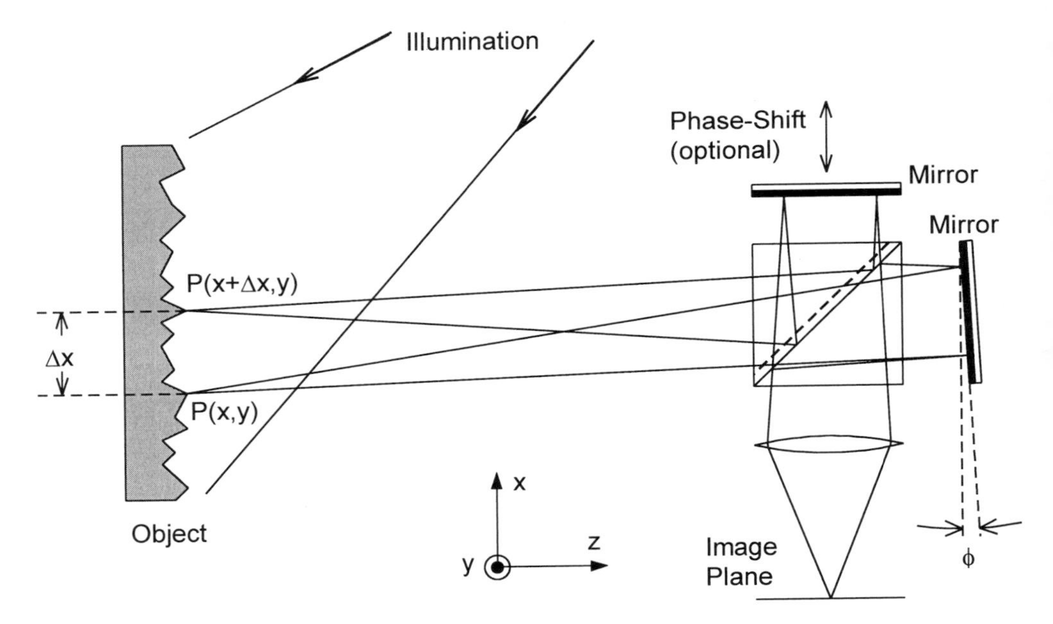

Figure 6.4: Arrangement for digital shearography

which is recorded, digitized, and stored. $\psi(x,y)$ is the randomly distributed phase difference $\psi(x,y) = \phi(x,y) - \phi(x+\Delta x, y)$. Deformation of the object leads to the wave fields

$$\begin{aligned} E_3(x,y) &= E_{01}(x,y)\, \mathrm{e}^{\mathrm{i}[\phi(x,y) + \Delta\phi(x,y)]} \\ E_4(x,y) &= E_{02}(x,y)\, \mathrm{e}^{\mathrm{i}[\phi(x+\Delta x,y) + \Delta\phi(x+\Delta x,y)]} \end{aligned} \tag{6.10}$$

whose superposition yields the speckle pattern

$$I_B(x,y) = I_1(x,y) + I_2(x,y) + 2\sqrt{I_1(x,y)\, I_2(x,y)} \cos[\psi(x,y) + \Delta\phi(x,y) - \Delta\phi(x+\Delta x,y)] \tag{6.11}$$

Pointwise subtraction gives

$$\begin{aligned} &(I_A - I_B)(x,y) \\ &= 2\sqrt{I_1\, I_2}\{\cos\psi(x,y) - \cos[\psi(x,y) + \Delta\phi(x,y) - \Delta\phi(x+\Delta x,y)]\} \\ &= 4\sqrt{I_1\, I_2} \sin[\psi(x,y) + \frac{\Delta\phi(x,y) - \Delta\phi(x+\Delta x,y)}{2}] \sin\frac{\Delta\phi(x,y) - \Delta\phi(x+\Delta x,y)}{2} \end{aligned} \tag{6.12}$$

Positive values may be obtained by taking the modulus or square. Again the term in the root is the background, the first sine-term is the stochastic speckle noise modulated by the second sine-term which stems from the deformation. The displacement vector field $\boldsymbol{d}(x,y)$ is in the argument of this sine by

$$\frac{\Delta\phi(x,y) - \Delta\phi(x+\Delta x,y)}{2} \tag{6.13}$$
$$= \frac{\pi}{\lambda}\{\boldsymbol{d}(x,y)\cdot[\boldsymbol{b}(x,y)-\boldsymbol{s}(x,y)] - \boldsymbol{d}(x+\Delta x,y)\cdot[\boldsymbol{b}(x+\Delta x,y)-\boldsymbol{s}(x+\Delta x,y)]\}$$
$$= \frac{\pi}{\lambda}[\boldsymbol{d}(x,y) - \boldsymbol{d}(x+\Delta x,y)]\cdot[\boldsymbol{b}(x,y)-\boldsymbol{s}(x,y)]$$
$$= \frac{\pi}{\lambda}\left[\frac{\boldsymbol{d}(x,y) - \boldsymbol{d}(x+\Delta x,y)}{\Delta x}\right]\cdot[\boldsymbol{b}(x,y)-\boldsymbol{s}(x,y)]\,\Delta x$$
$$\approx \frac{\partial \boldsymbol{d}(x,y)}{\partial x}\cdot\frac{\pi\Delta x}{\lambda}[\boldsymbol{b}(x,y)-\boldsymbol{s}(x,y)] \tag{6.14}$$

What produces the interference pattern is an approximation to the derivative of the displacement field in the direction of the image shearing, here the x-direction. For rigid body motions we have $\boldsymbol{d}(x,y) = const.$, implying $\partial\boldsymbol{d}(x,y)/\partial x = 0$. The sensitivity of the method can be adjusted by controlling the shearing. Besides the shearing in the x-direction we can shear in the y- or other angular directions. Also radial shearing, rotational shearing, inversion shear, or reversal shear can be performed with related optical arrangements [523].

6.4 Electro-Optic Holography

Electro-optic holography, also known as *electronic holography* or *TV-holography,* is a combination of *phase stepping* and digital speckle interferometry [54, 524, 525]. For static measurements n phase stepped speckle patterns are recorded in the unstressed and n phase stepped speckle patterns in the stressed state. The recorded intensities are, see (6.3)

$$\begin{aligned} I_n(x,y) &= I^{(ob)}(x,y) + I^{(ref)}(x,y) + 2\sqrt{I^{(ob)}(x,y)\,I^{(ref)}(x,y)}\cos[\psi(x,y)+\phi_{Rn}] \\ I'_n(x,y) &= I^{(ob)\prime}(x,y) + I^{(ref)}(x,y) \\ &\quad +2\sqrt{I^{(ob)\prime}(x,y)\,I^{(ref)}(x,y)}\cos[\psi(x,y)+\Delta\phi(x,y)+\phi_{Rn}] \end{aligned} \tag{6.15}$$

The notation is as in Sec. 6.2, ϕ_{Rn} are the phase shifts. While generally arbitrary phase shifts ϕ_{Rn} can be employed, see Sec. 4.5, the most used are $\phi_{R1} = 0^o$, $\phi_{R2} = 90^o$, $\phi_{R3} = 180^o$, and $\phi_{R4} = 270^o$. This results in

$$\begin{aligned} I_1(x,y) &= I^{(ob)}(x,y) + I^{(ref)}(x,y) + 2\sqrt{I^{(ob)}\,I^{(ref)}}\cos\psi(x,y) \\ I_2(x,y) &= I^{(ob)}(x,y) + I^{(ref)}(x,y) + 2\sqrt{I^{(ob)}\,I^{(ref)}}\sin\psi(x,y) \\ I_3(x,y) &= I^{(ob)}(x,y) + I^{(ref)}(x,y) - 2\sqrt{I^{(ob)}\,I^{(ref)}}\cos\psi(x,y) \\ I_4(x,y) &= I^{(ob)}(x,y) + I^{(ref)}(x,y) - 2\sqrt{I^{(ob)}\,I^{(ref)}}\sin\psi(x,y) \end{aligned} \tag{6.16}$$

and

$$I'_1(x,y) = I^{(ob)}(x,y) + I^{(ref)}(x,y) + 2\sqrt{I^{(ob)}\,I^{(ref)}}\cos[\psi(x,y)+\Delta\phi(x,y)]$$

$$
\begin{aligned}
I'_2(x,y) &= I^{(ob)}(x,y) + I^{(ref)}(x,y) + 2\sqrt{I^{(ob)}\, I^{(ref)}}\sin[\psi(x,y) + \Delta\phi(x,y)] \\
I'_3(x,y) &= I^{(ob)}(x,y) + I^{(ref)}(x,y) - 2\sqrt{I^{(ob)}\, I^{(ref)}}\cos[\psi(x,y) + \Delta\phi(x,y)] \\
I'_4(x,y) &= I^{(ob)}(x,y) + I^{(ref)}(x,y) - 2\sqrt{I^{(ob)}\, I^{(ref)}}\sin[\psi(x,y) + \Delta\phi(x,y)] \quad (6.17)
\end{aligned}
$$

These systems of equations can be solved, see (4.25), yielding $\psi(x,y)$ and $\psi(x,y) + \Delta\phi(x,y)$, whose difference is the interference phase distribution $\Delta\phi(x,y)$. The advantage over conventional DSPI-patterns becomes obvious, when $\sqrt{[(I_1 - I_3) + (I'_1 - I'_3)]^2 + [(I_2 - I_4) + (I'_2 - I'_4)]^2} = 8\sqrt{I^{(ob)}\, I^{(ref)}}\cos(\Delta\phi/2)$ is displayed.

The electro-optic holography also has been effectively applied to sinusoidally vibrating objects [158]. High quality time average interference patterns can be synthesized from phase stepped recordings. The argument of the Bessel function is determined with high accuracy, if the phase of the object or reference wave is modulated at the same frequency and phase as the object vibration [158].

In the described method the four phase shifted patterns were taken one after the other. To get several phase shifted patterns not sequentially in time but spatially separated, the in-line reference wave of digital speckle interferometry must be tilted, what is accomplished by shifting the focus of the reference wave little bit out of the focus of the camera lens. In [526] this shift was adjusted to produce phase shifts of 120° between subsequent pixels. The three phase shifted images now are coded in neighboring pixels. If the speckles are large enough - about three pixels per speckle - and if the phase to be measured does not vary too large over the speckle size, then the interference phase can be determined for each triple of neighboring pixels by $\Delta\phi = \arctan[\sqrt{3}(I_3 - I_2)/(2I_1 - I_2 - I_3)]$, see Table 4.1.

Appendix A

The Fourier Transform

The Fourier transform is an important tool used in the description of coherent optics and frequently applied in the evaluation of interference patterns. The treatment of diffraction effects, Appendix 2.4, strongly rests on the Fourier theory. Signal processing, especially two-dimensional signal processing is based on the Fourier transform. Fourier transform evaluation, Sec. 4.6.1, and spatial heterodyning, Sec. 4.6.5, are two methods for determining the interference phase distribution, which employ the Fourier transform and filtering in the spatial frequency domain. The main procedures of computer aided tomography, Appendix B, used here for the reconstruction of refractive index fields in phase objects, are based on the Fourier transform.

The objective of this appendix is to summarize the most important facts about the Fourier transform as far as they are used in this book. This remains far beyond being exhaustive, so the interested reader is referred to a number of excellent textbooks on one- and two-dimensional signal processing, where the details are introduced systematically, questions of existence of the Fourier transforms are discussed, proofs and examples are given [15, 176, 527, 528, 529, 530, 531, 532, 533, 534].

A.1 Definition of the Fourier transform

Let $f(x)$ be a one-dimensional complex-valued function. In practical applications the variable x stands for a *temporal* or a *spatial coordinate*. The *Fourier transform* of this $f(x)$ is defined as

$$\mathcal{F}\{f(x)\} = F(u) = \int_{-\infty}^{\infty} f(x) \mathrm{e}^{-\mathrm{i}2\pi ux} dx \tag{A.1}$$

The Fourier transform is a linear integral transformation that maps the complex function $f(x)$ to another complex function $F(u)$ of the variable u which stands for the *temporal frequency* or the *one-dimensional spatial frequency*. The *inverse Fourier transform* of $F(u)$ is defined as

$$\mathcal{F}^{-1}\{F(u)\} = \frac{1}{2\pi} \int_{-\infty}^{\infty} F(u) \mathrm{e}^{\mathrm{i}2\pi ux} du \tag{A.2}$$

Fourier's integral theorem states that

$$f(x) = \frac{1}{2\pi} \int_{-\infty}^{\infty} \left[\int_{-\infty}^{\infty} f(x) e^{-i2\pi ux} dx \right] e^{i2\pi ux} du \tag{A.3}$$

which means that the transformation is reciprocal

$$\mathcal{F}\{f(x)\} = F(u) \qquad \Longrightarrow \qquad \mathcal{F}^{-1}\{F(u)\} = f(x) \tag{A.4}$$

The two functions $f(x)$ and $F(u)$ together are called a *Fourier transform pair*. For any $f(x)$, if the Fourier transform exists, $F(u)$ is unique and vice versa.

Since the Fourier transform is an integral transformation, the question of existence of the integrals (A.1) and (A.2) must be addressed. But it is a notorious feature of the subject, that no direct and simple criterion is known, which is both sufficient and necessary in ensuring that a function belongs to an ordinary Fourier transform pair. A mathematically rigorous treatment of the Fourier transform is given in the framework of *generalized functions*, sometimes also named *distributions* [535].

Here we take a more pragmatic standpoint: Many of the functions we process, are digitized signals or images. These are necessarily truncated to finite duration and bounded. Thus they belong to those *transient functions* which go to zero for large positive and negative arguments rapidly enough that the integrals in (A.1) and (A.2) exist. Then the following corrollary states: If the integral of the absolute value of the function $f(x)$ exists

$$\int_{-\infty}^{\infty} |f(x)| dx < \infty \tag{A.5}$$

and it is either continuous or has only a finite number of discontinuities and these discontinuities are not infinite and furthermore it has at most a finite number of extrema in any finite interval, then its Fourier transform $F(u)$ exists for all values of u.

Periodic or *constant functions* do not belong to the class of transient functions, nevertheless they may be transformable. For their treatment as well as for other purposes the *Dirac delta* or *impulse* $\delta(x)$ is useful. It can be defined as the limit of a sequence of *rectangular functions* by

$$\mathrm{rect}(x) = \begin{cases} 1 & \text{for} \quad |x| < \frac{1}{2} \\ 0 & \text{elsewhere} \end{cases} \tag{A.6}$$

$$\delta(x) = \lim_{n \to \infty} \delta_n(x) = \lim_{n \to \infty} n\mathrm{rect}(nx) \tag{A.7}$$

The first components of the sequence $\{\delta_n(x)\}$ are illustrated in Fig. A.1 The limit of this sequence of functions is of infinite height but zero width in such a manner that the area is still unity

$$\int_{-\infty}^{\infty} \delta(x) dx = 1 \tag{A.8}$$

The impulse has the following *sampling property*

$$\int_{-\infty}^{\infty} f(x)\delta(x - x') dx = f(x') \tag{A.9}$$

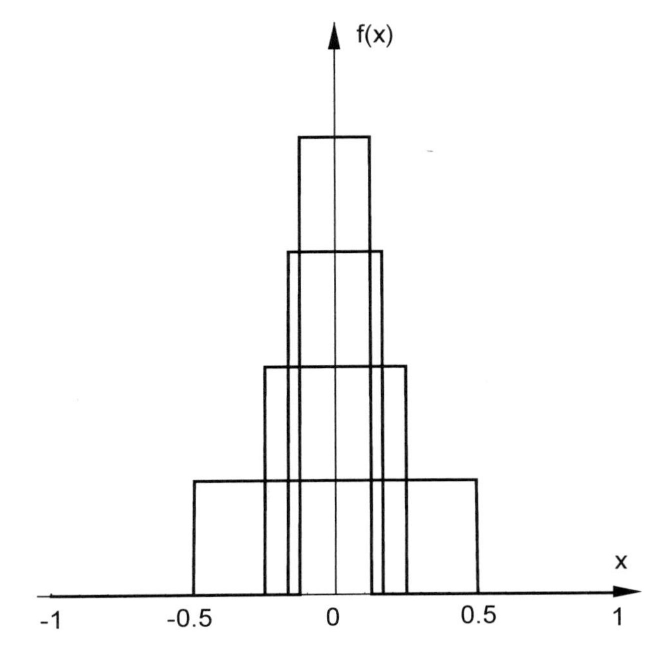

Figure A.1: Rectangular functions converging to the Dirac delta

where $\delta(x - x')$ denotes an impulse shifted to location $x = x'$. Based on these properties of the impulse it can be shown that

$$\mathcal{F}\{\cos(2\pi f x)\} = \frac{1}{2}\left[\delta(u - f) + \delta(u + f)\right] \tag{A.10}$$

$$\mathcal{F}\{\sin(2\pi f x)\} = \frac{\mathrm{i}}{2}\left[\delta(u + f) - \delta(u - f)\right] \tag{A.11}$$

$$\mathcal{F}\{1\} = \delta(u) \tag{A.12}$$

$$\mathcal{F}\{\delta(x)\} = 1 \tag{A.13}$$

Further useful Fourier transform pairs are given in Table A.1.

A.2 Fourier Representation

Reasons for representing a function, that describes the temporal fluctuation of a physical quantity or the spatial distribution of intensity in a two-dimensional image, in the temporal or spatial frequency domain are manifold. First it gives a new perspective and hopefully new insight to an otherwise difficult problem. But second, some processes like convolution, differentiation, or other filtering operations are much more easy to understand and performed more effectively in the spatial frequency domain. Although the transformation and the inverse transformation have to be carried out, this remains true in many applications.

In the following, four different varieties of the Fourier transform will be treated. The first is the *continuous Fourier transform* given in (A.1) and (A.2) which is mostly used

Table A.1: Useful Fourier transform pairs

Name	Function	Fourier transform
complex exponential	$e^{i2\pi fx}$	$\delta(u-f)$
Gaussian	$e^{-\pi x^2}$	$e^{-\pi u^2}$
rectangular pulse	$\Pi(x) = \begin{cases} 1 & \text{for } -\frac{1}{2} < x < \frac{1}{2} \\ \frac{1}{2} & \text{for } x = \pm\frac{1}{2} \\ 0 & \text{elsewhere} \end{cases}$	$\frac{\sin(\pi u)}{\pi u}$
triangular pulse	$\Lambda(x) = \begin{cases} 1-\lvert x\rvert & \text{for } \lvert x\rvert \le 1 \\ 0 & \text{for } \lvert x\rvert > 1 \end{cases}$	$\frac{\sin^2(\pi u)}{(\pi u)^2}$
unit step	$u(x) = \begin{cases} 1 & \text{for } x > 0 \\ \frac{1}{2} & \text{for } x = 0 \\ 0 & \text{for } x < 0 \end{cases}$	$\frac{1}{2}\left[\delta(u) - \frac{i}{\pi u}\right]$

in theoretical analysis. Given that with real world signals it is necessary to periodically sample the data, we are led to three other Fourier transforms which approximate either the time or frequency data as samples of the continuous functions.

Let a continuous function $f(x)$ be defined for $X_1 \le x \le X_2$. This function can be expressed as

$$f(x) = \sum_{k=-\infty}^{\infty} a_k e^{i2\pi k f_0 x} \tag{A.14}$$

where $f_0 = 1/X$ with $X = X_2 - X_1$. Here $f(x)$ is represented by an infinite linear combination of sines and cosines, the so called *sinusoids*, which oscillate with kf_0 cycles per unit of x.

$$e^{i2\pi k f_0 x} = \cos(2\pi k f_0 x) + i\sin(2\pi k f_0 x) \tag{A.15}$$

The frequencies of all sinusoids are integer multiples of the *fundamental frequency* f_0, the first four components of which are displayed in Fig. A.2. The coefficients a_k are called the *complex amplitude* of the k-th component and are obtained by

$$a_k = \frac{1}{X}\int_{X_1}^{X_2} f(x) e^{-i2\pi k \frac{x}{X}} dx \tag{A.16}$$

The representation (A.14) with the a_k defined by (A.16) is called the *Fourier series.*

The difference between the continuous Fourier transform and the Fourier series is that we have all frequencies between $-\infty$ and ∞ in the sinusoids of the continuous Fourier transform but only multiples of the fundamental frequency f_0 for the Fourier series. The series representation is contained in the continuous Fourier transform representation especially if $f(x)$ is assumed to be zero outside $[X_1, X_2]$, which restricts the range of integration in (A.1). The additional information in $F(u)$ for $u \ne kf_0$ in the continuous reconstruction (A.2) is necessary to constrain the values of the reconstructed $f(x) = \mathcal{F}^{-1}\{F(u)\}$ outside the interval $[X_1, X_2]$. If we reconstruct $f(x)$ only from the a_k by (A.14) we will of course

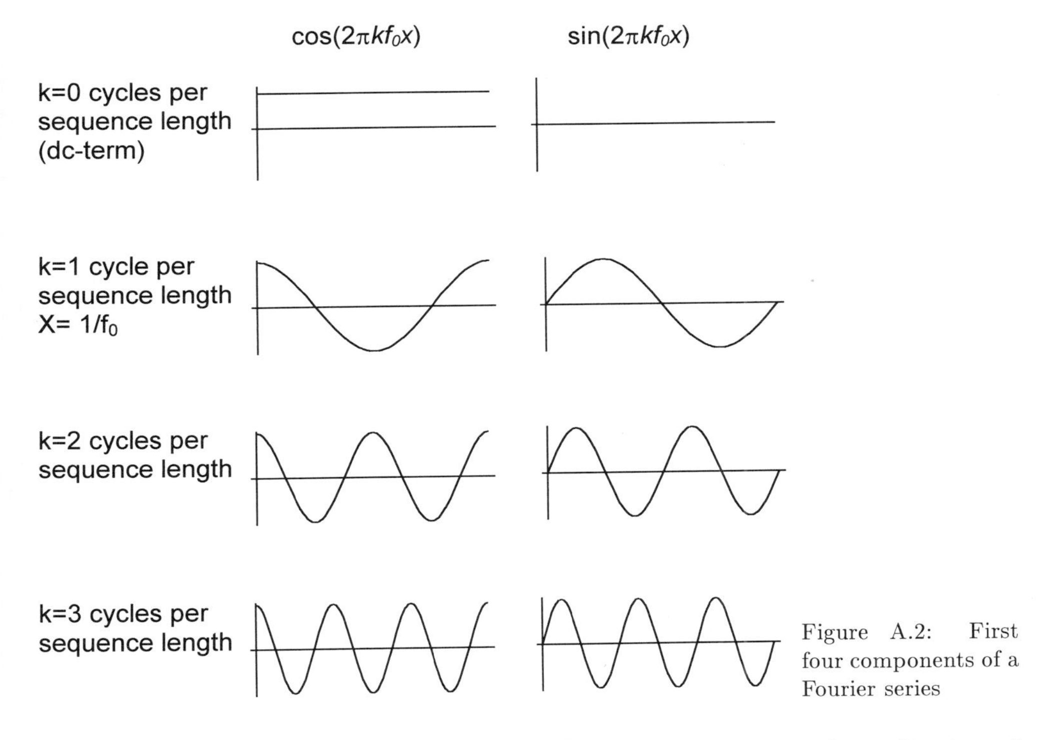

Figure A.2: First four components of a Fourier series

obtain the correct values of $f(x)$ within $[X_1, X_2]$, but we will get *periodic replications* of the original $f(x)$ outside the interval $[X_1, X_2]$. Contrary, using $F(u)$ and (A.2) for reconstruction, we will obtain $f(x)$ exactly within $[X_1, X_2]$ and the correct zero everywhere outside.

As in the continuous case a discrete function may also be given a frequency domain representation. Let $\{f_n\}$ be the discrete samples of a continuous function $f(x)$

$$f_n = f(nx_0) \qquad n = \ldots, -1, 0, 1, \ldots \tag{A.17}$$

Then the *discrete Fourier transform* is defined by

$$F(u) = \sum_{n=-\infty}^{\infty} f_n \mathrm{e}^{-\mathrm{i}2\pi unx_0} \tag{A.18}$$

From the discrete Fourier transform the function $f(t)$ can be recovered by

$$f(nx_0) = \frac{x_0}{2\pi} \int_{-\pi/x_0}^{\pi/x_0} F(u)\mathrm{e}^{\mathrm{i}2\pi unx_0} du \tag{A.19}$$

which gives the discrete function $f_n = f(nx_0)$ as an integral sum of sinusoids.

Although this discrete Fourier transform is useful for many theoretical discussions, for practical purposes the following *finite Fourier transform* is actually calculated. Let the discrete function

$$f(0), f(x_0), f(2x_0), \ldots, f((N-1)x_0) \tag{A.20}$$

be N elements long and be written as $\{f_0, f_1, f_2, \ldots, f_{N-1}\}$. Then the finite Fourier transform is defined as

$$F_k = \frac{1}{N} \sum_{n=0}^{N-1} f_n \mathrm{e}^{-\mathrm{i}2\pi \frac{kn}{N}} \qquad k = 0, 1, 2, \ldots, N-1 \tag{A.21}$$

The F_k's are samples of the continuous function $F(u)$ of (A.18) for $u = \frac{k}{Nx_0}$ with $k = 0, 1, \ldots, N-1$, as can be proven by rewriting (A.21) with these u's. This implies that if (A.21) is used to compute the frequency domain representation of a discrete function, a sampling interval of x_0 in the x-domain corresponds to a sampling interval of $1/(Nx_0)$ in the frequency domain.

The *inverse finite Fourier transform* is

$$f_n = \sum_{k=0}^{N-1} F_k \mathrm{e}^{\mathrm{i}2\pi \frac{kn}{N}} \tag{A.22}$$

Both (A.21) and (A.22) define sequences that are periodically replicated. Since $\mathrm{e}^{\mathrm{i}(2\pi/N)Nm} = 1$ for all integers m we see that

$$\begin{aligned} F_{Nm+i} &= F_i \quad \text{for all} \quad m \in \boldsymbol{Z} \\ \text{and} \quad f_{Nm+i} &= f_i \quad \text{for all} \quad m \in \boldsymbol{Z} \end{aligned} \tag{A.23}$$

The four types of Fourier transforms are summarized in Table A.2 [536].

Table A.2: Types of Fourier transforms

		Continuous Space/Time	Discrete Space/Time
Continuous Frequency	Name:	Continuous Fourier Transform	Discrete Fourier Transform
	Forward:	$F(u) = \int_{-\infty}^{\infty} f(x)\mathrm{e}^{-\mathrm{i}2\pi ux}\,dx$	$F(u) = \sum_{n=-\infty}^{\infty} f_n \mathrm{e}^{-\mathrm{i}2\pi unx_0}$
	Inverse:	$f(x) = \frac{1}{2\pi}\int_{-\infty}^{\infty} F(u)\mathrm{e}^{\mathrm{i}2\pi ux}\,du$	$f_n = \frac{x_0}{2\pi}\int_{-\pi/x_0}^{\pi/x_0} F(u)\mathrm{e}^{\mathrm{i}2\pi unx_0}\,du$
	Periodicity:	none	$F(u) = F(u + m(2\pi/x_0))$
Discrete Frequency	Name:	Fourier Series	Finite Fourier Transform
	Forward:	$F_k = \frac{1}{X}\int_{X_1}^{X_2} f(x)\mathrm{e}^{-\mathrm{i}2\pi k\frac{x}{X}}\,dx$	$F_k = \frac{1}{N}\sum_{n=0}^{N-1} f_n \mathrm{e}^{-\mathrm{i}2\pi\frac{kn}{N}}$
	Inverse:	$f(x) = \sum_{k=-\infty}^{\infty} F_k \mathrm{e}^{\mathrm{i}2\pi k\frac{x}{X}}$	$f_n = \sum_{k=0}^{N-1} F_k \mathrm{e}^{\mathrm{i}2\pi\frac{kn}{N}}$
	Periodicity:	$f(x) = f(x + mX)$	$f_i = f_{i+Nm}$ and $F_i = F_{i+Nm}$

A.3 Interpretation of the Fourier Transform

The Fourier transform of a real or complex function $f(x)$ generally is a complex function $F(u)$, which sometimes is called the *frequency spectrum* or if the spatial nature has to be emphasized, the *spatial frequency spectrum.* Normally one is not interested in how the values are distributed between real and imaginary part but in the modulus

$$|F(u)| = \sqrt{\mathrm{Re}^2\{F(u)\} + \mathrm{Im}^2\{F(u)\}} \tag{A.24}$$

called the *amplitude spectrum* and the *phase spectrum* $\phi(u)$ or shortly called *phase* defined by

$$\phi(u) = \arctan \frac{\mathrm{Im}\{F(u)\}}{\mathrm{Re}\{F(u)\}} \tag{A.25}$$

The value at frequency zero, $F(0)$, is always real, it stands for the *average* or *dc component* of the function, namely in the case of the finite Fourier transform we have

$$F_0 = \frac{1}{N} \sum_{n=0}^{N-1} f_n \tag{A.26}$$

The value F_1 stands for the sinusoid of 1 cycle per sequence length. The value F_k represents k cycles per sequence length, as long as $k \leq N/2$. For higher values we must recognize the periodicity property $F_{-k} = F_{N-k}$, so that for $k > N/2$ the F_k represents $-(N-k)$ cycles per sequence length. But since an output where the negative axis information follows the positive axis information is somewhat unnatural to look at, the output should be rearranged. Normal looking finite Fourier transform outputs with the dc component at the center between the positive and negative frequencies can be produced by multiplying the sequence f_n to be transformed with $(-1)^n$

$$f_n' = f_n(-1)^n \tag{A.27}$$

prior to calculating the finite Fourier transform.

A.4 Properties of the Fourier-Transform

There are a number of properties of the Fourier transform which are frequently used when processing data in the frequency domain. The most important of them are presented in the following without proof.

Generally a complex function $f(x)$ of a single real variable x has a Fourier transform that is also a complex function $F(u)$ of a real variable u. But several restricted classes of functions possess special *symmetry properties* under Fourier transform:

- An *even function* $f_e(x)$, that is $f_e(-x) = f_e(x)$ for all x, has an even Fourier transform.
- An *odd function* $f_o(x)$, that is $f_o(-x) = -f_o(x)$ for all x, has an odd Fourier transform.

- The transform of a real, even function is real and even.
- The transform of a real, odd function is imaginary and odd.
- The transform of an imaginary, even function is imaginary and even.
- The transform of an imaginary, odd function is real and odd.
- The transform of a real function is Hermitean, meaning $F(u) = F^*(-u)$.
- The transform of an imaginary function is anti-Hermitean, meaning $F(u) = -F^*(-u)$.
- The transform of a function with even real part and odd imaginary part is real.
- The transform of a function with odd real part and even imaginary part is imaginary.

The *linearity theorem* or *addition theorem* states

$$\mathcal{F}\{af(x) + bg(x)\} = a\mathcal{F}\{f(x)\} + b\mathcal{F}\{g(x)\} \tag{A.28}$$

The *shift theorem* describes the effect that moving the origin of a function has upon its transform

$$\mathcal{F}\{f(x-a)\} = \mathrm{e}^{-\mathrm{i}2\pi au}F(u) \tag{A.29}$$

with $F(u) = \mathcal{F}\{f(x)\}$. This means that shifting a function does not alter the amplitude of the Fourier transform but only introduces a phase shift proportional to both frequency and the amount of shift.

In linear system analysis the *convolution* of two functions $f(x)$ and $g(x)$ is defined as

$$f \star g(x) = \int_{-\infty}^{\infty} f(x')g(x-x')dx' \tag{A.30}$$

The *convolution theorem* states that convolution in one domain corresponds to multiplication in the other domain, in detail

$$\begin{aligned} \mathcal{F}\{f \star g(x)\} &= F(u)G(u) \\ \mathcal{F}^{-1}\{F(u)G(u)\} &= f(x) \star g(x) \end{aligned} \tag{A.31}$$

The *similarity theorem* describes the effect that a change of abscissa scale has on the Fourier transform

$$\mathcal{F}\{f(ax)\} = \frac{1}{|a|}F\left(\frac{u}{a}\right) \tag{A.32}$$

This means that if a function is broadened in the time or spatial domain, $a < 1$, its transform is contracted in the frequency domain and the amplitude is increased. Contrary if the function f is narrowed by an $a > 1$, the transform is broadened and the amplitude is flattened.

Differentiation in the temporal or spatial domain corresponds to a multiplication with a linear factor in the frequency domain

$$\mathcal{F}\{\frac{d}{dx}f(x)\} = \mathrm{i}2\pi u F(u) \tag{A.33}$$

For functions which are nonzero only over a finite portion of their domain, their *energy* E can be defined by

$$E = \int_{-\infty}^{\infty} |f(x)|^2 dx \tag{A.34}$$

provided the integral exists. *Rayleigh's theorem* states that

$$\int_{-\infty}^{\infty} |f(x)|^2 dx = \int_{-\infty}^{\infty} |F(u)|^2 du \tag{A.35}$$

which means that the transform carries the same energy as the original function.

Especially for the treatment of *random functions* the *autocorrelation function* $R(x')$ is used. It is defined by

$$R(x') = f(x) \star f(-x) = \int_{-\infty}^{\infty} f(x) f(x + x') dx \tag{A.36}$$

and is always even and has a maximum at $x' = 0$. The *autocorrelation theorem* states that

$$\int_{-\infty}^{\infty} R(x') dx' = \left[\int_{-\infty}^{\infty} f(x) dx\right]^2 \tag{A.37}$$

Every function has a unique autocorrelation function but the converse is not true. The autocorrelation theorem is closely related to the *Wiener-Khinchine theorem* which states

$$\mathcal{F}\{R(x)\} = \mathcal{F}\{f(x) \star f(-x)\} = |F(u)|^2 \tag{A.38}$$

A.5 The Sampling Theorem and Data Truncation Effects

In the definitions of the discrete and the finite Fourier transforms a sequence of numbers $\{f_n\}$ was used to approximate a continuous function $f(x)$. The question arises, how finely must the data be sampled to accurately represent the original signal. The answer is given by the *sampling theorem*: If a signal $f(x)$ has a Fourier transform $F(u)$ such that

$$F(u) = 0 \quad \text{for} \quad u \geq \frac{u_N}{2} \tag{A.39}$$

then the samples of f must be measured at a rate greater than u_N in order to ensure an exact reconstruction of $f(x)$ from the samples. That means if Δx is the interval between consecutive samples, it must be $2\pi/\Delta x > u_N$. The frequency u_N is known as the *Nyquist rate* or *Nyquist frequency* and represents the minimum frequency at which the data can be sampled without introducing errors. Functions $f(x)$ fulfilling (A.39) are called *band-limited functions*.

To consider the consequences of sampling at below the Nyquist rate, the sampled version $f_s(x)$ of a function $f(x)$ is expressed by a multiplication of the original continuous signal $f(x)$ with the sampling function $h(x)$ given by

$$h(x) = \sum_{n=-\infty}^{\infty} \delta(x - n\Delta x) \tag{A.40}$$

Since $h(x)$ is periodic, its Fourier transform is computed via the Fourier series to

$$H(u) = \frac{2\pi}{\Delta x} \sum_{n=-\infty}^{\infty} \delta\left(u - \frac{2\pi n}{\Delta x}\right) \tag{A.41}$$

Converting the multiplication into a convolution in the frequency domain we get the transform

$$F_s(u) = \frac{2\pi}{\Delta x} \sum_{n=-\infty}^{\infty} F\left(u - \frac{2\pi n}{\Delta x}\right) \tag{A.42}$$

which is a sum of shifted functions $F(u)$. The formula of the Fourier transform of a function f sampled at $n\Delta x$ is

$$F_s(u) = \frac{2\pi}{\Delta x} \sum_{n=-\infty}^{\infty} f(n\Delta x) \mathrm{e}^{-\mathrm{i}2\pi u n\Delta x} \tag{A.43}$$

Fig. A.3 shows the results of over- and undersampling. Fig. A.3a gives the spectrum $F(u)$, Fig. A.3b displays the result after sampling at a rate faster than the Nyquist rate, Fig. A.3c is the spectrum after sampling at exactly the Nyquist rate, and Fig. A.3d shows the overlap resulting from sampling at less than the Nyquist rate. An inverse Fourier transform of the spectrum in Fig. A.3d would produce an erroneous signal, the error is known as *aliasing* or as *moire pattern* in two-dimensional image processing.

In many applications we have only a finite number of samples over a finite time or space while the signal extends beyond the limits of this interval. Nevertheless we assume that all the significant transitions of the signal occur in this base interval. Let the signal $f_n = f(x_n)$ be defined for all n. The true discrete Fourier transform of this signal is

$$F(u) = \sum_{n=-\infty}^{\infty} f_n \mathrm{e}^{-\mathrm{i}2\pi u n x_0} \tag{A.44}$$

Suppose to take only an N-element transform meaning that of all the $x_n = nx_0$ we will retain only those going from $-(N/2-1)x_0$ to $(N/2)x_0$. It is assumed that N is even. The discrete Fourier transform of the *truncated data* is

$$\begin{aligned} F'(u) &= \sum_{n=-(N/2-1)}^{N/2} f_n \mathrm{e}^{-\mathrm{i}2\pi u n x_0} \\ &= \sum_{n=-\infty}^{\infty} f_n I_N(n) \mathrm{e}^{-\mathrm{i}2\pi u n x_0} \end{aligned} \tag{A.45}$$

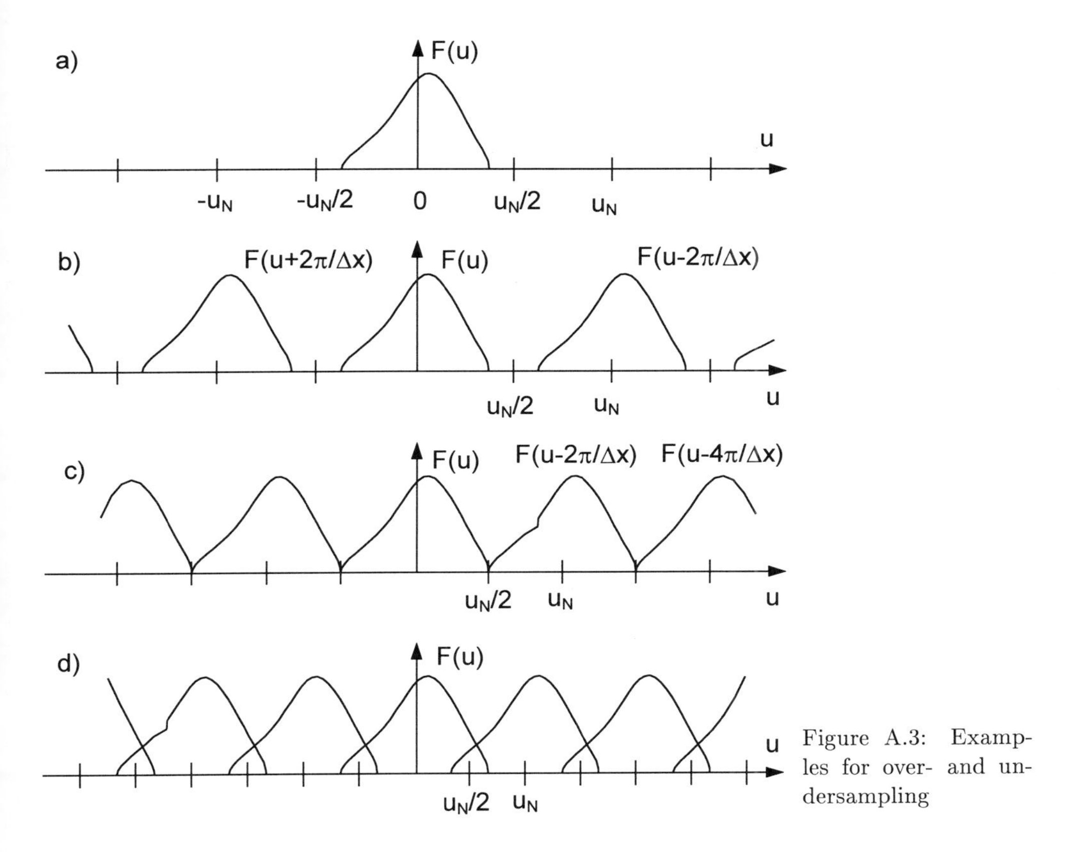

Figure A.3: Examples for over- and undersampling

where $I_N(n)$ is the function that is equal to 1 for n between $-(N/2-1)$ and $N/2$, and zero outside. The transform of $I_N(n)$ is

$$\begin{aligned} \mathcal{I}(u) &= \sum_{n=-(N/2-1)}^{N/2} \mathrm{e}^{-\mathrm{i}2\pi u n x_0} \\ &= \mathrm{e}^{-\mathrm{i}2\pi u \frac{x_0}{2}} \frac{\sin \frac{2\pi u N x_0}{2}}{\sin \frac{2\pi u x_0}{2}} \end{aligned} \tag{A.46}$$

By the convolution theorem we have

$$F'(u) = \frac{x_0}{2\pi} F(u) \star \mathcal{I}(u) \tag{A.47}$$

The function $\mathcal{I}(u)$ is displayed in Fig. A.4 and illustrates the nature of distortion introduced by *data truncation*.

Generally we can state that a function which vanishes outside a finite interval will yield a smooth transform with infinite extent: A function of bounded support cannot be band-limited and vice-versa.

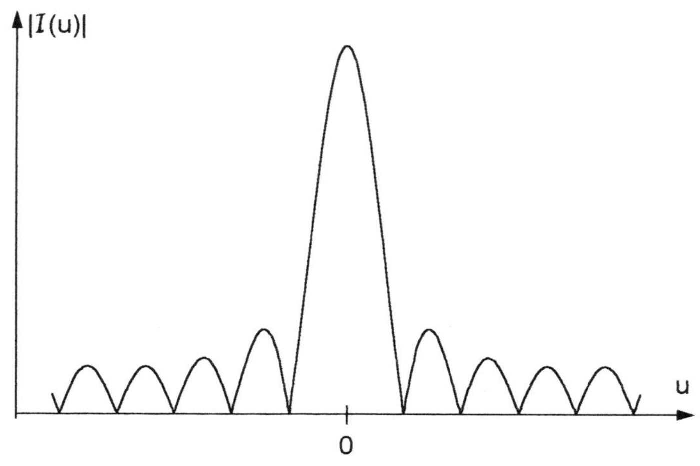

Figure A.4: Convolution function resulting from data truncation

Closely related to the truncation effect is the *leakage* effect. Consider a periodic continuous signal that is truncated to a finite interval. If the length of the interval is not an integer multiple of the period, a periodic continuation by putting the left end of the interval at the right end would produce discontinuities and would destroy the periodicity. The components of $f(t)$ not periodic over the interval appear to leak out into portions of the spectrum adjacent to the correct spatial value.

The truncation of a function to an interval can be represented by multiplication with a *rectangular window function* $w_r(x)$

$$w_r(x) = \begin{cases} 1 & : \text{ for } \quad x \in [x_1, x_2] \\ 0 & : \text{ elsewhere} \end{cases} \tag{A.48}$$

A method for reducing the error by leakage is the round off of the sharp corners of $w_r(x)$. Thus the function $f(x)$ is multiplied by a window $w(x)$ before calculating the finite Fourier transform. Often used is the so called *Hanning window* $w_H(x)$

$$w_H(x) = \frac{1}{2}\left(1 - \cos\frac{2\pi(x - x_1)}{x_2 - x_1}\right) \tag{A.49}$$

Its effect is shown in Fig. A.5. In Fig. A.5a we have a continuous signal and the corresponding amplitude spectrum. The leakage by the rectangular window is displayed in Fig. A.5b. In Fig. A.5c we recognize the significant reduction of leakage by the Hanning window. Other window functions are known, like the Hamming-, Bartlett-, cosine-taper-window, to name just a few.

The multiplication of a function $F(u)$ in the frequency domain with a window function $W(u)$ which is zero for $|u| > u_W$ also leads to consequences: The inverse transform of $[F \cdot W](u)$ is a band-limited approximation to $f(x)$, the inverse transform of the transform $F_s(u)$ of a sampled function $f(n\Delta x)$ multiplied with the window $[F_s \cdot W](u)$ gives an interpolated version of $f(n\Delta x)$. If the function $f(x)$ is band-limited and sampled with Δx fulfilling the sampling theorem, then a window $W(u)$ with $W(u) = 0$ for $|u| > 1/(2\Delta x)$

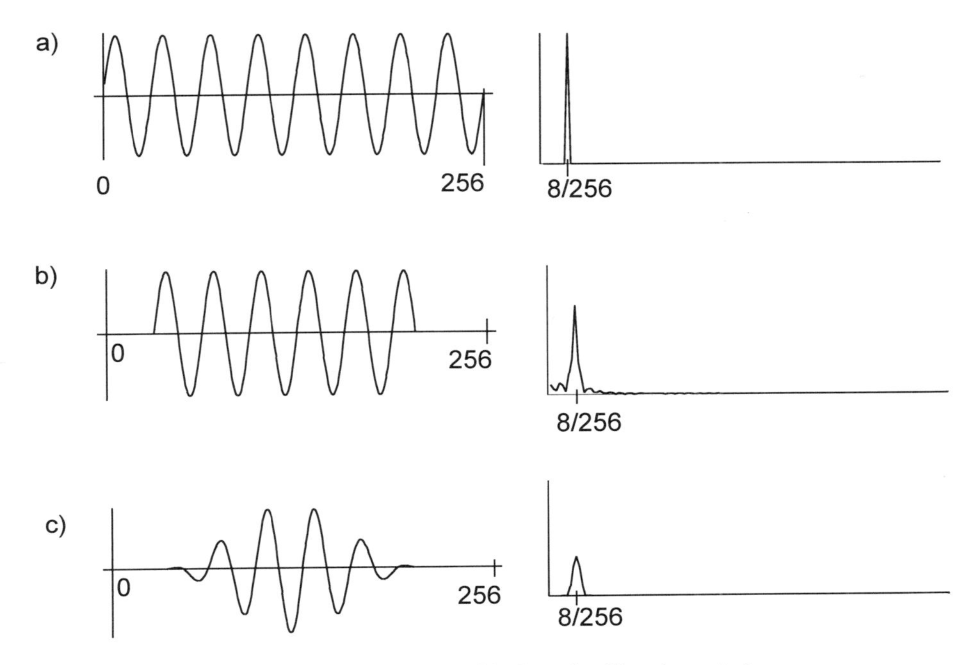

Figure A.5: Reduction of leakage by Hanning window

will cause no disturbing effects. Applying the same window to the transform of a function which is not band-limited will give rise to aliasing.

A.6 The Fast Fourier-Transform

The growing importance of the finite Fourier transform in practical applications of digital signal processing over the last decades is mainly based on the effective algorithm to compute the finite Fourier transform known as the *fast Fourier transform*, shortly *FFT*.

A simple measure of the amount of computations in (A.21) is the number of complex products. For the N elements of F_k there are N sums each with N products, so that N^2 products are needed to compute the whole set $\{F_k : k = 0, 1, \ldots, N-1\}$.

To simplify the notation, in the finite Fourier transform (A.21) let W_N represent the invariant part of the exponential term, i. e.

$$W_N = \mathrm{e}^{-\mathrm{i}\frac{2\pi}{N}} \tag{A.50}$$

so that (A.21) now becomes

$$F_k = \frac{1}{N} \sum_{n=0}^{N-1} f_n W_N^{kn} \qquad k = 0, 1, \ldots, N-1 \tag{A.51}$$

The cyclic character of the coefficients W_N^{kn} is illustrated in Fig. A.6 for $N = 8$. The

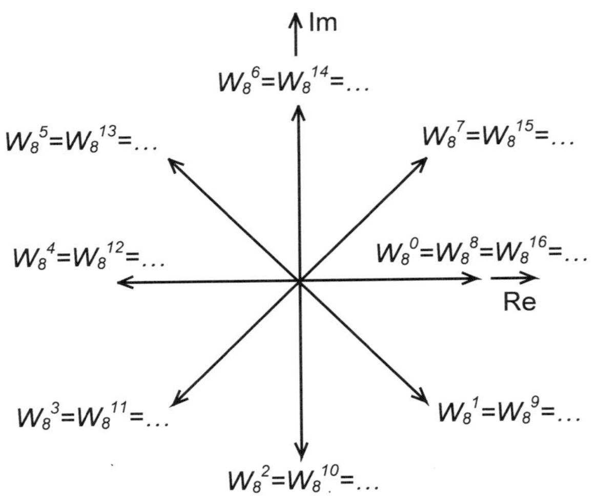

Figure A.6: Equivalence of different powers of W_N for $N = 8$

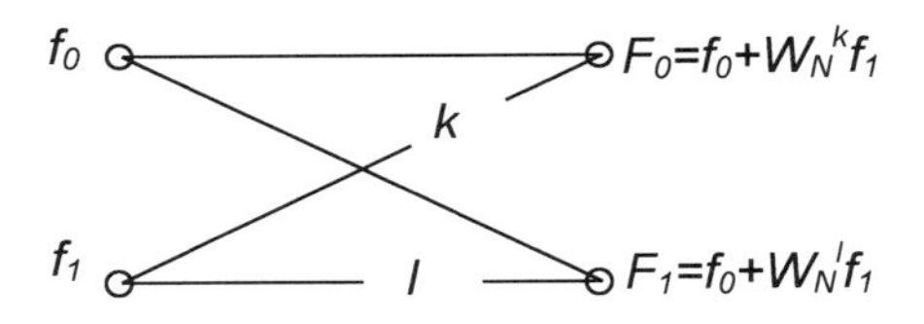

Figure A.7: Basic butterfly for FFT

redundancy shown in this representation, namely the equality of many different powers of W_N and half of all powers differing only by the sign, in the FFT algorithm is used in an intelligent way to save complex multiplications. Especially if N is a power of 2, the necessary computations are subdivided into basic transformations to the base $N = 2$. The corresponding signal-flow diagram for $N = 2$ often is called *butterfly*, Fig. A.7. One possible signal flow graph for calculation of an 8-point FFT is shown in Fig. A.8. A careful analysis of the *FFT algorithm* shows that now at most only $(N/2)\log_2 N$ complex products have to be calculated for an N-element Fourier transform, with N a power of 2. This number can be further reduced by recognizing that $W_N^0 = (1., 0.)$ in the algorithm. As an example take $N = 512$: Instead of $N^2 = 262,144$ complex multiplications now we need at most $(N/2)\log_2 N = 2,304$ complex multiplications. The advantage is obvious.

For practical implementations more savings are possible. The sines and cosines of W_N^{kn} should be taken from a look-up-table prepared once before the main computation. Nearly half of the computations can be saved when a real sequence is to be transformed by recognizing the vanishing imaginary parts and avoiding the multiplications with zero.

A short *FFT-subroutine* in FORTRAN, which may easily be translated into other programming languages is given here, Fig. A.9. It should help for first tests. More efficient programs should systematically avoid multiplications by zero and one, and should merge the multiplications by $\sqrt{2}$, or can be based on the Hartley transform. Detailed discussions of the FFT algorithms can be found in [537, 534].

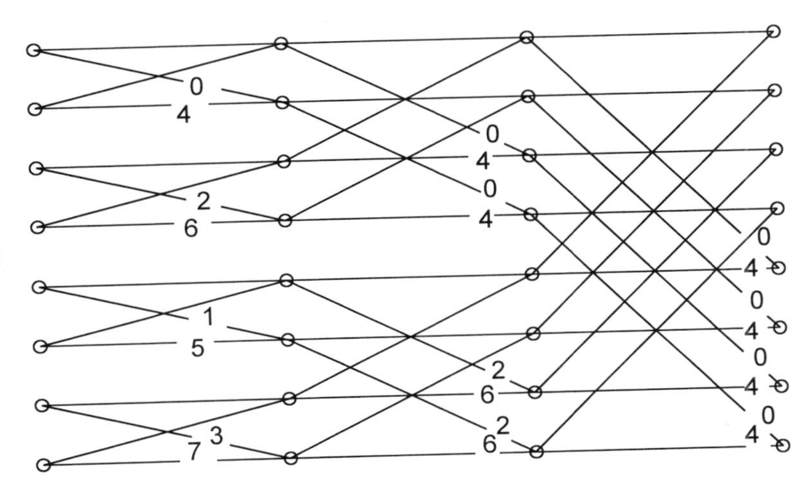

Figure A.8: Signal flow graph for FFT with $N = 8$

```
      SUBROUTINE FFT(FR,FI,K,IFLAG)
C - FAST FOURIER TRANSFORM
C - DATA IN FR (REAL) AND FI (IMAGINARY) ARRAYS
C - IF IFLAG=1 REPLACES (FR,FI) BY ITS DISCRETE FOURIER TRANSFORM
C - IF IFLAG=-1 BY ITS INVERSE DISCRETE TRANSFORM
C - NUMBER OF POINTS IS N=2**K; FR,FI DIMENSIONED IN MAIN PROGRAM
      DIMENSION FR(1),FI(1)
      N=2**K
      MR=0
      NM1=N-1
      DO 1 M=1,NM1
      L=N
 2    L=L/2
      IF(MR+L.GT.NM1) GOTO 2
      MR=MOD(MR,L)+L
      IF(MR.LE.M) GOTO 1
      TEMPR=FR(M+1)
      FR(M+1)=FR(MR+1)
      FR(MR+1)=TEMPR
      TEMPI=FI(M+1)
      FI(M+1)=FI(MR+1)
      FI(MR+1)=TEMPI
 1    CONTINUE
      L=1
 3    IF(L.GE.N) RETURN
      ISTEP=2*L
      EL=L
      DO 4 M=1,L
      A=IFLAG*3.1415926535*FLOAT(1-M)/EL
      WR=COS(A)
      WI=SIN(A)
      DO 4 I=M,N,ISTEP
      J=I+L
      TEMPR=WR*FR(J)-WI*FI(J)
      TEMPI=WR*FI(J)+WI*FR(J)
      FR(J)=FR(I)-TEMPR
      FI(J)=FI(I)-TEMPI
      FR(I)=FR(I)+TEMPR
      FI(I)=FI(I)+TEMPI
 4    CONTINUE
      L=ISTEP
      GOTO 3
      END
```

Figure A.9: FFT-program

A.7 Two-Dimensional Image Processing

For processing images like holographic interference patterns some signal processing concepts are extended to two dimensions in the following. A *picture* or *image* is nothing more than a two-dimensional real valued function $f(x, y)$ of two *spatial coordinates.* The values of this function can be interpreted as *gray-values,* so $f(x, y)$ gives the brightness distribution of, say, a black and white photograph.

Let L be an operation that maps an image f into another image $L[f]$. L is called *linear* if for all constants a, b and all images f, g

$$L[af + bg] = aL[f] + bL[g] \tag{A.52}$$

For the analysis of two-dimensional linear operations we need the concept of a *point source,* the two-dimensional equivalent to the delta impulse. Corresponding to the one-dimensional impulse response now we have the *point spread function* characterizing the operation L. Over the *two-dimensional rectangular function*

$$\text{rect}(x, y) = \begin{cases} 1 & : \text{ for } \quad |x| \leq \frac{1}{2} \quad \text{and} \quad |y| \leq \frac{1}{2} \\ 0 & : \text{ elsewhere} \end{cases} \tag{A.53}$$

and

$$\delta_n(x, y) = n^2 \text{rect}(nx, ny) \qquad n = 1, 2, \ldots \tag{A.54}$$

the point source δ is defined by

$$\delta(x, y) = \lim_{n \to \infty} \delta_n(x, y) \tag{A.55}$$

which has the properties

$$\int_{-\infty}^{\infty} \int_{-\infty}^{\infty} \delta(x, y) dx dy = 1 \tag{A.56}$$

and

$$\int_{-\infty}^{\infty} \int_{-\infty}^{\infty} f(x, y) \delta(x - a, y - b) dx dy = f(a, b) \tag{A.57}$$

Although the same notation is used for the impulse and the point source, no confusion should arise.

A linear operation L is called *shift invariant* if

$$L[f(x - a, y - b)] = L[f](x - a, y - b) \tag{A.58}$$

in other words, if the input f is shifted by (a, b) then the output $L[f]$ is also merely shifted by (a, b). Using the representation (A.57) and the linearity of L we get

$$\begin{aligned} L[f(x, y)] &= L\left[\int_{-\infty}^{\infty} \int_{-\infty}^{\infty} f(x', y') \delta(x' - x, y' - y) dx dy\right] \\ &= \int_{-\infty}^{\infty} \int_{-\infty}^{\infty} f(x', y') L[\delta(x' - x, y' - y)] dx dy \\ &= \int_{-\infty}^{\infty} \int_{-\infty}^{\infty} f(x', y') h_L(x' - x, y' - y) dx dy \end{aligned} \tag{A.59}$$

where the last equality uses the shift invariance of L, and h_L denotes the response of δ under L, the *point spread function.* The last expression in (A.59) defines the *two-dimensional convolution* $L[f] = f \star h_L = h_L \star f$. The *two-dimensional Fourier transform* $F(u,v)$ of the image $f(x,y)$ is defined by

$$F(u,v) = \int_{-\infty}^{\infty}\int_{-\infty}^{\infty} f(x,y)\mathrm{e}^{-\mathrm{i}2\pi(ux+vy)}dxdy \tag{A.60}$$

The two-dimensional Fourier transform can be considered as a one-dimensional transform with respect to, say, x followed by the one-dimensional transform with respect to y

$$F(u,v) = \int_{-\infty}^{\infty}\left[\int_{-\infty}^{\infty} f(x,y)\mathrm{e}^{-\mathrm{i}2\pi ux}dx\right]\mathrm{e}^{-\mathrm{i}2\pi vy}dy \tag{A.61}$$

This *separability* is used in the *two-dimensional FFT algorithm* where first all rows are replaced by their one-dimensional transforms and then all columns are transformed one-dimensionally or vice versa.

$$F(k,l) = \frac{1}{N^2}\sum_{m=0}^{N-1}\left[\sum_{n=0}^{N-1} f(n,m)W_N^{kn}\right]W_N^{lm} \tag{A.62}$$

with $k = 0,1,\ldots,N-1$, $l = 0,1,\ldots,N-1$ numbering the sample points in the spatial frequency domain.

The properties of the one-dimensional Fourier transform translate to two dimensions in a natural way, some additional properties due to the two dimensions come along. Some of these properties are summarized in Table A.3. Here $f(x,y)$, $f_1(x,y)$, $f_2(x,y)$ are functions in the spatial domain and $F(u,v)$, $F_1(u,v)$, $F_2(u,v)$ are the corresponding functions in the spatial frequency domain. a and b are scalar numbers. (r,Θ) are polar coordinates in the spatial, (p,ϕ) polar coordinates in the spatial frequency domain. α is an angular coordinate, x_0, y_0, u_0, v_0 are fixed spatial coordinates and spatial frequencies, resp. The transform $F(p)$ of the rotationally symmetric function $f(r)$ in line 7 of the table is called *Hankel transform.* It contains the zero-order *Bessel function* of the first kind J_0.

When a two-dimensional FFT of $f(x,y)$ is computed as described above, the zero peak at the spatial frequency $(0,0)$ will not occur at the center of the array, as one is used from the Fraunhofer diffraction patterns, but in the upper leftmost corner. A trick to force the frequency domain origin to approximately the center of the array (a precise center does not exist if N is an even number) is the multiplication of the data with $(-1)^{m+n}$ before the transform

$$f'(m,n) = f(m,n)(-1)^{m+n} \tag{A.63}$$

The *two-dimensional sampling theorem* states the following [533]: A function $f(x,y)$ whose Fourier transform $F(u,v)$ vanishes over all but a bounded region in the spatial frequency domain, can be reproduced everywhere from its values taken over a lattice of points $(m(\Delta x_1,\Delta y_1) + n(\Delta x_2,\Delta y_2))$, $m,n = 0,\pm1,\pm2,\ldots$ in the spatial domain provided the vectors $(\Delta x_1,\Delta y_1)$ and $(\Delta x_2,\Delta y_2)$ are small enough to ensure nonoverlapping of the spectrum $F(u,v)$ with its images on a periodic lattice of points $(l(\Delta u_1,\Delta v_1) + k(\Delta u_2,\Delta v_2))$,

Table A.3: Properties of the two-dimensional Fourier transform

Name	Function in the Spatial Domain	Transformed Function in the Spatial Frequency Domain
Linearity	$af_1(x,y)+bf_2(x,y)$	$aF_1(u,v)+bF_2(u,v)$
Scaling	$f(ax,by)$	$\frac{1}{\lvert ab\rvert}F\left(\frac{u}{a},\frac{v}{b}\right)$
Shifting	$f(x-x_0,y-y_0)$ $e^{i2\pi(u_0x+v_0y)}f(x,y)$	$e^{-i2\pi(ux_0+vy_0)}F(u,v)$ $F(u-u_0,v-v_0)$
Differentiation	$\left(\frac{\partial}{\partial x}\right)^m\left(\frac{\partial}{\partial y}\right)^n f(x,y)$	$(i2\pi u)^m(i2\pi v)^n F(u,v)$
Laplacian	$\nabla^2 f(x,y)=\left(\frac{\partial^2}{\partial x^2}+\frac{\partial^2}{\partial y^2}\right)f(x,y)$	$-4\pi^2(u^2+v^2)F(u,v)$
Rotation	$f(r,\Theta+\alpha)$	$F(p,\phi+\alpha)$
Rotational Symmetry	$f(r,\Theta)=f(r)$	$F(p,\phi)=F(p)$ $=2\pi\int_0^\infty rf(r)J_0(2\pi rp)dr$
Convolution	$f_1(x,y)\star f_2(x,y)$ $f_1(x,y)f_2(x,y)$	$F_1(u,v)F_2(u,v)$ $F_1(u,v)\star F_2(u,v)$
Separability	$f(x,y)=f_1(x)f_2(y)$	$F(u,v)=F_1(u)F_2(v)$
Parseval's Theorem	$\int_{-\infty}^{\infty}\int_{-\infty}^{\infty}f_1(x,y)f_2^*(x,y)dxdy=\int_{-\infty}^{\infty}\int_{-\infty}^{\infty}F_1(u,v)F_2^*(u,v)dudy$	
Conservation of Energy	$\int_{-\infty}^{\infty}\int_{-\infty}^{\infty}\lvert f(x,y)\rvert^2dxdy=\int_{-\infty}^{\infty}\int_{-\infty}^{\infty}\lvert F(u,v)\rvert^2dudy$	

$k,l=0,\pm1,\pm2,\ldots$ in the spatial frequency domain, where the vectors $(\Delta u_i,\Delta v_i)$, $i=1,2$, depend from the $(\Delta x_i,\Delta y_i)$, $i=1,2$, by

$$\Delta x_i\Delta u_j+\Delta y_i\Delta v_j=\begin{cases}0 & : \quad i\neq j\\ 1 & : \quad i=j\end{cases} \tag{A.64}$$

The definition of the sampling lattices over the two vectors reflects the fact that it is not necessary to sample in a rectangular grid. This is illustrated in Fig. A.10. Fig. A.10a displays the sampling points in the spatial domain, Fig. A.10b a boundary of the transform $F(u,v)$ of a band-limited two-dimensional function, and Fig. A.10c the nonoverlapping copies of $F(u,v)$.

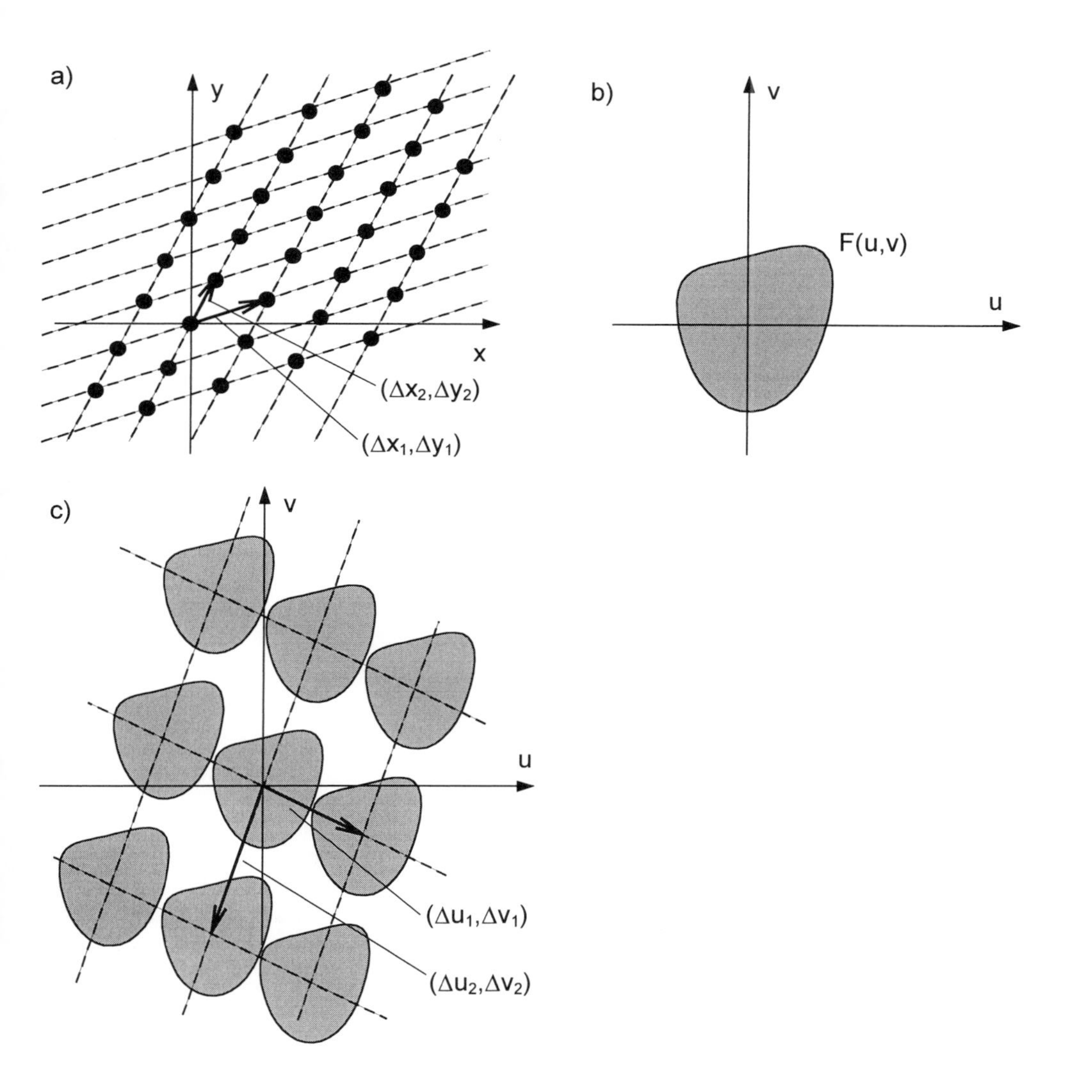

Figure A.10: Two-dimensional sampling theorem

Appendix B

Computer Aided Tomography

In holographic interferometry of refractive index fields the interference phase is given by the integral over the refractive index along the illuminating ray passing the measurement volume, (3.22). Therefore, except for very simple problems, we have to reconstruct the distribution of the refractive index from a sufficient number of projections, a problem first addressed by Radon [538]. Since the first use of this approach in medical diagnosis using X-rays, this so called field of *computer aided tomography* has emerged rapidly: signals from a diversity of sources from the whole electromagnetic spectrum are used [539], and the theoretical background of the computer evaluation algorithms has been refined.

Here only a very basic introduction to the main approaches to computer aided tomography should be given to constitute a background for the evaluation techniques used in holographic interferometric measurements at transparent refracting objects, see Sec. 5.8.

B.1 Mathematical Preliminaries

The measurement of the three-dimensional distribution of a physical quantity like the x-ray attenuation in human tissue or the *refractive index* in a transparent medium is simplified by the treatment of the measurement volume cut into two-dimensional plane slices. After evaluation many of such slices are stacked to build the three-dimensional result. So here we only have to consider a two-dimensional distribution of a physical quantity $f(x, y)$ in a single plane. Without loss of generality we can assume, that $f(x, y)$ is spatially bounded. Then by proper scaling and shifting we can ensure $f(x, y) = 0$ outside the unit circle Ω in the Cartesian coordinate system, Fig. B.1.

For the description of line integrals and projections the use of *polar coordinates* (t, ϕ) is advantageous, Fig. B.1.

$$f(x, y) = f(t\,\cos\phi, t\,\sin\phi) \tag{B.1}$$

A line in the plane now can be described by the two parameters s and θ: s is the signed distance of the line to the origin of the coordinate system and θ is the angle between the line and the y-axis, Fig. B.1. The projections of $f(x, y)$ along lines are called the *Radon*

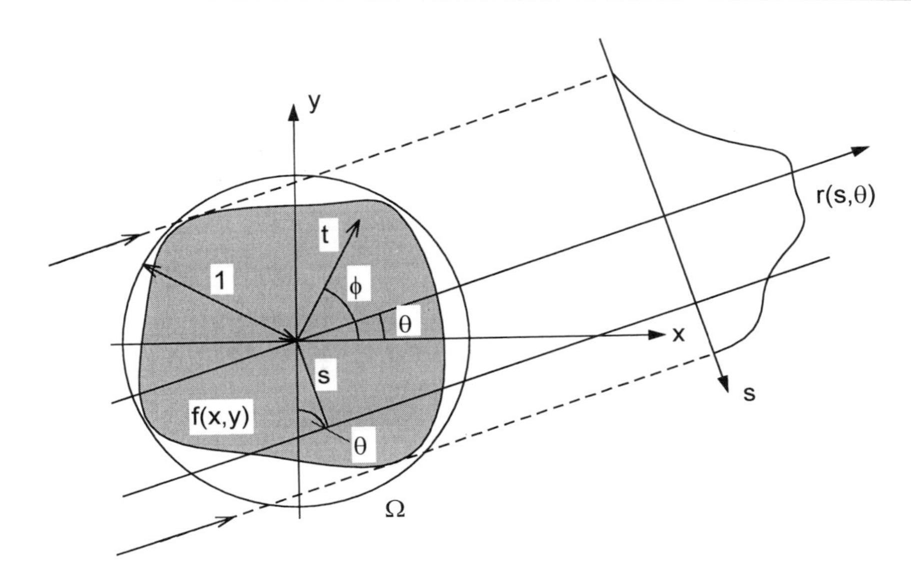

Figure B.1: Projection through a refractive index field

transforms $r(s, \theta)$, defined as

$$r(s, \theta) = \int_{-T}^{T} f(s \cos\theta - t \sin\theta, s \sin\theta + t \cos\theta) dt \tag{B.2}$$

or employing the delta function

$$r(s, \theta) = \int_{-\infty}^{\infty} \int_{-\infty}^{\infty} f(x, y) \delta(x \cos\theta + y \sin\theta - s) dx\, dy \tag{B.3}$$

The T of (B.2) can be set to $T(s) = (1 - s^2)^{1/2}$ since by assumption $f(s, \theta) = 0$ for $|s| > 1$, which is outside the unit-circle.

(s, θ) and $(-s, \theta + \pi)$ represent the same line in the plane, therefore this is true for the projections

$$r(s, \theta) = r(-s, \theta + \pi) \tag{B.4}$$

In the sequel some reconstruction methods will be introduced. This outline will be restricted on parallel projections, all rays establishing the projections are parallel. The case of fan beam projection is refered to the literature [536].

What is measured in practical experiments are the estimated projections $r(s, \theta)$ for discrete values of s and θ. From these projections the distribution $f(x, y)$ has to be reconstructed, which should be a two-dimensional array of numbers, each representing the physical quantity to be measured in an elementary cell.

Algorithms become more easy, if r is uniformly sampled in s and θ. Therefore assume a set of projections in N angular directions separated by $\Delta\theta$, each of these consisting of M equidistant beams separated by Δs, Fig. B.2.

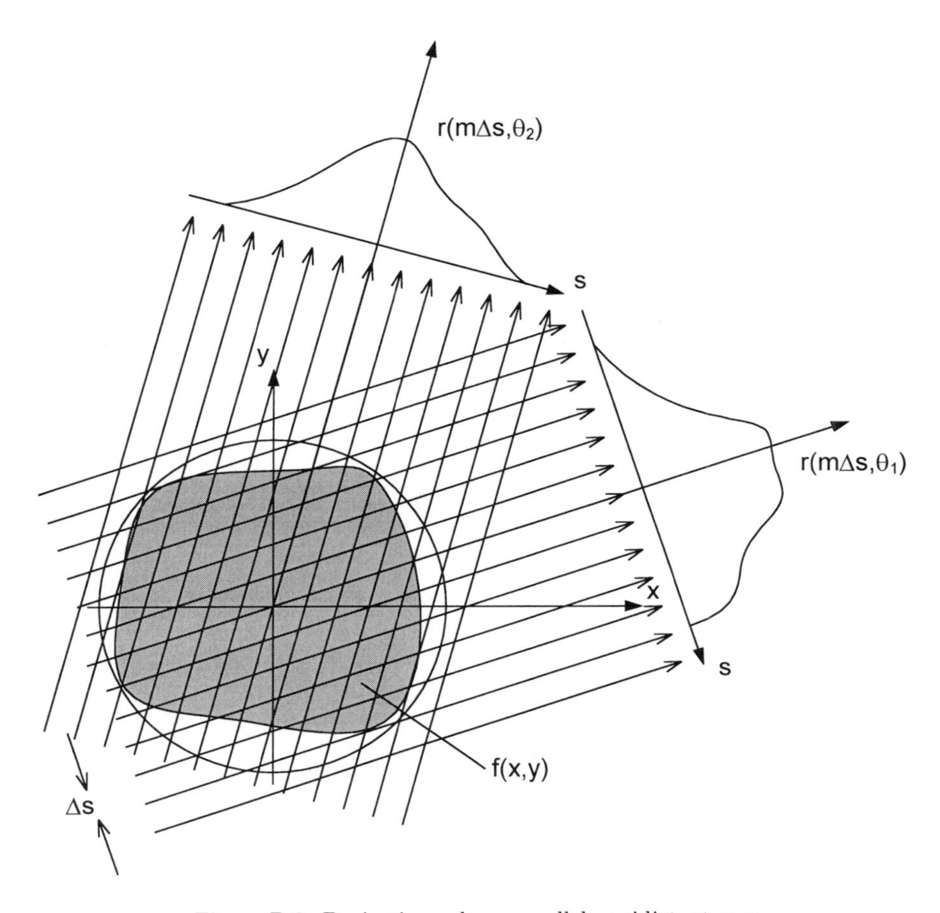

Figure B.2: Projections along parallel equidistant rays

B.2 The Generalized Projection Theorem

Let $f(x, y)$ be a function as in Sec. B.1 and $r(s, \theta)$ its Radon-transform. Let further $w(s)$ be any function of a single variable such that the following integrals exist. Then for all angles θ we have

$$\int_{-1}^{1} r(s,\theta)w(s)ds = \iint_{\Omega} f(x,y)w(x\,\cos\theta + y\,\sin\theta)dx\,dy \tag{B.5}$$

This theorem is proved by replacing $r(s, \theta)$ by its definition (B.2) and a change of the integration variables from the rotating coordinates (s, t) to fixed coordinates (x, y):

$$\begin{aligned} \int_{-1}^{1} r(s,\theta)w(s)ds &= \int_{-1}^{1}\int_{-T}^{T} f(s\,\cos\theta - t\,\sin\theta, s\,\sin\theta + t\,\cos\theta)dt\,w(s)\,ds \\ &= \iint_{\Omega} f(x,y)w(x\,\cos\theta + y\,\sin\theta)dx\,dy \end{aligned} \tag{B.6}$$

The change of variables is $x = s\cos\theta - t\sin\theta$, $y = s\sin\theta + t\cos\theta$, or conversely $s = x\cos\theta + y\sin\theta$, $t = y\cos\theta - x\sin\theta$, and $ds\,dt = dx\,dy$ since the Jacobian is 1.

This *generalized projection theorem* states the equivalence of an operation on the projection r with a related operation on the object f itself.

By taking $w(s) = \exp(-2\pi i\rho s)$ we obtain the *projection theorem for Fourier transforms*, also called *Fourier slice theorem* [536]:

$$R(\rho, \theta) = F(\rho\cos\theta, \rho\sin\theta) \tag{B.7}$$

with $R(\rho,\theta) = \mathcal{F}\{r(s,\theta)\}$ and $F(u,v) = \mathcal{F}\{f\}(u,v)$ This theorem states that the Fourier transform of a parallel projection of $f(x,y)$, taken at a fixed angle θ, gives a one-dimensional slice of the Fourier transform $F(u,v)$, subtending an angle θ with the u-axis. In Fig. B.3 the Fourier transform of the projection $r(s,\theta)$ gives the values of $F(u,v)$ along

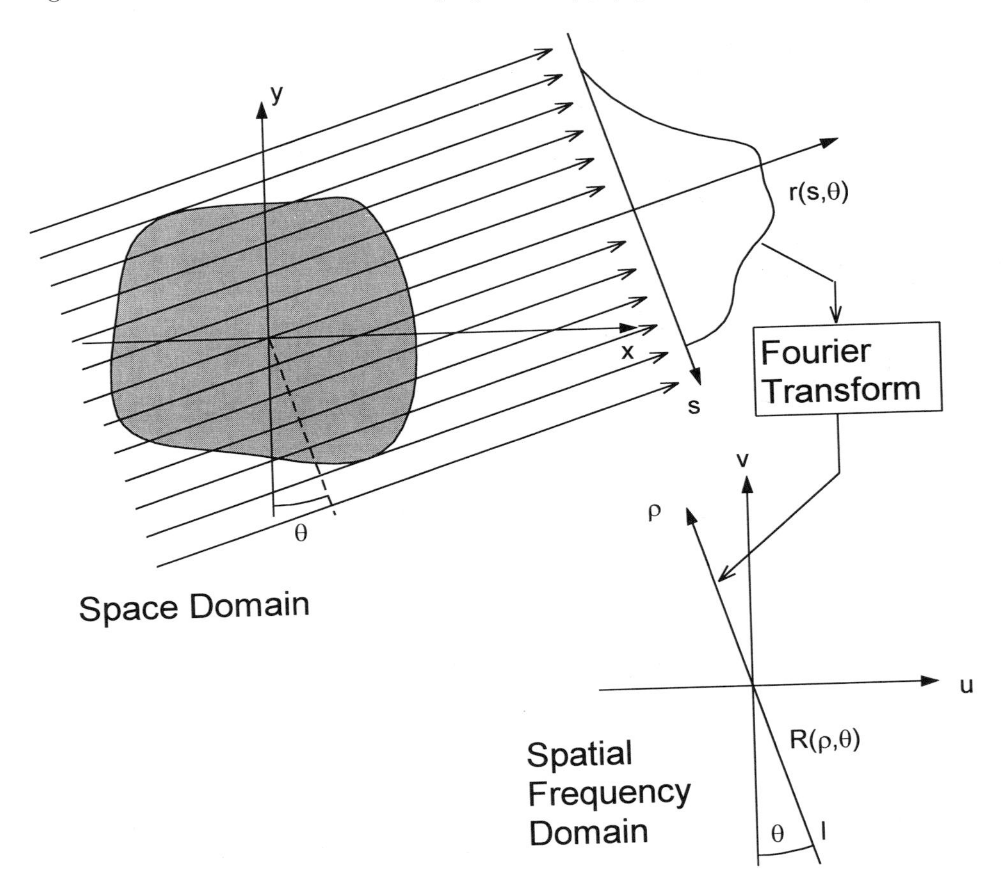

Figure B.3: Projection theorem for Fourier transforms

line l. The angular coordinate in the spatial frequency domain, expressed by Θ, here coincides with θ.

A direct approach to reconstruction based on the projection theorem for Fourier transforms would consist of collecting enough slices in the frequency domain, which are produced by transforming the projections along a sufficient number of angles, and to compute the inverse transform of the compiled two-dimensional frequency spectrum. But the inverse transform requires rectangularly sampled data, while the slices yield samples along radial lines, Fig. B.4. To obtain data in a square grid, a sort of nearest neighbor, linear,

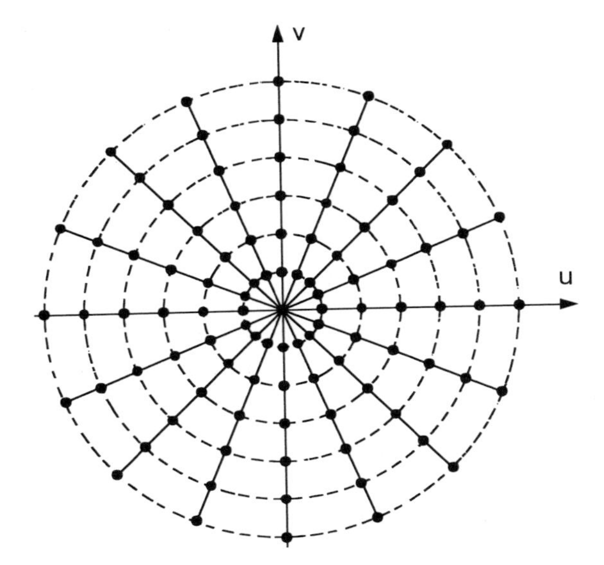

Figure B.4: Slices in the frequency domain

or bilinear interpolation must be performed. Since the data points become sparser as they are farther away from the center, Fig. B.4, the interpolation error increases. That means high frequency components of the reconstructed image are more likely to be degraded than are low frequency items. Altogether this approach has only theoretical value, in practice the filtered backprojection method, which also is based on the projection theorem for Fourier transforms, is the one most frequently applied.

B.3 Reconstruction by Filtered Backprojection

The function $f(x, y)$ to be reconstructed can be written as the inverse Fourier transform of its transform $F(u, v)$, which expressed in polar coordinates in the frequency domain is

$$\begin{aligned} f(x,y) &= \int_{-\infty}^{\infty}\int_{-\infty}^{\infty} F(u,v)\, \mathrm{e}^{\mathrm{i}2\pi(ux+vy)} du\, dv \\ &= \int_{0}^{2\pi}\int_{0}^{\infty} F(\rho,\Theta)\, \mathrm{e}^{\mathrm{i}2\pi(\rho x\, \cos\Theta + \rho y\, \sin\Theta)} \rho\, d\rho\, d\Theta \end{aligned} \tag{B.8}$$

with $u = \rho\, \cos\Theta$, $v = \rho\, \sin\Theta$, and $du\, dv = \rho\, d\rho\, d\Theta$. The integral is split into two and employing the property $F(\rho, \Theta + \pi) = F(-\rho, \Theta)$ we get

$$\begin{aligned} f(x,y) &= \int_0^\pi \int_0^\infty F(\rho,\Theta)\,\mathrm{e}^{\mathrm{i}2\pi\rho(x\,\cos\Theta + y\,\sin\Theta)}\rho\,d\rho\,d\Theta \\ &+ \int_0^\pi \int_0^\infty F(\rho,\Theta+\pi)\,\mathrm{e}^{\mathrm{i}2\pi\rho(x\,\cos(\Theta+\pi) + y\,\sin(\Theta+\pi))}\rho\,d\rho\,d\Theta \\ &= \int_0^\pi \left\{\int_{-\infty}^\infty F(\rho,\Theta)|\rho|\,\mathrm{e}^{\mathrm{i}2\pi\rho t}d\rho\right\}\,d\Theta \end{aligned} \tag{B.9}$$

with the abbreviation $t = x\,\cos\Theta + y\,\sin\Theta$.

The Fourier transform for each angle Θ is the corresponding transformed projection, so we get

$$f(x,y) = \int_0^\pi \left\{\int_{-\infty}^\infty R(\rho,\Theta)|\rho|\,\mathrm{e}^{\mathrm{i}2\pi\rho t}d\rho\right\}\,d\Theta \tag{B.10}$$

If we introduce the quantity

$$q_\Theta(t) = \int_{-\infty}^\infty R(\rho,\Theta)|\rho|\,\mathrm{e}^{\mathrm{i}2\pi\rho t}d\rho \tag{B.11}$$

(B.10) reads

$$f(x,y) = \int_0^\pi q_\Theta(x\,\cos\Theta + y\,\sin\Theta)\,d\Theta \tag{B.12}$$

These last two equations (B.11) and (B.12) are the heart of the *reconstruction by filtered backprojection*. (B.11) represents a filtering operation applied to the projection $R(\rho,\Theta)$, where the frequency response of the filter is given by $|\rho|$, so we name $q_\Theta(t)$ the *filtered projection*. The filtered projections $q_\Theta(t)$ for different angles Θ then are backprojected in the spatial domain to form an estimate of $f(x,y)$, which is the content of (B.12). This can be imagined with the help of Fig. B.5. For a given angle Θ_i the filtered projection contributes its value $q_{\Theta_i}(t)$ to all points (x,y) in the image plane, for which $t = x\,\cos\Theta_i + y\,\sin\Theta_i$. Thus the filtered projection for t is smeared back over all points along the projection line l of Fig. B.5. Since here we assume a circular area of interest in the image plane, $q_{\Theta_i}(t)$ is smeared over all points between P_1 and P_2. Altogether this is done for all the $q_{\Theta_i}(t)$ over all t as well as for all angles Θ_i.

B.4 Practical Implementation of Filtered Backprojection

In practice the energy contained in the spectra above a certain frequency is negligible, that means for practical purposes the projection may be considered as bandlimited. If ρ_{max} is higher than the highest significant frequency component in each projection, the projections can be sampled error-free at intervals of $\Delta t = 1/(2\rho_{max})$. We assume further that the projection data are zero for large $|t|$, then the M samples of a projection are

$$r(m\Delta t,\Theta) \qquad \text{with} \qquad m = -M/2,\ldots,+M/2-1 \tag{B.13}$$

Now the FFT algorithm can be used to calculate the Fourier transforms $R(\rho,\Theta)$ of the projections

$$R(\rho,\Theta) \approx R\left(m\frac{2\rho_{max}}{M},\Theta\right) = \frac{1}{2\rho_{max}}\sum_{k=-M/2}^{M/2-1} r\left(\frac{k}{2\rho_{max}},\Theta\right)\mathrm{e}^{-\mathrm{i}2\pi\frac{mk}{M}} \tag{B.14}$$

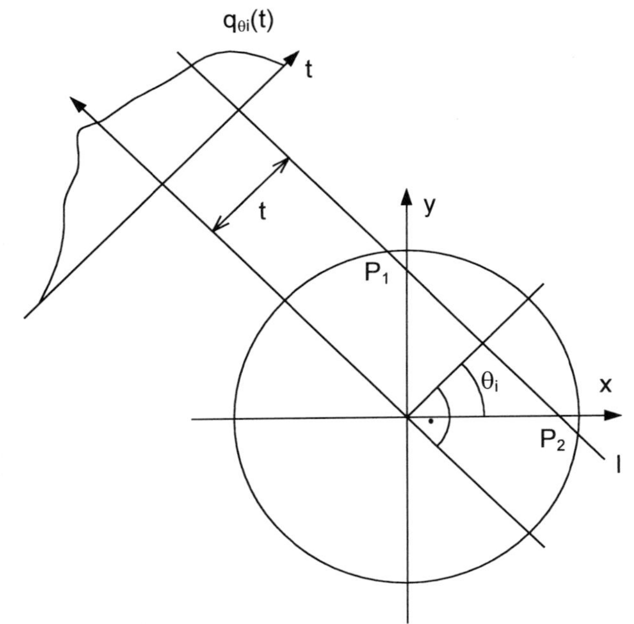

Figure B.5: Backprojection of the filtered projection

This projection now has to be filtered according to (B.11)

$$q_\Theta(t) \approx \frac{2\rho_{max}}{M} \sum_{m=-M/2}^{M/2-1} R\left(m\frac{2\rho_{max}}{M}, \Theta\right) \left|m\frac{2\rho_{max}}{M}\right| \mathrm{e}^{\mathrm{i}2\pi(2\rho_{max}/M)t} \tag{B.15}$$

The multiplication in the frequency domain before taking the inverse transform is equivalent to a convolution in the spatial, the t-domain

$$q_\Theta(t) = \int r(s,\theta)h(t-s)\,ds \tag{B.16}$$

where $h(t)$ is the inverse Fourier transform of the function $|\rho|$ multiplied with a window function in the frequency domain. A simple window function $B(\rho)$ only reflects the bandlimitness, so that the convolution kernel is the inverse transform of

$$H(\rho) = |\rho|B(\rho) \tag{B.17}$$

with $B(\rho) = 1$ for $|\rho| < \rho_{max}$ and zero otherwise. The impulse response of $H(\rho)$ is

$$\begin{aligned} h(t) &= \int_{-\infty}^{\infty} H(\rho)\mathrm{e}^{\mathrm{i}2\pi\rho t}d\rho \\ &= \frac{1}{2\Delta t}\frac{\sin 2\pi t/(2\Delta t)}{2\pi t/(2\Delta t)} - \frac{1}{4(\Delta t)^2}\frac{\sin \pi t/(2\Delta t)}{\pi t/(2\Delta t)} \end{aligned} \tag{B.18}$$

The projection data are measured with the sampling interval Δt, consequently we only need $h(t)$ at the sampled points. These values are

$$h(m\Delta t) = \begin{cases} \frac{1}{4(\Delta t)^2} & n = 0 \\ 0 & n \quad \text{even} \\ -\frac{1}{m^2\pi^2(\Delta t)^2} & n \quad \text{odd} \end{cases} \tag{B.19}$$

This is the well known *Ramachandran-Lakshminarayanan-kernel.* Another frequently used kernel is the *Shepp-Logan-kernel* [540]

$$h(m\Delta t) = \frac{-2}{\pi^2(\Delta t)^2(4m^2 - 1)} \tag{B.20}$$

Unfortunately the values of $x \cos\theta_i + y \sin\theta_i$ not always correspond to the sampled points $m\Delta t$. Therefore the filtered projection $q_\Theta(t)$ must be interpolated between the sampling points. Linear interpolation in most applications is adequate.

Altogether a practical implementation of the filtered backprojection method consists of the following steps:

- Choice of parameters: These are the number M of sampling points in each projection, the number N of angular projections, the sampling interval Δt, the angular separation $\Delta\theta = \pi/N$, the cutoff frequency $\rho_{max} = 1/(2\Delta t)$, the number of pixels $K \times L$ in the reconstructed image, the pixel distances Δx and Δy in the reconstructed image, the interpolation procedure (e. g. piecewise linear), the type of convolution kernel (e. g. Shepp-Logan).
- Projection data input: These are N vectors each having M components.
- Convolution of each projection with the kernel: This is performed in the spatial domain or by multiplication in the frequency domain.
- Interpolation of the filtered projections
- Backprojection
- Display of results.

A very simple example is shown in Fig. B.6. The function $f(x, y)$ to be reconstructed from four projections in the directions $\theta = 0^o$, 45°, 90°, and 135° is, Fig. B.6a

$$f(x,y) = \begin{cases} 1 & 0 \le x \le .25 \quad \text{and} \quad 0 \le y \le .25 \\ .5 & .25 \le x \le .5 \quad \text{and} \quad 0 \le y \le .5 \\ .5 & .25 \le y \le .5 \quad \text{and} \quad 0 \le x \le .25 \\ 0 & \text{elsewhere} \end{cases} \tag{B.21}$$

The four projections are shown in Fig. B.6c to f, the Shepp-Logan-kernel to be applied to the projections in Fig. B.6b. The filtered and interpolated projections are given in Fig. B.6g to j. The resulting $f(x, y)$ after backprojection is displayed in pseudo 3D in Fig. B.6k and in gray-values in Fig. B.6l. It must be admitted that this example is far from practice because four projections are never sufficient.

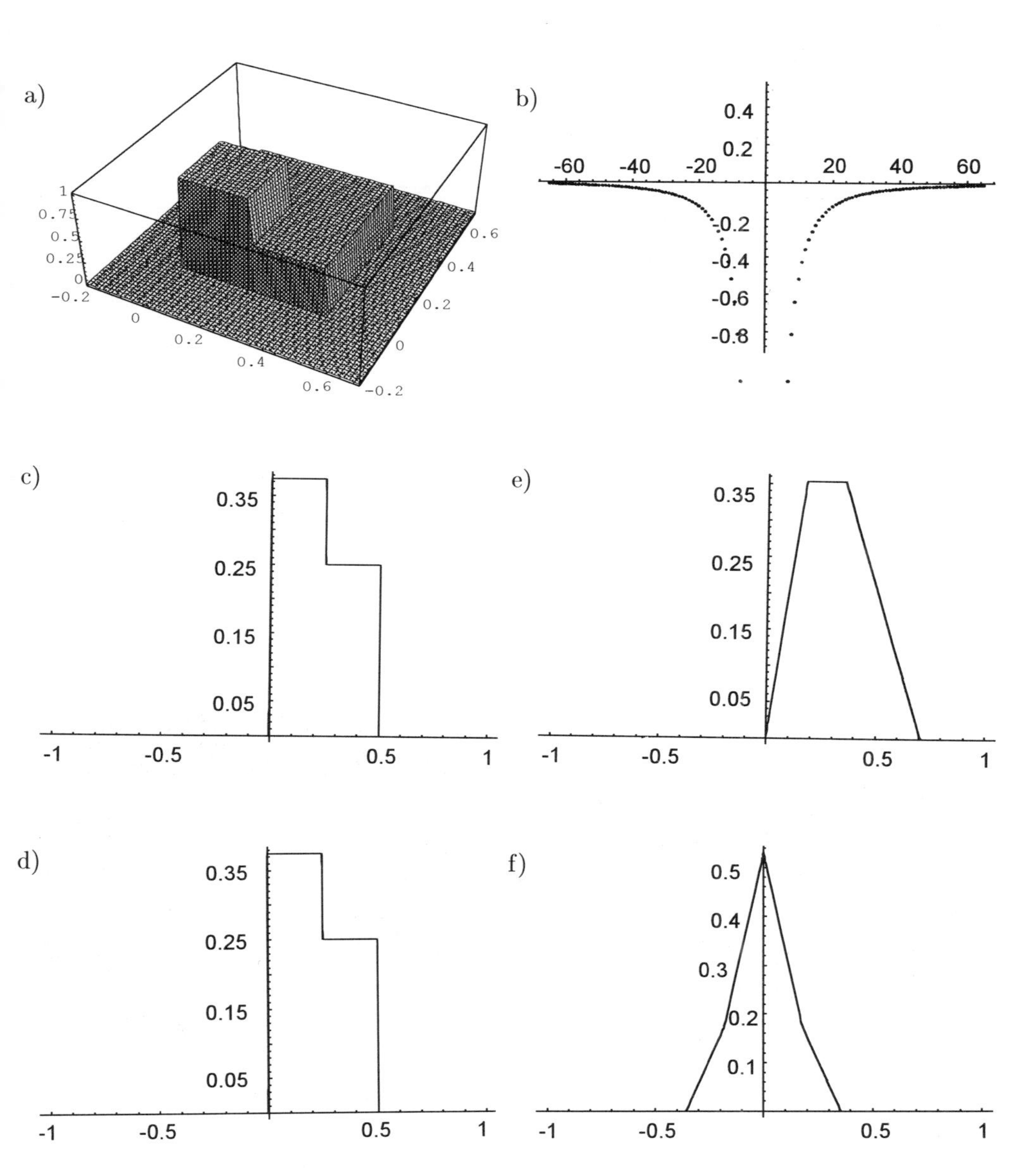

Figure B.6: Numerical example for reconstruction by filtered backprojection, (a) 2D test function $f(x, y)$, (b) Shepp-Logan-convolution kernel, (c - f) Projections of $f(x, y)$, (c) $\theta = 0^o$, (d) $\theta = 45^o$, (e) $\theta = 90^o$, (f) $\theta = 135^o$

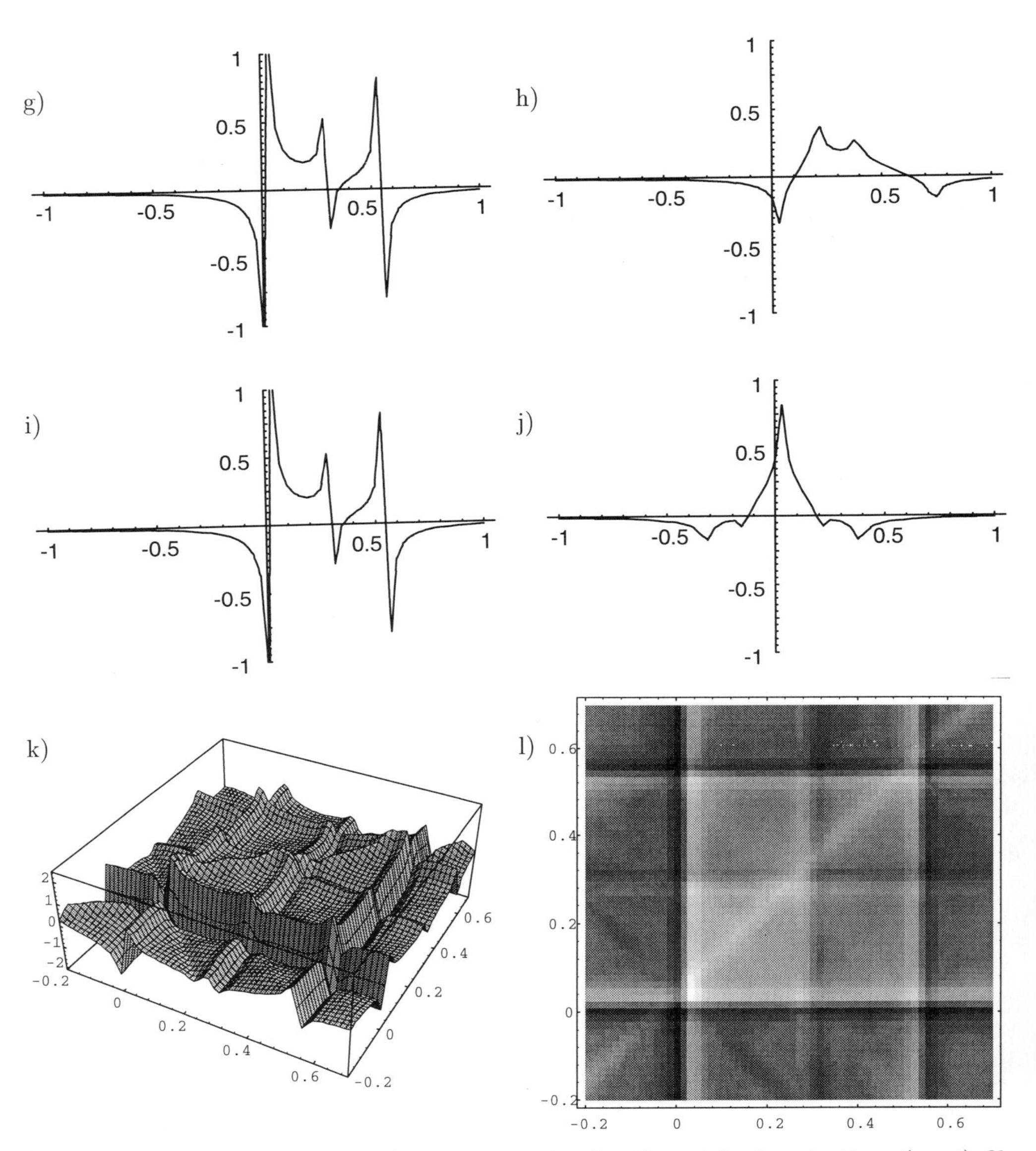

Figure B.6 Numerical example for reconstruction by filtered backprojection, (g - j) filtered and interpolated projections, (k, l) reconstructed function, (k) pseudo 3D display, (l) gray-scale display

B.5 Algebraic Reconstruction Techniques

The *algebraic reconstruction techniques*, also known as *series expansion reconstruction* methods principally differ from the transformation based methods as the filtered back-projection reconstruction. While in the transform methods the problem is treated as a continuous one until it is discretized for computational implementation, the algebraic reconstructions are discretized from the beginning [541, 542, 543, 544]. The following short outline again only deals in two dimensions.

The interesting area is partitioned into a Cartesian grid of square cells, the pixels, numbered consecutively from 1 to N, Fig. B.7. It is assumed that the function $f(x, y)$ to

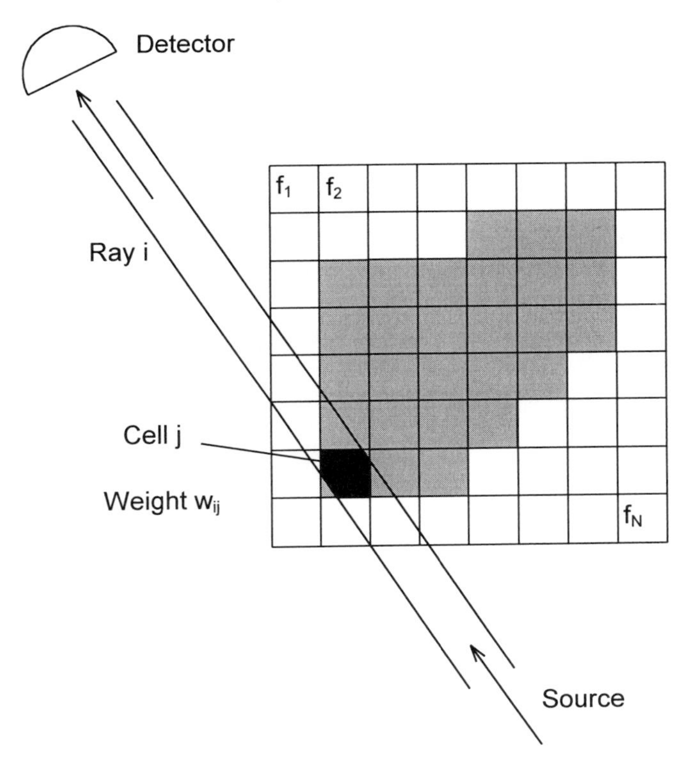

Figure B.7: Geometry for algebraic reconstruction

be reconstructed is constant in each cell, f_j being its value in cell j. Let M rays probe the area, one of these rays being indicated in Fig. B.7. The rays now have a finite thickness. The different portions each ray i intercepts with each cell j are quantified by w_{ij}. After running through the grid the integrated signal p_i of ray i is the sum of the values f_j of all subtended pixels weighted with the area w_{ij} the ray covers of the cell:

$$p_i = \sum_{j=1}^{N} w_{ij} f_j \qquad i = 1, \ldots, M \tag{B.22}$$

This system of linear equations can be written in matrix form $\boldsymbol{p} = \boldsymbol{W} \cdot \boldsymbol{f}$. Since for each ray only the intercepted pixels yield a $w_{ij} \neq 0$, $\boldsymbol{W}$ is a sparsely occupied matrix, but its size frequently is of the order $10^6 \times 10^6$.

A solution by direct matrix inversion is not feasible because of the size of the problem, sometimes the system is underdetermined with less projections than cells, often the system is overdetermined and inconsistent, and generally all projections are measured with limited accuracy. Instead one has to look for appropriate iterative solutions.

The primary algebraic reconstruction technique is based on Kaczmarz' algorithm for solving a system of linear equations [541]: Beginning with an arbitrary initial guess, the solutions are iteratively refined by taking into account the measurements along one ray during one iteration step. Convergence of the iteration is forced by introduction of relaxation parameters.

Another approach replaces the equalities (B.22) by inequalities

$$p_i - \varepsilon_i \leq \sum_{j=1}^{N} w_{ij} f_j \leq p_i - \varepsilon_i \qquad i = 1, \ldots, M \tag{B.23}$$

where the tolerances ε_i reflect the limited measurement accuracy. The iteration looks for a vector in the intersection of all hyperslabs defined by the inequalities (B.23).

Other methods use the concepts of entropy optimization, quadratic optimization, least squares regularization or statistical techniques [541, 536].

This short outline should be finished by the observation that still today most commercial equipment works with transform methods. But due to the versatility and flexibility of the algebraic reconstruction techniques and the enormous increase in computer speed these methods have good prospects for the future.

Appendix C

Bessel Functions

Bessel functions arise in solving differential equations for systems with cylindrical symmetry. The *Bessel functions* $J_n(z)$ and $Y_n(z)$ are linearly independent solutions to the differential equation

$$z^2 \frac{d^2y}{dz^2} + z\frac{dy}{dz} + (z^2 - n^2)y = 0 \tag{C.1}$$

$J_n(z)$ is called the *Bessel function of the first kind* , $Y_n(z)$ is referred to as the *Bessel function of the second kind.* For integer n, the $J_n(z)$ are regular at $z = 0$, the $Y_n(z)$ have a logarithmic divergence at $z = 0$.

Alternatively the Bessel function of the first kind can be defined over the integral

$$J_n(z) = \frac{1}{2\pi}\int_0^{2\pi} \cos(z \sin t - nt)dt \qquad z \in C, \qquad n = 0, 1, 2, \ldots \tag{C.2}$$

or as the power series

$$\begin{aligned} J_n(z) &= \frac{z^n}{2^n 0!n!} - \frac{z^{n+2}}{2^{n+2}1!(n+1)!} + \frac{z^{n+4}}{2^{n+4}2!(n+2)!} - \frac{z^{n+6}}{2^{n+6}3!(n+3)!} + - \ldots \\ &= \sum_{i=0}^{\infty} \frac{(-1)^i z^{n+2i}}{2^{n+2i}i!(n+i)!} \qquad \text{for} \qquad |z| < \infty \end{aligned} \tag{C.3}$$

Since the absolute values of the coefficients of this power series decrease very rapidly, this representation is useful for a practical calculation even for large $|z|$.

By proper combination of the power series components and using the Euler formula one obtains the useful formula

$$\sum_{n=-\infty}^{\infty} J_n(z)e^{in\phi} = e^{iz\sin\phi} \tag{C.4}$$

We see in (C.3) that for real x $J_n(x)$ is also real and an even function for even n and an odd function if n is odd. Fig. C.1 shows the first four real Bessel functions $J_0(x), \ldots, J_3(x)$. All real Bessel functions are bounded by the functions $\pm\sqrt{2/(\pi x)}$ which also are shown in Fig. C.1. We recognize the damped oscillation of all curves as well as the distribution of the zeroes becoming more and more regular with increasing x.

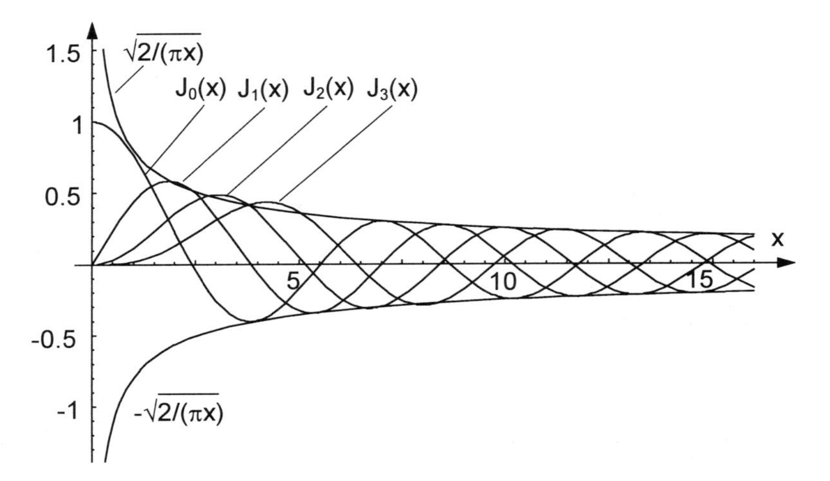

Figure C.1: Real Bessel functions of the first kind

The most interesting Bessel function in holographic interferometry is the real Bessel function of the first kind and zero order $\mathrm{J}_0(x)$ because it describes the intensity distribution resulting from the *time average method* for analysing harmonically vibrating objects. The zeroes b_m, $\mathrm{J}_0(b_m) = 0, \quad m = 1, 2, \ldots$, can be approximated for sufficiently large m by

$$b_m^* = (m - \frac{1}{4})\pi \tag{C.5}$$

Accordingly an approximating function to $\mathrm{J}_0(x)$ for large x is

$$\mathrm{J}_0^*(x) = \sqrt{\frac{2}{\pi x}} \cos(x - \frac{\pi}{4}) \tag{C.6}$$

For higher order n we have

$$\mathrm{J}_n^*(x) = \sqrt{\frac{2}{\pi x}} \cos(x - \frac{2n+1}{4}\pi) \tag{C.7}$$

A table of the first 30 zeroes of $\mathrm{J}_0(x)$ together with the approximating b_m^* is given with a precision of five digits behind the decimal point in Table C.1.

At this point it should be mentioned that the n-th order Bessel functions of the first kind also appear in the Zernike description of wavefronts in optical testing: A data spectrum $D(u, v)$ is described by a sum of Zernike terms

$$Z_{m,n}(r, \Theta) = \mathrm{R}_{m,n}(r) e^{im\Theta}, \qquad m, n \in N, \quad n > 0, \quad m < n, \quad m + n \text{ even} \tag{C.8}$$

where the $\mathrm{R}_{m,n}(r)$ are *Zernike polynomials.* The Zernike terms form a complete orthogonal set of functions over the unit circle. The Fourier transforms of the $Z_{m,n}(r, \Theta)$ are

$$\mathcal{F}\{Z_{m,n}(r, \Theta)\} = \mathcal{Z}_{m,n}(\rho, \theta) = (-1)^n A_n(\rho) \mathrm{e}^{im\theta} \tag{C.9}$$

Table C.1: Zeroes of the real Bessel function of the first kind and zero order

m	b_m	b_m^*	m	b_m	b_m^*
1	2.40483	2.35619	16	49.48261	49.48008
2	5.52008	5.49779	17	52.62405	52.62168
3	8.65373	8.63938	18	55.76551	55.76327
4	11.79153	11.78097	19	58.90698	58.90486
5	14.93092	14.92257	20	62.04847	62.04645
6	18.07106	18.06416	21	65.18996	65.18805
7	21.21164	21.20575	22	68.33147	68.32964
8	24.35247	24.34734	23	71.47298	71.47123
9	27.49348	27.48894	24	74.61450	74.61283
10	30.63461	30.63053	25	77.75603	77.75442
11	33.77582	33.77212	26	80.89756	80.89601
12	36.91710	36.91371	27	84.03909	84.03760
13	40.05843	40.05531	28	87.18063	87.17920
14	43.19979	43.19690	29	90.32217	90.32078
15	46.34119	46.33849	30	93.46372	93.46238

where

$$A_n(\rho) = \frac{1}{\rho} \mathrm{J}_{n+1}(2\pi\rho) \tag{C.10}$$

with (ρ, θ) being the spatial frequency coordinates in polar form [265].

Bibliography

[1] D. Gabor. A new microscopic principle. *Nature*, **161**, 777–778, 1948.

[2] D. Gabor. Microscopy by reconstructed wavefronts. *Proc. Royal Society A*, **197**, 454–487, 1949.

[3] D. Gabor. Microscopy by reconstructed wavefronts: II. *Proc. Phys. Society B*, **64**, 449–469, 1951.

[4] G. L. Rogers. Experiments in diffraction microscopy. *Proc. Roy. Soc. Edinb.*, **63A**, 193–221, 1952.

[5] H. M. A. El-Sum and P. Kirkpatrick. Microscopy by reconstructed wavefronts. *Phys. Rev.*, **85**, 763, 1952.

[6] A. W. Lohmann. Optische Einseitenbandübertragung angewandt auf das Gabor-Mikroskop. *Optica Acta*, **3**, 97–99, 1956.

[7] E. N. Leith and J. Upatnieks. Reconstructed wavefronts and communication theory. *Journ. Opt. Soc. Amer.*, **52**, 1123–1130, 1962.

[8] E. N. Leith and J. Upatnieks. Wavefront reconstruction with continuous-tone objects. *Journ. Opt. Soc. Amer.*, **53**, 1377–1381, 1963.

[9] E. N. Leith and J. Upatnieks. Wavefront reconstruction with diffused illumination and three-dimensional objects. *Journ. Opt. Soc. Amer.*, **54**, 1295–1301, 1964.

[10] R. L. Powell and K. A. Stetson. Interferometric vibration analysis by wavefront reconstruction. *Journ. Opt. Soc. Amer.*, **55**, 1593–1508, 1965.

[11] P. Hariharan. Laser interferometry: current trends and future prospects. In R. J. Pryputniewicz, ed., *Laser Interferometry IV: Computer-Aided Interferometry, Proc. of Soc. Photo-Opt. Instr. Eng.*, **1553**, 2–11, 1991.

[12] H. Rottenkolber and W. Jüptner. Holographic interferometry in the next decade. In R. J. Pryputniewicz, ed., *Laser Interferometry: Quantitative Analysis of Interferograms, Proc. of Soc. Photo-Opt. Instr. Eng.*, **1162**, 2–15, 1989.

[13] H. J. Tiziani. Optical methods for precision measurements. *Optical and Quantum Electronics*, **21**, 253–282, 1989.

[14] J. Durnin. Exact solutions for nondiffracting beams. I. The scalar theory. *Journ. Opt. Soc. Amer. A*, **4**, 651–654, 1987.

[15] J. W. Goodman. *Introduction to Fourier Optics.* McGraw-Hill, 1968.

[16] W. Lauterborn, T. Kurz, and M. Wiesenfeldt. *Coherent Optics.* (In German). Springer, 1993.

[17] J. W. Goodman. Statistical properties of laser speckle patterns. In J. C. Dainty, ed., *Laser Speckle and Related Phenomena, Springer Series Topics in Applied Physics*, **9**, 9–75, 1975.

[18] K. J. Gasvik. *Optical Metrology*. J. Wiley and Sons, 1987.

[19] E. B. Champagne. Non-paraxial imaging, magnification and aberration properties in holography. *Journ. Opt. Soc. Amer.*, **57**, 51–55, 1967.

[20] R. W. Meier. Magnification and third-order aberrations in holography. *Journ. Opt. Soc. Amer.*, **55**, 987–992, 1965.

[21] J. F. Miles. Imaging and magnification properties in holography. *Optica Acta*, **19**, 165–186, 1972.

[22] D. B. Neumann. Geometrical relationships between the original object and the two images of a hologram reconstruction. *Journ. Opt. Soc. Amer.*, **56**, 858–861, 1966.

[23] W. Schumann and M. Dubas. On the motion of holographic images caused by movements of the reconstruction light source, with the aim of application to deformation analysis. *Optik*, **46**, 377–392, 1976.

[24] C. M. Vest. *Holographic Interferometry*. J. Wiley and Sons, 1979.

[25] P. Hariharan. *Optical Holography: Principles, Techniques and Applications*. Cambridge University Press, 1984.

[26] P. Hariharan. Basic principles. In P. K. Rastogi, ed., *Holographic Interferometry*, 7–32. Springer Series in Optical Sciences, **68**, 1994.

[27] G. Lai and T. Yatagai. Dual-reference holographic interferometry with a double pulsed laser. In F.-P. Chiang, ed., *International Conference on Photomechanics and Speckle Metrology, Proc. of Soc. Photo-Opt. Instr. Eng.*, **814**, 346–351, 1987.

[28] D. Paoletti, S. Amadesi, and A. D'Altorio. A fringe control method for real time HNDT. *Opt. Comm.*, **49**, 98–102, 1984.

[29] J. N. Latta. Computer-based analysis of hologram imagery and aberrations. I. hologram types and their nonchromatic aberrations. *Appl. Opt.*, **10**, 599–608, 1971.

[30] J. N. Latta. Computer-based analysis of hologram imagery and aberrations. II. aberrations induced by a wavelength shift. *Appl. Opt.*, **10**, 609–618, 1971.

[31] J. N. Latta. Computer-based analysis of holography using ray tracing. *Appl. Opt.*, **10**, 2698–2710, 1971.

[32] J. F. Miles. Evaluation of the wavefront aberration in holography. *Optica Acta*, **20**, 19–31, 1973.

[33] G. W. Stroke and A. E. Labeyrie. White light reconstruction of holographic images using the Lippmann-Bragg diffraction effect. *Phys. Lett.*, **20**, 368–370, 1966.

[34] Yu. I. Ostrovsky, M. M. Butusov, and G. V. Ostrovskaya. *Interferometry by Holography*. Springer-Verlag, 1980.

[35] R. T. Pitlak and R. Page. Pulsed lasers for holographic interferometry. *Opt. Eng.*, **24**, 639–644, 1985.

[36] A. J. Decker. Holographic interferometry with an injection seeded Nd:YAG laser and two reference beams. *Appl. Opt.*, **29**, 2697–2700, 1990.

[37] G. Hüttmann, W. H. Lauterborn, and E. Schmitz. Holography with a frequency-doubled Nd:YAG laser. In W. Jüptner, ed., *Holography Techniques and Applications*, *Proc. of Soc. Photo-Opt. Instr. Eng.*, **1026**, 14–21, 1988.

[38] L. Crawforth, C.-K. Lee, and A. C. Munce. Application of pulsed laser holographic interferometry to the study of magnetic disc drive component motions. In K. Stetson and R. Pryputniewicz, eds., *International Conference on Hologram Interferometry and Speckle Metrology*, 404–412, 1990.

[39] P. Hariharan. High-precision, digital, phase-stepping interferometry with laser diode. In R. J. Pryputniewicz, ed., *Laser Interferometry: Quantitative Analysis of Interferograms*, *Proc. of Soc. Photo-Opt. Instr. Eng.*, **1162**, 86–91, 1989.

[40] Y. Ishii, J. Chen, and K. Murata. Digital phase-measuring interferometry with a tunable laser diode. *Opt. Lett.*, **12**, 233–235, 1987.

[41] R. J. Parker. A quarter century of thermoplastic holography. In K. Stetson and R. Pryputniewicz, eds., *International Conference on Hologram Interferometry and Speckle Metrology*, 217–224, 1990.

[42] J. P. Herriau, A. Delboulbe, and J. P. Huignard. Non destructive testing using real time holographic interferometry in BSO crystals. In W. F. Fagan, ed., *Industrial Applications of Laser Technology*, *Proc. of Soc. Photo-Opt. Instr. Eng.*, **398**, 123–129, 1983.

[43] H. J. Tiziani. Real-time metrology with BSO crystals. *Optica Acta*, **29**, 463–470, 1982.

[44] X. Wang, R. Magnusson, and A. Haji-Sheikh. Real-time interferometry with photorefractive reference holograms. *Appl. Opt.*, **32**, 1983–1986, 1993.

[45] J. P. Bentley. *Principles of Measurement Systems, 2nd edition*, **chap. 15**, Optical measurement systems, 364–405. Longman Scientific and Technical, 1988.

[46] G. O. Reynolds, J. B. DeVelis, G. B. Parrent, and B. J. Thompson. *The New Physical Optics Notebook: Tutorials in Fourier Optics.* SPIE Optical Engineering Press, 1989.

[47] J. A. Gilbert, D. R. Matthys, and Ch. M. Hendren. Displacement analysis of the interior walls of a pipe using panoramic holo-interferometry. In F. P. Chiang, ed., *Second International Conference on Photomechanics and Speckle Metrology*, *Proc. of Soc. Photo-Opt. Instr. Eng.*, **1554**, 128–134, 1991.

[48] N. Abramson. *The Making and Evaluation of Holograms.* Academic Press, 1981.

[49] B. E. Jones. Optical fibre sensors and systems for industry. *Journ. Phys. E: Sci. Instr.*, **18**, 770–782, 1985.

[50] F. Solitro, L. Gatti, F. Bedarida, and L. Michetti. Multidirectional holographic interferometer (MHOI) with fiber optics for study of crystal growth in microgravity. In R. J. Pryputniewicz, ed., *Laser Interferometry: Quantitative Analysis of Interferograms*, *Proc. of Soc. Photo-Opt. Instr. Eng.*, **1162**, 62–65, 1989.

[51] J. A. Gilbert, T. D. Dudderar, M. E. Schultz, and A. J. Boehnlein. The monomode fiber - a new tool for holographic interferometry. *Exp. Mech.*, **23**, 190–195, 1983.

[52] J. A. Gilbert, T. D. Dudderar, and A. Nose. Remote deformation field measurement through different media using fiber optics. *Opt. Eng.*, **24**, 628–631, 1985.

[53] S. Pflüger, R. Noll, J. Hertzberg, V. Sturm, and W. Meesters. Pulsed holography with elongated Q-switch pulse, (in German). In VDI-Technology Centre, ed., *Holographic-interferometric metrology*, Laser-Research and Laser-Technology, 131–145, 1995.

[54] K. Creath. Phase-shifting holographic interferometry. In P. K. Rastogi, ed., *Holographic Interferometry*, 109–150. Springer Series in Optical Sciences, **68**, 1994.

[55] G. E. Sommargren. Double exposure holographic interferometry using commonpath reference waves. *Appl. Opt.*, **16**, 1736–1741, 1977.

[56] M. Kujawinska and D. W. Robinson. Automatic fringe pattern analysis for holographic measurement of transient event. In W. Jüptner, ed., *Holography Techniques and Applications, Proc. of Soc. Photo-Opt. Instr. Eng.*, **1026**, 93–103, 1988.

[57] M. Kujawinska and D. W. Robinson. Multichannel phase-stepped holographic interferometry. *Appl. Opt.*, **27**, 312–320, 1988.

[58] C. R. Mercer and G. Beheim. Fiber optic phase stepping system for interferometry. *Appl. Opt.*, **30**, 729–734, 1991.

[59] M. Takeda and M. Kitoh. Spatio-temporal frequency-multiplex heterodyne interferometry. In R. J. Pryputniewicz, ed., *Laser Interferometry IV: Computer-Aided Interferometry, Proc. of Soc. Photo-Opt. Instr. Eng.*, **1553**, 1991.

[60] R. W. Larson, J. S. Zelenka, and E. L. Johansen. Microwave radar imagery. In *Engineering Applications of Holography*, Proc. of Soc. Photo-Opt. Instr. Eng., 14, 1972.

[61] K. Suzuki and B. P. Hildebrand. Holographic interferometry with acoustic waves. In N. Booth, ed., *Acoustical Holography*, **6**, 577–595. Plenum Press, New York, 1975.

[62] J. E. Sollid and J. B. Swint. A determination of the optimum beam ratio to produce maximum contrast photographic reconstructions from double-exposure holographic interferograms. *Appl. Opt.*, **9**, 2717–2719, 1970.

[63] J. E. Sollid. Holographic interferometry applied to measurements of small static displacements of diffusely reflecting surfaces. *Appl. Opt.*, **8**, 1587–1595, 1969.

[64] R. Pawluczyk and Z. Kraska. Diffuse illumination in holographic double-aperture interferometry. *Appl. Opt.*, **24**, 3072–3078, 1985.

[65] J. D. Trolinger. The holography of phase objects. In N. A. Massie, ed., *Interferometric Metrology, Critical Reviews, Proc. of Soc. Photo-Opt. Instr. Eng.*, 128–139, **816**, 1987.

[66] R. E. Brooks, L. O. Heflinger, and R. F. Wuerker. Interferometry with a holographically reconstructed comparison beam. *Appl. Phys. Lett.*, **7**, 248–249, 1965.

[67] L. O. Heflinger, R. F. Wuerker, and R. E. Brooks. Holographic interferometry. *Journ. Appl. Phys.*, **37**, 642–649, 1966.

[68] N. L. Hecht, J. E. Minardi, D. Lewis, and R. L. Fusek. Quantitative theory for predicting fringe pattern formation in holographic interferometry. *Appl. Opt.*, **12**, 2665–2676, 1973.

[69] J. Janta and M. Miler. Model interferogram as an aid for holographic interferometry. *J. Optics*, **8**, 301–307, 1977.

[70] R. Höfling and W. Osten. Displacement measurement by image processed speckle patterns. *J. Mod. Opt.*, **34**, 607–617, 1987.

[71] Th. Kreis, W. Jüptner, and R. Biedermann. Neural network approach to holographic nondestructive testing. *Appl. Opt.*, **34**, 1407–1415, 1995.

[72] G. S. Rightley, L. K. Matthews, and G. P. Mulholland. Holographic analysis and experimental error. In K. Stetson and R. Pryputniewicz, eds., *International Conference on Hologram Interferometry and Speckle Metrology*, 343–350, 1990.

[73] W. Jüptner, K. Ringer, and H. Welling. The evaluation of interference fringes for holographic strain and translation measurement, (in German). *Optik*, **38**, 437–448, 1973.

[74] N. Abramson. The holo-diagram: A practical device for making and evaluating holograms. *Appl. Opt.*, **8**, 1235–1240, 1969.

[75] N. Abramson. The holo-diagram II: A practical device for information retrieval in hologram interferometry. *Appl. Opt.*, **9**, 97–101, 1970.

[76] N. Abramson. The holo-diagram III: A practical device for predicting fringe patterns in hologram interferometry. *Appl. Opt.*, **9**, 2311–2320, 1970.

[77] N. Abramson. The holo-diagram IV: A practical device for simulating fringe patterns in hologram interferometry. *Appl. Opt.*, **10**, 2155–2161, 1971.

[78] N. Abramson. The holo-diagram V: A device for practical interpreting of hologram fringes. *Appl. Opt.*, **11**, 1143–1147, 1972.

[79] S. Toyooka. Holographic interferometry with increased sensitivity for diffusely reflecting objects. *Appl. Opt.*, **16**, 1054–1057, 1977.

[80] R. Dändliker, E. Marom, and F. M. Mottier. Two-reference-beam holographic interferometry. *Journ. Opt. Soc. Amer.*, **66**, 23–30, 1976.

[81] R. Dändliker. Two-reference-beam holographic interferometry. In P. K. Rastogi, ed., *Holographic Interferometry*, 75–108. Springer Series in Optical Sciences, Vol. 68, 1994.

[82] R. Dändliker, R. Thalmann, and J.-F. Willemin. Fringe interpolation by two-reference-beam holographic interferometry: reducing sensitivity to hologram misalignment. *Opt. Comm.*, **42**, 301–306, 1982.

[83] S. Walles. On the concept of homologous rays in holographic interferometry of diffusely reflecting surfaces. *Optica Acta*, **17**, 899–913, 1970.

[84] K. A. Stetson. Use of fringe vectors in hologram interferometry to determine fringe localization. *Journ. Opt. Soc. Amer.*, **66**, 626–627, 1976.

[85] P. K. Rastogi. Techniques to measure displacements, derivatives and surface shapes. extension to comparative holography. In P. K. Rastogi, ed., *Holographic Interferometry*, 213–292. Springer Series in Optical Sciences, **68**, 1994.

[86] R. Jones and C. Wykes. *Holographic and Speckle Interferometry*. Cambridge University Press, second edition, 1989.

[87] J. Blanco-Garcia, J. L. Fernandez, and M. Perez-Amor. Fringe localization control in holographic interferometry. *Appl. Opt.*, **31**, 488–496, 1992.

[88] K. A. Stetson. The argument of the fringe function in hologram interferometry of general deformations. *Optik*, **31**, 576–591, 1970.

[89] K. A. Haines and B. P. Hildebrand. Interferometric measurements on diffuse surfaces by holographic techniques. *IEEE Trans.*, **IM 15**, 149–161, 1966.

[90] K. A. Haines and B. P. Hildebrand. Surface-deformation measurement using the wavefront reconstruction technique. *Appl. Opt.*, **5**, 595–602, 1966.

[91] N. E. Molin and K. A. Stetson. Fringe localization in hologram interferometry of mutually independent and dependent rotations around orthogonal, non-intersecting axes. *Optik*, **33**, 399–422, 1970.

[92] N. E. Molin and K. A. Stetson. Measurement of fringe loci and localization in hologram interferometry for pivot motion, in-plane rotation and in-plane translation. Part I. *Optik*, **31**, 157–177, 1970.

[93] N. E. Molin and K. A. Stetson. Measurement of fringe loci and localization in hologram interferometry for pivot motion, in-plane rotation and in-plane translation. Part II. *Optik*, **31**, 281–291, 1970.

[94] K. A. Stetson. A rigorous treatment of the fringes of hologram interferometry. *Optik*, **29**, 386–400, 1969.

[95] K. A. Stetson. Fringe interpolation for hologram interferometry of rigid-body motions and homogeneous deformations. *Journ. Opt. Soc. Amer.*, **64**, 1–10, 1974.

[96] K. A. Stetson. Fringe vectors and observed fringe vectors in hologram interferometry. *Appl. Opt.*, **14**, 272–273, 1975.

[97] K. A. Stetson. Holographic strain analysis by fringe localization planes. *Journ. Opt. Soc. Amer.*, **66**, 627, 1976.

[98] P. M. Boone and R. Verbiest. Applications of hologram interferometry and translation measurement. *Optica Acta*, **16**, 555–567, 1969.

[99] M. A. Machado Gama. Fringe localization and visibility in hologram and classical broad source interferometry. *Opt. Comm.*, **8**, 362–365, 1973.

[100] T. Tsuruta, N. Shiotake, and Y. Itoh. Formation and localization of holographically produced interference fringes. *Optica Acta*, **16**, 723–733, 1969.

[101] W. T. Welford. Fringe visibility and localization in hologram interferometry. *Opt. Comm.*, **1**, 123–125, 1969.

[102] W. T. Welford. Fringe visibility and localization in hologram interferometry with parallel displacement. *Opt. Comm.*, **1**, 311–314, 1970.

[103] P. K. Rastogi. Visualization and measurement of slope and curvature fields using holographic interferometry: an application to flaw detection. *Journal of Modern Optics*, **38**, 1251–1263, 1991.

[104] E. Schnack and P. A. Klumpp. Shearographic and holographic defect detection for composite materials. In R. J. Pryputniewicz, ed., *Laser Interferometry IV: Computer-Aided Interferometry*, *Proc. of Soc. Photo-Opt. Instr. Eng.*, **1553**, 332–348, 1991.

[105] H. Steinbichler, S. Leidenbach, J. Engelsberger, E.-H. Nösekabel, and J. Sun. Optical measuring systems help to find constructional weak spots. In R. J. Pryputniewicz, ed., *Laser Interferometry IV: Computer-Aided Interferometry*, *Proc. of Soc. Photo-Opt. Instr. Eng.*, **1553**, 240–249, 1991.

[106] G. Birnbaum and C. M. Vest. Holographic nondestructive evaluation: status and future. *Intern. Advances in Nondestructive Testing*, **9**, 257–282, 1983.

[107] S. Amadesi, A. D'Altorio, and D. Paoletti. Single-two hologram interferometry: a combined method for dynamic tests on painted wooden statues. *J. Optics*, **14**, 243–246, 1983.

[108] G. M. Brown. Computer automated holometry for automotive applications. In W. F. Fagan, ed., *Industrial Optoelectronic Measurement Systems using Coherent Light*, *Proc. of Soc. Photo-Opt. Instr. Eng.*, **863**, 213–222, 1987.

[109] G. Cavaccini, A. Ciliberto, L. D'Antonio, P. Ferraro, A. Ortona, and C. Sabatino. Holographic interferometry used for structural analysis of bonded structures in aerospace industry. In K. Stetson and R. Pryputniewicz, eds., *International Conference on Hologram Interferometry and Speckle Metrology*, 379–384, 1990.

[110] J. Gryzagoridis. Holographic non-destructive testing of composites. *Opt. Laser Technol.*, **21**, 113–116, 1989.

[111] Y. Y. Hung. Shearography versus holography in nondestructive evaluation of tires and composites. In F.-P. Chiang, ed., *International Conference on Photomechanics and Speckle Metrology*, *Proc. of Soc. Photo-Opt. Instr. Eng.*, **814**, 433–442, 1987.

[112] H. Kasprzak, H. Podbielska, and G. von Bally. Human tibia rigidity examined in bending and torsion loading by using double-exposure holographic interferometry. In W. Jüptner, ed., *Holography Techniques and Applications*, *Proc. of Soc. Photo-Opt. Instr. Eng.*, **1026**, 196–201, 1988.

[113] H. G. Leis. Vibration analysis of an 8-cylinder V-engine by time-averaged holographic interferometry. In W. F. Fagan, ed., *Industrial Applications of Laser Technology*, *Proc. of Soc. Photo-Opt. Instr. Eng.*, **398**, 90–94, 1983.

[114] J. W. Newman. Production and field inspection of advanced materials with holographic and speckle interferometry. In F.-P. Chiang, ed., *International Conference on Photomechanics and Speckle Metrology*, *Proc. of Soc. Photo-Opt. Instr. Eng.*, **814**, 421–427, 1987.

[115] B. Ovryn. Holographic Interferometry. *CRC Critical Reviews in Biomedical Engineering*, **16**, 269–322, 1989.

[116] P. Paulet. Holographic measuring of the deformations of various internal combustion engine parts. In W. F. Fagan, ed., *Industrial Applications of Laser Technology*, *Proc. of Soc. Photo-Opt. Instr. Eng.*, **398**, 30–34, 1983.

[117] C. A. Sciammarella and B. Subbaraman. Application of holographic interferometry to the NDE of composites. In SEM, ed., *Optical Methods in Composites*, Proc. of the 1986 SEM Fall Conference on Experimental Mechanics, 11–16, 1986.

[118] P. Zanetta, G. P. Solomos, M. Zürn, and A. C. Lucia. Holographic detection of defects in composites. *Opt. Laser Technol.*, **25**, 97–102, 1993.

[119] J. E. Sollid and K. A. Stetson. Strains from holographic data. *Exp. Mech.*, **18**, 208–214, 1978.

[120] K. D. Hinsch. Holographic interferometry of surface deformations of transparent fluids. *Appl. Opt.*, **17**, 3101–3107, 1978.

[121] C. M. E. Holden, S. C. J. Parker, and P. J. Bryanston-Cross. Quantitative three-dimensional holographic interferometry for flow field analysis. *Opt. Lasers in Eng.*, **19**, 285–298, 1993.

[122] T. A. W. M. Lanen. Digital holographic interferometry in flow research. *Opt. Comm.*, **79**, 386–396, 1990.

[123] T. A. W. M. Lanen, P. G. Bakker, and P. J. Bryanston-Cross. Digital holographic interferometry in high-speed flow research. *Exp. Fluids*, **13**, 56–62, 1992.

[124] T. A. W. M. Lanen, C. Nebbeling, and J. L. van Ingen. Phase-stepping holographic interferometry in studying transparent density fields around 2-D objects of arbitrary shape. *Opt. Comm.*, **76**, 268–275, 1990.

[125] J. D. Trolinger, N. Brock, P. DeBarber, J. Hsu, and J. Millerd. Recent developments in optical flow diagnostics. In AIAA, ed., *Proc. of 32nd Aerospace Sciences Meeting and Exhibit*, 1–21, 1994.

[126] J. D. Trolinger. The new methods in holographic flow diagnostics. In K. Stetson and R. Pryputniewicz, eds., *International Conference on Hologram Interferometry and Speckle Metrology*, 488–493, 1990.

[127] R. D. Matulka and D. J. Collins. Determination of three-dimensional density fields from holographic interferograms. *Journ. Appl. Phys.*, **42**, 1109–1119, 1971.

[128] M. J. Ehrlich and J. W. Wagner. Investigation of transient acoustic wave and shock wave propagation in bulk materials using high speed pulsed holographic interferometry and computer assisted optical tomography. In K. Stetson and R. Pryputniewicz, eds., *International Conference on Hologram Interferometry and Speckle Metrology*, 422–428, 1990.

[129] R. J. Parker. Extraction of 3-d flow data from transonic flow holograms. In W. F. Fagan, ed., *Industrial Optoelectronic Measurement Systems using Coherent Light*, *Proc. of Soc. Photo-Opt. Instr. Eng.*, **863**, 78–85, 1987.

[130] R. J. Parker and M. Reeves. Holographic flow visualization in rotating turbomachinery. In K. Stetson and R. Pryputniewicz, eds., *International Conference on Hologram Interferometry and Speckle Metrology*, 500–507, 1990.

[131] M. Watanabe, A. Abe, R. T. Casey, and K. Takayama. Holographic interferometric observation of shock wave phenomena. In R. J. Pryputniewicz, ed., *Laser Interferometry IV: Computer-Aided Interferometry*, *Proc. of Soc. Photo-Opt. Instr. Eng.*, **1553**, 418–426, 1991.

[132] D. W. Watt and C. M. Vest. Digital interferometry for flow visualization. *Exp. Fluids*, **5**, 401–306, 1987.

[133] D. W. Watt and C. M. Vest. Turbulent flow visualization by interferometric integral imaging and computed tomography. *Exp. Fluids*, **8**, 301–311, 1990.

[134] J. S. Steckenrider, M. J. Ehrlich, and J. W. Wagner. Pulsed holographic recording of very high speed transient events. In F. P. Chiang, ed., *Second International Conference on Photomechanics and Speckle Metrology*, *Proc. of Soc. Photo-Opt. Instr. Eng.*, **1554**, 106–112, 1991.

[135] B. Lu, X. Yang, H. Abendroth, H. Eggers, and E. Ziolkowski. Measurement of a three-dimensional temperature field applying ESPI and CT techniques. *Opt. Comm.*, **69**, 6–10, 1988.

[136] W. Merzkirch. Calculation of fluid density from holographic interferograms. In P. Smigielski, ed., *Deuxieme Colloque Franco-Allemand sur les Applications de l'Holographie*, 195–207, 1989.

[137] N. Rubayi, J. Yeakle, and M. A. Wright. Holographic investigation of various stressing techniques for detecting flaws in composite laminates. In SEM, ed., *Optical Methods in Composites*, Proc. of the 1986 SEM Fall Conference on Experimental Mechanics, 1–10, 1986.

[138] D. L. Mader. Holographic interferometry on pipes: precision interpretation by least-squares fitting. *Appl. Opt.*, **24**, 3784–3790, 1985.

[139] P. Del Vo and M. L. Rizzi. Vibrational testing of an X-ray concentrator by holographic interferometry. In F.-P. Chiang, ed., *International Conference on Photomechanics and Speckle Metrology, Proc. of Soc. Photo-Opt. Instr. Eng.*, **814**, 357–364, 1987.

[140] H.-A. Crostack, E. H. Meyer, and K.-J. Pohl. Holographic interferometric ultrasound imaging - technique and application in nondestructive testing. In W. Jüptner, ed., *Holography Techniques and Applications, Proc. of Soc. Photo-Opt. Instr. Eng.*, **1026**, 123–132, 1988.

[141] K.-J. Pohl and H.-A. Crostack. Holographic visualization of laser-induced ultrasonic Rayleigh waves. In K. Stetson and R. Pryputniewicz, eds., *International Conference on Hologram Interferometry and Speckle Metrology*, 449–456, 1990.

[142] P. Gren and R. Olsson. Deformation during impact on an orthotropic composite plate. In K. Stetson and R. Pryputniewicz, eds., *International Conference on Hologram Interferometry and Speckle Metrology*, 429–434, 1990.

[143] D. Fraile, F. Gascon, and A. Varade. Fringe pattern in holographic interferometry for thermal expansion characterization of anisotropic bodies. *Appl. Opt.*, **31**, 7371–7374, 1992.

[144] D. W. Watt, T. S. Gross, and S. D. Hening. Three-illumination-beam phase-shifted holographic interferometry study of thermally induced displacements on a printed wiring board. *Appl. Opt.*, **30**, 1617–1623, 1991.

[145] A. Choudry. Digital holographic interferometry of convective heat transport. *Appl. Opt.*, **20**, 1240–1244, 1981.

[146] A. Choudry. Automated fringe reduction techniques. In N. A. Massie, ed., *Interferometric Metrology, Critical Reviews, Proc. of Soc. Photo-Opt. Instr. Eng.*, **816**, 49–55, 1987.

[147] A. Choudry, H. Dekker, and D. Enard. Automated interferometric evaluation of optical components at the European Southern Observatory (ESO). In W. F. Fagan, ed., *Industrial Applications of Laser Technology, Proc. of Soc. Photo-Opt. Instr. Eng.*, **398**, 66–72, 1983.

[148] A. C. Gillies. Image processing approach to fringe patterns. *Opt. Eng.*, **27**, 861–866, 1988.

[149] F. Ginesu and F. Bertolino. Numerical analysis of fringe patterns for structural engineering problems. In R. J. Pryputniewicz, *Laser Interferometry IV: Computer-Aided Interferometry, Proc. of Soc. Photo-Opt. Instr. Eng.*, **1553**, 313–324, 1991.

[150] Th. Kreis. Computer aided evaluation of fringe patterns. *Opt. Lasers in Eng.*, **19**, 221–240, 1993.

[151] M. Kujawinska. The architecture of a multipurpose fringe pattern analysis system. *Opt. Lasers in Eng.*, **19**, 261–268, 1993.

[152] G. E. Maddux. Video/computer techniques for static and dynamic experimental mechanics. In R. J. Pryputniewicz, ed., *Industrial Laser Interferometry, Proc. of Soc. Photo-Opt. Instr. Eng.*, **746**, 52–57, 1987.

[153] R. J. Pryputniewicz. Review of methods for automatic analysis for fringes in hologram interferometry. In N. A. Massie, ed., *Interferometric Metrology, Critical Reviews, Proc. of Soc. Photo-Opt. Instr. Eng.*, **816**, 140–148, 1987.

[154] G. T. Reid. Automatic fringe pattern analysis: A review. *Opt. Lasers in Eng.*, **7**, 37–68, 1986.

[155] D. W. Robinson and G. T. Reid, eds.. *Interferogram Analysis: Digital Fringe Pattern Measurement Techniques.* Institute of Physics Publ., Bristol and Philadelphia, 1993.

[156] C. A. Sciammarella and G. Bhat. Computer assisted techniques to evaluate fringe patterns. In R. J. Pryputniewicz, ed., *Laser Interferometry IV: Computer-Aided Interferometry, Proc. of Soc. Photo-Opt. Instr. Eng.*, **1553**, 252–262, 1991.

[157] J. S. Sirkis, Y.-M. Chen, H. Singh, and A. Y. Cheng. Computerized optical fringe pattern analysis in photomechanics: a review. *Opt. Eng.*, **31**, 305–314, 1992.

[158] R. J. Pryputniewicz. Quantitative determination of displacements and strains from holograms. In P. K. Rastogi, ed., *Holographic Interferometry*, 33–74. Springer Series in Optical Sciences, **68**, 1994.

[159] N. Abramson. The rose of error or the importance of sign. In *Technical Digest of Topical Meeting On Hologram Interferometry and Speckle Metrology.* Opt. Soc. Amer., TuB4-1–TuB4-4, 1980.

[160] Th. Kreis. Computer-aided evaluation of holographic interferograms. In P. K. Rastogi, ed., *Holographic Interferometry*, 151–212. Springer Series in Optical Sciences, **68**, 1994.

[161] D. R. Matthys, J. A. Gilbert, T. D. Dudderar, and K. W. Koenig. A windowing technique for the automated analysis of holo-interferograms. *Opt. Lasers in Eng.*, **8**, 123–136, 1988.

[162] P. D. Plotkowski, M. Y. Y. Hung, J. D. Hovanesian, and G. Gerhardt. Improved fringe carrier techniques for unambiguous determination of holographically recorded displacements. *Opt. Eng.*, **24**, 754–756, 1985.

[163] J. Petkovsek and K. Rankel. Measurement of three-dimensional displacement by four small holograms. In P. Meyrueis and M. Grosmann, eds., *2nd European Congress on Optics applied to Metrology, Proc. of Soc. Photo-Opt. Instr. Eng.*, **210**, 173–177, 1979.

[164] N. Eichhorn and W. Osten. An algorithm for the fast derivation of line structures from interferograms. *Journal of Modern Optics*, **35**, 1717–1725, 1988.

[165] W. Osten. *Digital Processing and Evaluation of Interference Images (in German).* Akademie-Verlag, 1991.

[166] W. Luth. *Isolation of moving objects in digital image sequences, (in German).* Academy of Sciences of the GDR, 1989.

[167] K. Creath. Submicron linewidth measurement using an interferometric optical profiler. *Proc. of Soc. Photo-Opt. Instr. Eng.*, **1464**, 1991.

[168] K. Creath. Holographic contour and deformation measurement using a 1.4 million element detector array. *Appl. Opt.*, **28**, 2170–2175, 1989.

[169] K. Creath. Phase-measurement interferometry: Beware these errors. In R. J. Pryputniewicz, Laser Interferometry IV: Computer-Aided Interferometry, *Proc. of Soc. Photo-Opt. Instr. Eng.*, **1553**, 213–220, 1991.

[170] W. Osten, R. Höfling, and J. Saedler. Two computer-aided methods for data reduction from interferograms. In W. F. Fagan, ed., *Industrial Optoelectronic Measurement Systems using Coherent Light, Proc. of Soc. Photo-Opt. Instr. Eng.*, **863**, 105–113, 1987.

[171] W. R. J. Funnell. Image processing applied to the interactive analysis of interferometric fringes. *Appl. Opt.*, **20**, 3245–3250, 1981.

[172] W. Osten, J. Saedler, and H. Rottenkolber. Interpretation of interferometric fringe patterns using digital image processing, (in German). *Technisches Messen tm*, **54**, 285–290, 1988.

[173] T. Yatagai. Intensity based analysis methods. In D. W. Robinson and G. T. Reid, eds., *Interferogram Analysis: Digital Fringe Pattern Measurement Techniques*, 72–93, Bristol and Philadelphia, Institute of Physics Publ. 1993.

[174] Th. Kreis and H. Kreitlow. Quantitative evaluation of holographic interference patterns under image processing aspects. In P. Meyrueis and M. Grosmann, eds., *2nd European Congress on Optics applied to Metrology*, *Proc. of Soc. Photo-Opt. Instr. Eng.*, **210**, 196–202, 1979.

[175] A. K. Jain and C. R. Christensen. Digital processing of images in speckle noise. In W. H. Carter, ed., *Applications of Speckle Phenomena*, *Proc. of Soc. Photo-Opt. Instr. Eng.*, **243**, 46–50, 1980.

[176] J. S. Lim and H. Nawab. Techniques for speckle noise removal. *Opt. Eng.*, **20**, 472–480, 1981.

[177] F. A. Sadjadi. Perspectives on techniques for enhancing speckled imagery. *Opt. Eng.*, **29**, 25–30, 1990.

[178] T. R. Crimmins. Geometric filter for speckle reduction. *Appl. Opt.*, **24**, 1438–1443, 1985.

[179] T. R. Crimmins. Geometric filter for reducing speckle. *Opt. Eng.*, **25**, 651–654, 1986.

[180] E. Bieber and W. Osten. Improvement of speckled fringe patterns by Wiener filtration. In Z. Jaroszewicz and M. Pluta, eds., *Interferometry 89*, *Proc. of Soc. Photo-Opt. Instr. Eng.*, **1121**, 393–399, 1989.

[181] Y. Katzir, I. Glaser, A. A. Friesem, and B. Sharon. On-line acquisition and analysis for holographic nondestructive evaluation. *Opt. Eng.*, **21**, 1016–1021, 1982.

[182] F. Becker, G. E. A. Meier, and H. Wegner. Automatic evaluation of interferograms. In A. G. Tescher, ed., *Applications of Digital Image Processing*, *Proc. of Soc. Photo-Opt. Instr. Eng.*, **359**, 386–393, 1982.

[183] F. Becker and Y. H. Yu. Digital fringe reduction technique applied to the measurement of three-dimensional transonic flow fields. *Opt. Eng.*, **24**, 429–434, 1985.

[184] J. Budzinski. SNOP: a method for skeletonization of a fringe pattern along a fringe direction. *Appl. Opt.*, **31**, 3109–3113, 1992.

[185] S. Nakadate, N. Magome, T. Honda, and J. Tsujiuchi. Hybrid holographic interferometer for measuring three-dimensional deformations. *Opt. Eng.*, **20**, 246–252, 1981.

[186] T. Yatagai, S. Nakadate, M. Idesawa, and H. Saito. Automatic fringe analysis using digital image processing techniques. *Opt. Eng.*, **21**, 432–435, 1982.

[187] E. A. Mnatsakanyan and S. V. Nefyodov. Skeletonizing of interferometric images for finding maximum and minimum centres of fringes. In K. Stetson and R. Pryputniewicz, eds., *International Conference on Hologram Interferometry and Speckle Metrology*, 351–355, 1990.

[188] H. Winter, S. Unger, and W. Osten. The application of adaptive and anisotropic filteringfor the extraction of fringe pattern skeletons. In W. Osten, R. J. Pryputniewicz, G. T. Reid, and H. Rottenkolber, eds., *Fringe '89, Automatic Processing of Fringe Patterns*, Physical Research, 158–166. Akademie-Verlag Berlin, 1989.

[189] J. A. Aparicio, J. L. Molpeceres, A. M. de Frutos, C. de Castro, S. Caceres, and F. A. Frechoso. Improved algorithm for the analysis of holographic interferograms. *Opt. Eng.*, **32**, 963–969, 1993.

[190] F. Bertolino and F. Ginesu. Semiautomatic approach of grating techniques. *Opt. Lasers in Eng.*, **19**, 313–323, 1993.

[191] A. E. Ennos, D. W. Robinson, and D. C. Williams. Automatic fringe analysis in holographic interferometry. *Optica Acta*, **32**, 135–145, 1985.

[192] S. Krishnaswamy. Algorithm for computer tracing of interference fringes. *Appl. Opt.*, **30**, 1624–1628, 1991.

[193] R. Nübel. Computer-aided evaluation method for interferograms. *Exp. Fluids*, **12**, 166–172, 1992.

[194] G. W. Johnson, D. C. Leiner, and D. T. Moore. Phase-locked interferometry. *Opt. Eng.*, **18**, 46–52, 1979.

[195] G. A. Mastin and D. C. Ghiglia. Digital extraction of interference fringe contours. *Appl. Opt.*, **24**, 1727–1728, 1985.

[196] V. Srinivasan, S.-T. Yeo, and P. Chaturvedi. Fringe processing and analysis with a neural network. *Opt. Eng.*, **33**, 1166–1171, 1994.

[197] K. H. Womack, K. L. Underwood, and D. Forbes. Microprocessor-based video interferogram analysis system. In *Minicomputers and Microprocessors in Optical Systems, Proc. of Soc. Photo-Opt. Instr. Eng.*, **230**, 168–179, 1980.

[198] A. Colin and W. Osten. Automatic support for consistent labelling of skeletonized fringe patterns. *J. Mod. Opt.*, **42**, 945–954, 1995.

[199] J. B. Schemm and C. M. Vest. Fringe pattern recognition and interpolation using nonlinear regression analysis. *Appl. Opt.*, **22**, 2850–2853, 1983.

[200] U. Mieth and W. Osten. Three methods for the interpolation of phase values between fringe pattern skeletons. In W. Osten, R. J. Pryputniewicz, G. T. Reid, and H. Rottenkolber, eds., *Fringe '89, Automatic Processing of Fringe Patterns*, Physical Research, 118–123. Akademie-Verlag Berlin, 1989.

[201] R. Dändliker, B. Ineichen, and F. M. Mottier. High resolution hologram interferometry by electronic phase measurement. *Opt. Comm.*, **9**, 412–416, 1973.

[202] P. V. Farrell, G. S. Springer, and C. M. Vest. Heterodyne holographic interferometry: concentration and temperature measurements in gas mixtures. *Appl. Opt.*, **21**, 1624–1627, 1982.

[203] R. J. Pryputniewicz. Heterodyne holography, applications in studies of small components. *Opt. Eng.*, **24**, 849–854, 1985.

[204] R. Thalmann and R. Dändliker. Strain measurement by heterodyne holographic interferometry. *Appl. Opt.*, **26**, 1964–1971, 1987.

[205] Th. Kreis, J. Geldmacher, and R. Biedermann. *Theoretical and experimental investigations of the accuracy of diverse methods for evaluating interference patterns.* (In German). Report on DFG-Project Kr953/2-1 edition, 1993.

[206] Th. Kreis, J. Geldmacher, and W. Jüptner. A comparison of interference phase determination methods with respect to achievable accuracy. In W. Jüptner and W. Osten, eds., *Fringe '93*, 51–59, 1993.

[207] B. Breuckmann and W. Thieme. Computer-aided analysis of holographic interferograms using the phase-shift method. *Appl. Opt.*, **24**, 2145–2149, 1985.

[208] R. Dändliker and R. Thalmann. Heterodyne and quasi-heterodyne holographic interferometry. *Opt. Eng.*, **24**, 824–831, 1985.

[209] P. Hariharan. Quasi-heterodyne hologram interferometry. *Opt. Eng.*, **24**, 632–638, 1985.

[210] P. Hariharan, B. F. Oreb, and N. Brown. A digital phase-measurement system for real-time holographic interferometry. *Opt. Comm.*, **41**, 393–396, 1982.

[211] R. Thalmann and R. Dändliker. Automated evaluation of 3-D displacement and strain by quasi-heterodyne holographic interferometry. In W. F. Fagan, ed., *Optics in Engineering Measurement, Proc. of Soc. Photo-Opt. Instr. Eng.*, **599**, 141–148, 1985.

[212] O. Y. Kwon and D. M. Shough. Multichannel grating phase-shift interferometers. In W. F. Fagan, ed., *Optics in Engineering Measurements, Proc. of Soc. Photo-Opt. Instr. Eng.*, **599**, 273, 1985.

[213] J. E. Greivenkamp. Generalized data reduction for heterodyne interferometry. *Opt. Eng.*, **23**, 350–352, 1984.

[214] P. L. Wizinowich. Phase shifting interferometry in the presence of vibration: a new algorithm and system. *Appl. Opt.*, **29**, 3271–3279, 1990.

[215] D. P. Towers, P. J. Bryanston-Cross, and C. E. Towers. The automatic quantitative analysis of phase stepped interferograms. In K. Stetson and R. Pryputniewicz, eds., *International Conference on Hologram Interferometry and Speckle Metrology*, 480–483, 1990.

[216] P. Carre. Installation et utilisation du comparateur photoelectrique et interferentiel du Bureau International des Poids et Mesures. *Metrologia*, **2**, 13–23, 1966.

[217] W. Jüptner, Th. Kreis, and H. Kreitlow. Automatic evaluation of holographic interferograms by reference beam phase shifting. In W. F. Fagan, ed., *Industrial Applications of Laser Technology, Proc. of Soc. Photo-Opt. Instr. Eng.*, **398**, 22–29, 1983.

[218] V. A. Deason and G. D. Lassahn. Novel approaches to automated reduction of phase shifted interferograms. In R. J. Pryputniewicz, *Laser Interferometry IV: Computer-Aided Interferometry, Proc. of Soc. Photo-Opt. Instr. Eng.*, **1553**, 569–582, 1991.

[219] X.-Y. Su, A. M. Zarubin, and G. von Bally. Modulation analysis of phase-shifted holographic interferograms. *Opt. Comm.*, **105**, 379–387, 1994.

[220] M. Chang, Ch.-P. Hu, P. Lam, and J. C. Wyant. High precision deformation measurement by digital phase shifting interferometry. *Appl. Opt.*, **24**, 3780–3783, 1985.

[221] Th. Kreis. Quantitative evaluation of interference patterns. In W. F. Fagan, ed., *Industrial Optoelectronic Measurement Systems using Coherent Light, Proc. of Soc. Photo-Opt. Instr. Eng.*, **863**, 68–77, 1987.

[222] C. P. Brophy. Effect of intensity error correlation on the computed phase of phase shifting interferometry. *Journ. Opt. Soc. Amer. A*, **7**, 537–541, 1990.

[223] Y. Y. Cheng and J. C. Wyant. Phase shifter calibration in phase-shifting interferometry. *Appl. Opt.*, **24**, 3049–3052, 1985.

[224] K. Freischlad and C. L. Koliopoulos. Fourier description of digital phase-measuring interferometry. *Journ. Opt. Soc. Amer. A*, **7**, 542–551, 1990.

[225] J. Schwider, R. Burow, K.-E. Elssner, J. Grzanna, R. Spolaczyk, and K. Merkel. Digital wave-front measuring interferometry: some systematic error sources. *Appl. Opt.*, **22**, 3421–3432, 1983.

[226] J. van Wingerden, H. J. Frankena, and C. Smorenburg. Linear approximation for measurement errors in phase shifting interferometry. *Appl. Opt.*, **30**, 2718–2729, 1991.

[227] K. Kinnstaetter, A. W. Lohmann, J. Schwider, and N. Streibl. Accuracy of phase shifting interferometry. *Appl. Opt.*, **27**, 5082–5089, 1988.

[228] P. J. de Groot. Derivation of algorithms for phase-shifting interferometry using the concept of a data-sampling window. *Appl. Opt.*, **34**, 4723–4730, 1995.

[229] P. J. de Groot. Vibration in phase-shifting interferometry. *Journ. Opt. Soc. Amer. A*, **12**, 354–365, 1995.

[230] K. G. Larkin and B. F. Oreb. Design and assessment of symmetrical phase-shifting algorithms. *Journ. Opt. Soc. Amer. A*, **9**, 1740–1748, 1992.

[231] D. R. Burton and M. J. Lalor. Managing some of the problems of Fourier fringe analysis. In G. T. Reid, ed., *Fringe Pattern Analysis, Proc. of Soc. Photo-Opt. Instr. Eng.*, **1163**, 149–160, 1989.

[232] Th. Kreis. Digital holographic interference-phase measurement using the Fourier-transform method. *Journ. Opt. Soc. Amer. A*, **3**, 847–855, 1986.

[233] Th. Kreis. Fourier-transform evaluation of holographic interference patterns. In F.-P. Chiang, ed., *International Conference on Photomechanics and Speckle Metrology, Proc. of Soc. Photo-Opt. Instr. Eng.*, **814**, 365–371, 1987.

[234] Th. Kreis and W. Jüptner. Fourier-transform evaluation of interference patterns: the role of filtering in the spatial-frequency domain. In R. J. Pryputniewicz, ed., *Laser Interferometry: Quantitative Analysis of Interferograms, Proc. of Soc. Photo-Opt. Instr. Eng.*, **1162**, 116–125, 1989.

[235] Th. Kreis and W. Jüptner. Fourier-transform evaluation of interference patterns: demodulation and sign-ambiguity. In R. J. Pryputniewicz, ed., *Laser Interferometry IV: Computer-Aided Interferometry, Proc. of Soc. Photo-Opt. Instr. Eng.*, **1553**, 263–273, 1991.

[236] A. A. Malcolm and D. R. Burton. The relationship between Fourier fringe analysis and the FFT. In R. J. Pryputniewicz, ed., *Laser Interferometry IV: Computer-Aided Interferometry, Proc. of Soc. Photo-Opt. Instr. Eng.*, **1553**, 286–297, 1991.

[237] Q.-S. Ru, T. Honda, J. Tsujiuchi, and N. Ohyama. Fringe analysis by using 2-D-Fresnel transforms. *Opt. Comm.*, **66**, 21–24, 1988.

[238] Q.-S. Ru, N. Ohyama, T. Honda, and J. Tsujiuchi. Constant radial shearing interferometry with circular gratings. *Appl. Opt.*, **28**, 3350–3353, 1989.

[239] M. Takeda, H. Ina, and S. Kobayashi. Fourier-transform method of fringe-pattern analysis for computer-based topography and interferometry. *Journ. Opt. Soc. Amer.*, **72**, 156–160, 1982.

[240] Th. Kreis. Automatic evaluation of interference patterns. In W. Jüptner, ed., *Holography Techniques and Applications, Proc. of Soc. Photo-Opt. Instr. Eng.*, **1026**, 80–89, 1988.

[241] K. E. Perry Jr. and J. McKelvie. A comparison of phase shifting and Fourier methods in the analysis of discontinuous fringe patterns. *Opt. Lasers in Eng.*, **19**, 269–284, 1993.

[242] P. Zanetta, D. Albrecht, G. Schirripa Spagnolo, and D. Paoletti. Application of fast fourier transform techniques to the quantitative analysis of holographic and TV-holographic interferograms. *Optik*, **97**, 47–52, 1994.

[243] N. Ohyama, S. Kinoshita, A. Cornejo-Rodriguez, T. Honda, and J. Tsujiuchi. Accuracy of phase determination with unequal reference phase shift. *Journ. Opt. Soc. Amer. A*, **5**, 2019–2025, 1988.

[244] G. Lai and T. Yatagai. Generalized phase-shifting interferometry. *Journ. Opt. Soc. Amer. A*, **8**, 822–827, 1991.

[245] D. J. Bone, H.-A. Bachor, and R. J. Sandeman. Fringe-pattern analysis using 2-D Fourier transform. *Appl. Opt.*, **25**, 1653–1660, 1986.

[246] P. J. Bryanston-Cross, C. Quan, and T. R. Judge. Application of the FFT method for the quantitative extraction of information from high-resolution interferometric and photoelastic data. *Opt. Laser Technol.*, **26**, 147–155, 1994.

[247] D. R. Burton and M. J. Lalor. The precision measurement of engineering form by computer analysis of optically generated contours. In D. W. Braggins, ed., *Industrial Inspection*, of *Proc. of Soc. Photo-Opt. Instr. Eng.*, **1010**, 17–24, 1989.

[248] C. Gorecki. Interferogram analysis using a Fourier transform method for automatic 3D surface measurement. *Pure Appl. Opt.*, **1**, 103–110, 1992.

[249] J. Gu and F. Chen. Fast Fourier transform, iteration, and least-squares-fit demodulation image processing for analysis of single-carrier fringe pattern. *Journ. Opt. Soc. Amer. A*, **12**, 2159–2164, 1995.

[250] K. A. Nugent. Interferogram analysis using an accurate fully automatic algorithm. *Appl. Opt.*, **24**, 3101–3105, 1985.

[251] D. Paoletti and G. Schirripa Spagnolo. Fourier transform for sandwich holograms evaluation. *J. Optics*, **25**, 17–23, 1994.

[252] R. Preater and R. Swain. Fourier transform fringe analysis of ESPI fringes from rotating components. In G. M. Brown, K. G. Harding, and H. Stahl, eds., *Industrial Applications of Optical Inspection, Metrology, and Sensing*, *Proc. of Soc. Photo-Opt. Instr. Eng.*, **1821**, 82–100, 1992.

[253] C. Roddier and F. Roddier. Interferogram analysis using Fourier transform techniques. *Appl. Opt.*, **26**, 1668–1673, 1987.

[254] M. Takeda. Spatial-carrier fringe-pattern analysis and its applications to precision interferometry and profilometry: An overview. *Industrial Metrology*, **1**, 79–99, 1990.

[255] M. Takeda. Spatial carrier heterodyne techniques for precision interferometry and profilometry: An overview. In Z. Jaroszewicz and M. Pluta, eds., *Interferometry '89*, *Proc. of Soc. Photo-Opt. Instr. Eng.*, **1121**, 73–88, 1989.

[256] M. Takeda and Z. Tung. Subfringe holographic interferometry by computer-based spatial-carrier fringe-pattern analysis. *J. Optics*, **16**, 127–131, 1985.

[257] S. Toyooka, H. Nishida, and J. Takezaki. Automatic analysis of holographic and shearographic fringes to measure flexural strains in plates. *Opt. Eng.*, **28**, 55–60, 1989.

[258] G. O. Reynolds, D. A. Servaes, L. Ramos-Izquierdo, J. B. DeVelis, D. C. Peirce, P. D. Hilton, and R. A. Mayville. Holographic fringe linearization interferometry for defect detection. *Opt. Eng.*, **24**, 757–768, 1985.

[259] D. R. Matthys, T. D. Dudderar, and J. A. Gilbert. Automated analysis of holointerferograms for the determination of surface displacement. *Exp. Mech.*, **28**, 86–91, 1988.

[260] J. D. Trolinger. Application of generalized phase control during reconstruction to flow visualization holography. *Appl. Opt.*, **18**, 766–774, 1979.

[261] D. Tentori and D. Salazar. Hologram interferometry: carrier fringes. *Appl. Opt.*, **30**, 5157–5158, 1991.

[262] P. Long, D. Hsu, and B. Wang. Cylindrical fringe carrier technique in holographic interferometry. *Opt. Eng.*, **27**, 867–869, 1988.

[263] W. W. Macy Jr. Two-dimensional fringe-pattern analysis. *Appl. Opt.*, **22**, 3898–3901, 1983.

[264] P. L. Ransom and J. V. Kokal. Interferogram analysis by a modified sinusoid fitting technique. *Appl. Opt.*, **25**, 4199–4204, 1986.

[265] K. H. Womack. Frequency domain description of interferogram analysis. *Opt. Eng.*, **23**, 396–400, 1984.

[266] K. H. Womack. Interferometric phase measurement using spatial synchronuous detection. *Opt. Eng.*, **23**, 391–395, 1984.

[267] U. Schnars. Direct phase determination in hologram interferometry with use of digitally recorded holograms. *Journ. Opt. Soc. Amer. A*, **11**, 2011–2015, 1994.

[268] L. P. Yaroslavskii and N. S. Merzlyakov. *Methods of Digital Holography.* Consultants Bureau, New York, 1980.

[269] U. Schnars and W. Jüptner. Direct recording of holograms by a CCD target and numerical reconstruction. *Appl. Opt.*, **33**, 179–181, 1994.

[270] U. Schnars, Th. Kreis, and W. Jüptner. CCD-recording and numerical reconstruction of holograms and holographic interferograms. In M. Kujawinska, R. J. Pryputniewicz, and M. Takeda, eds., *Interferometry VII: Techniques and Analysis*, *Proc. of Soc. Photo-Opt. Instr. Eng.*, **2544**, 57–63, 1995.

[271] U. Schnars, Th. Kreis, and W. Jüptner. Digital recording and numerical reconstruction of holograms: Reduction of the spatial frequency spectrum. *Opt. Eng.*, 1996.

[272] U. Schnars and W. Jüptner. Digital recording and reconstruction of holograms in hologram interferometry and shearography. *Appl. Opt.*, **33**, 4373–4377, 1994.

[273] M. J. Landry and C. M. Wise. Automatic data reduction of certain holographic interferograms. *Appl. Opt.*, **12**, 2320–2327, 1973.

[274] E. B. Aleksandrov and A. M. Bonch-Bruevich. Investigation of surface strains by the hologram technique. *Sov. Phys. Tech. Phys.*, **12**, 258–265, 1967.

[275] A. E. Ennos. Measurement of in-plane surface strain by hologram interferometry. *Journ. Phys. E: Scient. Instr.*, **1**, 731–734, 1968.

[276] L. Ek and K. Biedermann. Analysis of a system for hologram interferometry with a continuously scanning reconstruction beam. *Appl. Opt.*, **16**, 2535–2542, 1977.

[277] V. Fossati Bellani and A. Sona. Measurement of three-dimensional displacements by scanning a double-exposure hologram. *Appl. Opt.*, **13**, 1337–1341, 1974.

[278] L. Ek and K. Biedermann. Implementation of hologram interferometry with a continuously scanning reconstruction beam. *Appl. Opt.*, **17**, 1727–1732, 1978.

[279] R. Peralta-Fabi. Measurements of microdisplacements by holographic image processing. In W. F. Fagan, ed., *Industrial Applications of Laser Technology, Proc. of Soc. Photo-Opt. Instr. Eng.*, **398**, 169–173, 1983.

[280] H. Rytz. Holographic measurement of three dimensional displacements by means of a holographic plate. In W. F. Fagan, ed., *Industrial Applications of Laser Technology, Proc. of Soc. Photo-Opt. Instr. Eng.*, **398**, 53–58, 1983.

[281] J. M. Huntley and J. R. Buckland. Characterization of sources of 2π phase discontinuity in speckle interferograms. *Journ. Opt. Soc. Amer. A*, **12**, 1990–1996, 1995.

[282] W. Osten and R. Höfling. The inverse modulo process in automatic fringe analysis - problems and approaches. In K. Stetson and R. Pryputniewicz, eds., *International Conference on Hologram Interferometry and Speckle Metrology*, 301–309, 1990.

[283] D. R. Burton and M. J. Lalor. Multichannel Fourier fringe analysis as an aid to automatic phase unwrapping. *Appl. Opt.*, **33**, 2939–2948, 1994.

[284] D. J. Bone. Fourier fringe analysis: the two-dimensional phase unwrapping problem. *Appl. Opt.*, **30**, 3627–3632, 1991.

[285] K. Itoh. Analysis of the phase unwrapping algorithm. *Appl. Opt.*, **21**, 2470, 1982.

[286] J. M. Tribolet. A new phase unwrapping algorithm. *IEEE Trans. on Acoust., Speech, and Signal Proc.*, **25**, 170–177, 1977.

[287] D. W. Robinson. Phase unwrapping methods. In D. W. Robinson and G. T. Reid, eds., *Interferogram Analysis: Digital Fringe Pattern Measurement Techniques*, 194–229, Institute of Physics Publ., Bristol and Philadelphia, 1993.

[288] M. Takeda. Fringe formula for projection type moire topography. *Opt. Lasers in Eng.*, **3**, 45–52, 1982.

[289] D. W. Robinson and D. C. Williams. Digital phase stepping speckle interferometry. *Opt. Comm.*, **57**, 26–30, 1986.

[290] H. A. Vrooman and A. A. M. Maas. Image processing in digital speckle interferometry. In H. Halliwell, ed., *Fringe Analysis 89*, FASIG, 1989.

[291] K. Andresen and Q. Yu. Robust phase unwrapping by spin filtering using a phase direction map. In W. Jüptner and W. Osten, eds., *Fringe '93*, 154–156, 1993.

[292] P. J. Bryanston-Cross and C. Quan. Examples of automatic phase unwrapping applied to interferometric and photoelastic images. In W. Jüptner and W. Osten, eds., *Fringe '93*, 121–135, 1993.

[293] D. G. Ghiglia and L. A. Romero. Robust two-dimensional weighted and unweighted phase unwrapping that uses fast transforms and iterative methods. *Journ. Opt. Soc. Amer. A*, **11**, 107–117, 1994.

[294] T. R. Judge, Ch. Quan, and P. J. Bryanston-Cross. Holographic deformation measurements by Fourier transform technique with automatic phase unwrapping. *Opt. Eng.*, **31**, 533–543, 1992.

[295] J. A. Quiroga and E. Bernabeu. Phase-unwrapping algorithm for noisy phase-map processing. *Appl. Opt.*, **33**, 6725–6731, 1994.

[296] J. Schörner, A. Ettemeyer, U. Neupert, H. Rottenkolber, C. Winter, and P. Obermeier. New approaches in interpreting holographic images. *Opt. Lasers in Eng.*, **14**, 283–291, 1991.

[297] R. Cusack, J. M. Huntley, and H. T. Goldrein. Improved noise-immune phase-unwrapping algorithm. *Appl. Opt.*, **34**, 781–789, 1995.

[298] J. M. Huntley. Noise-immune phase unwrapping algorithm. *Appl. Opt.*, **28**, 3268–3270, 1989.

[299] J. J. Gierloff. Phase unwrapping by regions. In R. E. Fischer and W. J. Smith, eds., *Current Developments in Optical Engineering II, Proc. of Soc. Photo-Opt. Instr. Eng.*, **818**, 2–9, 1987.

[300] O. Y. Kwon. Advanced wavefront sensing at lockheed. In N. A. Massie, ed., *Interferometric Metrology, Proc. of Soc. Photo-Opt. Instr. Eng.*, **816**, 196–211, 1987.

[301] D. Winter, R. Ritter, and H. Sadewasser. Evaluation of modulo 2π phase-images with regional discontinuities: area based unwrapping. In W. Jüptner and W. Osten, eds., *Fringe '93*, 157–159, 1993.

[302] P. Stephenson, D. R. Burton, and M. J. Lalor. Data validation techniques in a tiled phase unwrapping algorithm. *Opt. Eng.*, **33**, 3703–3708, 1994.

[303] D. P. Towers, T. Judge, and P. J. Bryanston-Cross. A quasi heterodyne holographic technique and automatic algorithms for phase unwrapping. In G. T. Reid, ed., *Fringe Pattern Analysis, Proc. of Soc. Photo-Opt. Instr. Eng.*, **1163**, 95–119, 1989.

[304] D. P. Towers, T. R. Judge, and P. J. Bryanston-Cross. Automatic interferogram analysis techniques applied to quasi-heterodyne holography and ESPI. *Opt. Lasers in Eng.*, **14**, 239–281, 1991.

[305] D. G. Ghiglia, G. A. Mastin, and L. A. Romero. Cellular-automata method for phase unwrapping. *Journ. Opt. Soc. Amer. A*, **4**, 267–280, 1987.

[306] A. Spik and D. W. Robinson. Investigation of the cellular automata method for phase unwrapping and its implementation on an array processor. *Opt. Lasers in Eng.*, **14**, 25–37, 1991.

[307] R. J. Green and J. G. Walker. Phase unwrapping using a priori knowledge about the band limits of a function. In D. W. Braggins, ed., *Industrial Inspection, Proc. of Soc. Photo-Opt. Instr. Eng.*, **1010**, 36–43, 1989.

[308] R. J. Green, J. G. Walker, and D. W. Robinson. Investigation of the Fourier-transform method of fringe pattern analysis. *Opt. Lasers in Eng.*, **8**, 29–44, 1988.

[309] Th. Kreis, R. Biedermann, and W. Jüptner. Evaluation of holographic interference patterns by artificial neural networks. In M. Kujawinska, R. J. Pryputniewicz, and M. Takeda, eds., *Interferometry VII: Techniques and Analysis, Proc. of Soc. Photo-Opt. Instr. Eng.*, **2544**, 11–24, 1995.

[310] M. Takeda, K. Nagatome, and Y. Watanabe. Phase unwrapping by neural network. In W. Jüptner and W. Osten, eds., *Fringe '93*, 136–141, 1993.

[311] J. M. Huntley, R. Cusack, and H. Saldner. New phase unwrapping algorithms. In W. Jüptner and W. Osten, eds., *Fringe '93*, 148–153, 1993.

[312] J. M. Huntley and H. Saldner. Temporal phase-unwrapping algorithm for automated interferogram analysis. *Appl. Opt.*, **32**, 3047–3052, 1993.

[313] J. S. Lim. *Two-Dimensional Signal and Image Processing.* Prentice-Hall Intern., Inc., 1990.

[314] J. Ebbeni. Measurement of mechanical deformations from holographic interferograms. In P. Smigielski, ed., *Deuxieme Colloque Franco-Allemand sur les Applications de l'Holographie*, 1–18, 1989.

[315] Z. Füzessy and N. Abramson. Measurement of 3-D displacement: sandwich holography and regulated path length interferometry. *Appl. Opt.*, **21**, 260–264, 1982.

[316] E. Müller, V. Hrdliczka, and D. E. Cuche. Computer-based evaluation of holographic interferograms. In W. F. Fagan, ed., *Industrial Applications of Laser Technology*, *Proc. of Soc. Photo-Opt. Instr. Eng.*, **398**, 46–52, 1983.

[317] R. J. Pryputniewicz and W. W. Bowley. Techniques of holographic displacement measurement: an experimental comparison. *Appl. Opt.*, **17**, 1748–1756, 1978.

[318] K. A. Stetson. Use of sensivity vector variations to determine absolute displacements in double exposure hologram interferometry. *Appl. Opt.*, **29**, 502–504, 1990.

[319] Z. Füzessy. Application of double-pulse holography for the investigations of machines and systems. In G. Frankowski, N. Abramson, and Z. Füzessy, eds., *Application of Metrological Laser Methods in Machines and Systems*, *Physical Research*, **15**, 75–107. Akademie Verlag, 1991.

[320] Z. Füzessy and F. Gyimesi. Difference holographic interferometry: displacement measurement. *Opt. Eng.*, **23**, 780–783, 1984.

[321] Z. Füzessy and P. Wesolowski. Simplified static holographic evaluation method omitting the zero-order fringe. *Opt. Eng.*, **24**, 1023–1025, 1985.

[322] W. Jüptner, J. Geldmacher, Th. Bischof, and Th. Kreis. Measurement of the deformation of a pressure vessel above a weld point. In R. J. Pryputniewicz, G. M. Brown, and W. Jüptner, eds., *Interferometry: Applications*, *Proc. of Soc. Photo-Opt. Instr. Eng.*, **1756**, 98–105, 1992.

[323] Th. Kreis and W. Jüptner. Determination of defects by combination of holographic interferometry and finite-element-method, (in German). *VDI-Berichte*, **631**, 139–151, 1987.

[324] M. Medhat. Computer analysis of interferograms by fitting quadratic forms. *Journal of Modern Optics*, **38**, 121–128, 1991.

[325] J. D. Trolinger, D. C. Weber, G. C. Pardoen, G. T. Gunnarsson, and W. F. Fagan. Application of long-range holography in earthquake engineering. *Opt. Eng.*, **30**, 1315–1319, 1991.

[326] W. Jüptner and H. Kreitlow. Holographic recording of non-vibration protected objects (in German). In W. Waidelich, ed., *Laser 77 Opto-Electronics*, 420–425. ipc science and technology press, 1977.

[327] D. B. Neumann and H. W. Rose. Improvement of recorded holographic fringes by feedback control. *Appl. Opt.*, **6**, 1097–1104, 1967.

[328] H. W. Rose and H. D. Pruett. Stabilization of holographic fringes by FM feedback. *Appl. Opt.*, **7**, 87–89, 1968.

[329] H. Kreitlow, Th. Kreis, and W. Jüptner. Holographic interferometry with reference beams modulated by the object motion. *Appl. Opt.*, **26**, 4256–4262, 1987.

[330] F. M. Mottier. Holography of randomly moving objects. *Appl. Phys. Lett.*, **15**, 44–45, 1969.

[331] J. P. Waters. Object motion compensation by speckle reference beam interferometry. *Appl. Opt.*, **11**, 630–636, 1972.

[332] D. B. Neumann and R. C. Penn. Object motion compensation using reflection holography. *Journ. Opt. Soc. Amer.*, **62**, 1373, 1972.

[333] N. Abramson. Sandwich hologram interferometry: A new dimension in holographic comparison. *Appl. Opt.*, **13**, 2019–2025, 1974.

[334] N. Abramson. Sandwich hologram interferometry 2: Some practical calculations. *Appl. Opt.*, **14**, 981–984, 1975.

[335] N. Abramson. Sandwich hologram interferometry 4: Holographic studies of two milling machines. *Appl. Opt.*, **16**, 2521–2531, 1977.

[336] N. Abramson and H. Bjelkhagen. Industrial holographic measurements. *Appl. Opt.*, **11**, 2792–2796, 1973.

[337] N. Abramson and H. Bjelkhagen. Pulsed sandwich holography 2: Practical application. *Appl. Opt.*, **17**, 187–191, 1978.

[338] N. Abramson and H. Bjelkhagen. Sandwich hologram interferometry 5: Measurement of in-plane displacement and compensation for rigid body motion. *Appl. Opt.*, **18**, 2870–2880, 1979.

[339] N. Abramson, H. Bjelkhagen, and P. Skande. Sandwich holography for storing information interferometrically with a high degree of accuracy. *Appl. Opt.*, **18**, 2017–2021, 1979.

[340] H. Bjelkhagen. Pulsed sandwich holography. *Appl. Opt.*, **16**, 1727–1731, 1977.

[341] G. Lai and T. Yatagai. Dual-reference holographic interferometry with a double pulsed laser. *Appl. Opt.*, **27**, 3855–3858, 1988.

[342] A. Stimpfling and P. Smigielski. New method for compensating and measuring any motion of three-dimensional objects in holographic interferometry. *Opt. Eng.*, **24**, 821–823, 1985.

[343] P. Smigielski. New possibilities of holographic interferometry. In W. Jüptner, ed., *Holography Techniques and Applications*, *Proc. of Soc. Photo-Opt. Instr. Eng.*, **1026**, 90–92, 1988.

[344] B. Lutz and W. Schumann. Approach to extend the domain of visibility of recovered modified fringes when holographic interferometry is applied to large deformations. *Opt. Eng.*, **34**, 1879–1886, 1995.

[345] W. Schumann. Holographic interferometry applied to the case of large deformations. *Journ. Opt. Soc. Amer. A*, **6**, 1738–1747, 1989.

[346] S. Toyooka. Holographic interferometer, proof against external vibrations. *Optica Acta*, **29**, 861–865, 1982.

[347] P. Cielo. *Optical Techniques for Industrial Inspection.* Academic Press, 1988.

[348] R. J. Pryputniewicz. Determination of the sensitivity vectors directly from holograms. *Journ. Opt. Soc. Amer.*, **67**, 1351–1353, 1977.

[349] R. J. Pryputniewicz and K. A. Stetson. Determination of sensitivity vectors in hologram interferometry from two known rotations of the object. *Appl. Opt.*, **19**, 2201–2205, 1980.

[350] D. Vogel, V. Grosser, W. Osten, J. Vogel, and R. Höfling. Holographic 3d-measurement technique based on a digital image processing system. In W. Jüptner and W. Osten, eds., *Fringe '89*, 33–42, 1989.

[351] R. Dändliker and R. Thalmann. Determination of 3-D displacement and strain by holographic interferometry for non-plane objects. In W. F. Fagan, ed., *Industrial Applications of Laser Technology*, *Proc. of Soc. Photo-Opt. Instr. Eng.*, **398**, 11–16, 1983.

[352] S. K. Dhir and J. P. Sikora. An improved method for obtaining the general displacement field from a holographic interferogram. *Exp. Mech.*, **12**, 323–327, 1972.

[353] P. W. King. Holographic interferometry technique utilizing two plates and relative fringe orders for measuring micro-displacements. *Appl. Opt.*, **13**, 231–233, 1974.

[354] C. A. Sciammarella and T. Y. Chang. Holographic interferometry applied to the solution of a shell problem. *Exp. Mech.*, **14**, 217–224, 1974.

[355] C. A. Sciammarella and J. A. Gilbert. Strain analysis of a disk subjected to diametral compression by means of holographic interferometry. *Appl. Opt.*, **12**, 1951–1956, 1973.

[356] W. Schumann, J.-P. Zürcher, and D. Cuche. *Holography and Deformation Analysis.* Springer-Verlag, 1985.

[357] G. Schönebeck. New holographic means to exactly determine coefficients of elasticity. In W. F. Fagan, ed., *Industrial Applications of Laser Technology*, *Proc. of Soc. Photo-Opt. Instr. Eng.*, 130–136, **398**, 1983.

[358] G. Schönebeck. Holography and torsional problems. In W. F. Fagan, ed., *Industrial Optoelectronic Measurement Systems using Coherent Light*, *Proc. of Soc. Photo-Opt. Instr. Eng.*, **863**, 173–177, 1987.

[359] D. Cuche and W. Schumann. Fringe modification with amplification in holographic interferometry and application of this to determine strain and rotation. In W. F. Fagan, ed., *Industrial Applications of Laser Technology*, *Proc. of Soc. Photo-Opt. Instr. Eng.*, **398**, 35–45, 1983.

[360] C. A. Sciammarella and R. Narayanan. The determination of the components of the strain tensor in holographic interferometry. *Exp. Mech.*, **24**, 257–264, 1984.

[361] K. A. Stetson. The relationship between strain and derivatives of observed displacement in coherent optical metrology. *Exp. Mech.*, **7**, 273–275, 1981.

[362] L. H. Taylor and G. B. Brandt. An error analysis of holographic strains determined by cubic splines. *Exp. Mech.*, 543–548, 1972.

[363] D. E. Cuche. Determination of the Poisson's ratio by the holographic moire technique. In W. Jüptner, ed., *Holography Techniques and Applications*, *Proc. of Soc. Photo-Opt. Instr. Eng.*, **1026**, 165–170, 1988.

[364] R. J. Pryputniewicz and K. A. Stetson. Holographic strain analysis: extension of fringe vector method to include perspective. *Appl. Opt.*, **15**, 725–728, 1976.

[365] R. J. Pryputniewicz. Holographic strain analysis: an experimental implementation of the fringe-vector theory. *Appl. Opt.*, **17**, 3613–3618, 1978.

[366] K. A. Stetson. Homogeneous deformations: determination by fringe vectors in hologram interferometry. *Appl. Opt.*, **14**, 2256–2259, 1975.

[367] G. M. Brown. Fringe analysis for automotive applications. *Opt. Lasers in Eng.*, **19**, 203–220, 1993.

[368] W. Jüptner and Th. Kreis, eds.. *An External Interface for Processing 3-D Holographic and X-Ray Images.* Research Reports ESPRIT. Springer-Verlag, 1989.

[369] A. S. Kobayashi. Hybrid experimental-numerical stress analysis. *Exp. Mech.*, **23**, 338–347, 1983.

[370] M. M. Ratnam and W. T. Evans. Comparison of measurement of piston deformation using holographic interferometry and finite elements. *Exp. Mech.*, **12**, 336–342, 1993.

[371] J. M. Weathers, W. A. Foster, W. F. Swinson, and J. L. Turner. Integration of laser-speckle and finite-element techniques of stress analysis. *Exp. Mech.*, **3**, 60–65, 1985.

[372] K. Oh and R. J. Pryputniewicz. Application of electro-optic holography in the study of cantilever plate vibration with concentrated masses. In K. Stetson and R. Pryputniewicz, eds., *International Conference on Hologram Interferometry and Speckle Metrology*, 245–253, 1990.

[373] H. Borner, M. Schulz, J. Villain, and H. Steinbichler. Application of holographic interferometry supported by FEM-calculations during the development of a new assembly technique. In W. Jüptner, ed., *Holography Techniques and Applications*, *Proc. of Soc. Photo-Opt. Instr. Eng.*, **1026**, 171–175, 1988.

[374] M. A. Caponero, A. De Angelis, V. R. Filetti, and S. Gammella. Structural analysis of an aircraft turbine blade prototype by use of holographic interferometry. In R. J. Pryputniewicz, G. M. Brown, and W. Jüptner, eds., *Interferometry VI: Applications*, *Proc. of Soc. Photo-Opt. Instr. Eng.*, **2004**, 150–161, 1993.

[375] X. Bohineust, V. Linet, and F. Dupuy. Development and application of holographic modal decomposition techniques to acoustic analysis of vehicle structures. In R. J. Pryputniewicz, G. M. Brown, and W. Jüptner, eds., *Interferometry VI: Applications*, *Proc. of Soc. Photo-Opt. Instr. Eng.*, **2004**, 118–129, 1993.

[376] R. J. Pryputniewicz. Holographic and finite element studies of vibrating beams. In W. F. Fagan, ed., *Optics in Engineering Measurement*, *Proc. of Soc. Photo-Opt. Instr. Eng.*, **599**, 54–62, 1985.

[377] Th. Bischof and W. Jüptner. Investigation of the stress distribution in intact bonds by holographic interferometry and finite element method. In R. J. Pryputniewicz, ed., *Laser Interferometry IV: Computer-Aided Interferometry*, *Proc. of Soc. Photo-Opt. Instr. Eng.*, **1553**, 326–331, 1991.

[378] W. Jüptner, Th. Kreis, J. Geldmacher, and Th. Bischof. Detection of defects in adhesion bonds using holographic interferometry, (in German). *Qualität und Zuverlässigkeit*, **36**, 417–423, 1991.

[379] J. Balas, J. Sladek, and M. Drzik. Stress analysis by combination of holographic interferometry and boundary-integral method. *Exp. Mech.*, **83**, 196–202, 1983.

[380] C. A. Sciammarella. Contributions of interferometry to the field of fracture mechanics. In R. J. Pryputniewicz, G. M. Brown, and W. Jüptner, eds., *Interferometry VI: Applications*, *Proc. of Soc. Photo-Opt. Instr. Eng.*, **2004**, 204–214, 1993.

[381] J. M. Huntley and L. R. Benckert. Measurement of dynamic crack tip displacement field by speckle photography and interferometry. *Opt. Lasers in Eng.*, **19**, 299–312, 1993.

[382] L. W. Meyer, W. Jüptner, and H.-D. Steffens. Fracture toughness investigations using holographic interferometry. In W. Waidelich, ed., *Laser 75 Opto-Electronics*, 203–205. IPC Science and Technology Press, 1975.

[383] A. J. Moore and J. R. Tyrer. The evaluation of fracture mechanics parameters from electronic speckle pattern interferometric fringe patterns. *Opt. Lasers in Eng.*, **19**, 325–336, 1993.

[384] W. Jüptner, K. Grünewald, R. Zirn, and H. Kreitlow. Measurement of the stress intensity factor k_i in large specimens by means of holographic interferometry. In D. Vukicevic, ed., *Holographic Data Nondestructive Testing*, *Proc. of Soc. Photo-Opt. Instr. Eng.*, **370**, 62–65, 1982.

[385] P. Will, W. Totzauer, and B. Michel. Generalized J-integral of fracture mechanics from holographic data. *Phys. Stat. Sol. (a)*, **95**, K113–K116, 1986.

[386] K. A. Stetson and I. R. Harrison. Computer-aided holographic vibration analysis for vectorial displacements of bladed disks. *Appl. Opt.*, **17**, 1733–1738, 1978.

[387] K. A. Stetson. Holographic vibration analysis. In R. K. Erf, ed., *Holographic Non-Destuctive Testing*, 181–220. Academic Press, 1974.

[388] J. Janta and M. Miler. Time average holographic interferometry of damped oscillations. *Optik*, **36**, 185–195, 1972.

[389] P. C. Gupta and K. Singh. Time-average hologram interferometry of periodic, non-cosinusoidal vibrations. *Appl. Phys.*, **6**, 233–240, 1975.

[390] P. C. Gupta and K. Singh. Characteristic fringe function for time-average holography of periodic nonsinusoidal vibrations. *Appl. Opt.*, **14**, 129–133, 1975.

[391] P. C. Gupta and K. Singh. Hologram interferometry of vibrations represented by the square of a Jacobian elliptic function. *Nouv. Rev. Opt.*, **7**, 95–100, 1976.

[392] P. Hariharan, B. F. Oreb, and C. H. Freund. Stroboscopic holographic interferometry: measurements of vector components of a vibration. *Appl. Opt.*, **26**, 3899–3903, 1987.

[393] P. Sajenko and C. D. Johnson. Stroboscopic holographic interferometry. *Appl. Phys. Lett.*, **13**, 22–24, 1968.

[394] S. Nakadate, H. Saito, and T. Nakajima. Vibration measurement using phase-shifting stroboscopic interferometry. *Optica Acta*, **33**, 1295–1309, 1986.

[395] B. Ineichen and J. Mastner. Vibration analysis by stroboscopic two-reference-beam heterodyne holographic interferometry. In P. Meyrueis and M. Grosmann, eds., *2nd European Congress on Optics applied to Metrology*, *Proc. of Soc. Photo-Opt. Instr. Eng.*, **210**, 207–212, 1979.

[396] K. A. Stetson. Method of vibration measurements in heterodyne interferometry. *Opt. Lett.*, **7**, 233–234, 1982.

[397] P. Hariharan. Application of holographic subtraction to time-average hologram interferometry of vibrating objects. *Appl. Opt.*, **12**, 143–146, 1973.

[398] R. J. Pryputniewicz. Time-average holography in vibration analysis. *Opt. Eng.*, **24**, 843–848, 1985.

[399] R. Tonin and D. A. Bies. Analysis of 3-D vibrations from time-averaged holograms. *Appl. Opt.*, **17**, 3713–3721, 1978.

[400] R. Tonin and D. A. Bies. General theory of time-averaged holography for the study of three-dimensional vibrations at a single frequency. *Journ. Opt. Soc. Amer.*, **68**, 924–931, 1978.

[401] Y. Xu, C. M. Vest, and J. D. Murray. Holographic interferometry used to demonstrate a theory of pattern formation in animal coats. *Appl. Opt.*, **22**, 3479–3483, 1983.

[402] K. Antropius. Fundamentals of time-average holographic interferometry and its applications to vibration measurements. In G. Frankowski, N. Abramson, and Z. Füzessy, eds., *Application of Metrological Laser Methods in Machines and Systems, Physical Research*, **15**, 167–194. Akademie Verlag, 1991.

[403] C. S. Vikram. Study of vibrations. In P. K. Rastogi, ed., *Holographic Interferometry*, 293–318. Springer Series in Optical Sciences, **68**, 1994.

[404] M. Ueda, S. Miida, and T. Sato. Signal-to-noise ratio and smallest detectable vibration amplitude in frequency-translated holography - an analysis. *Appl. Opt.*, **15**, 2690–2694, 1976.

[405] C. C. Aleksoff. Temporally modulated holography. *Appl. Opt.*, **10**, 1329–1341, 1971.

[406] J. W. Goodman. Temporal filtering properties of holograms. *Appl. Opt.*, **6**, 857–859, 1967.

[407] N. Takai, M. Yamada, and T. Idogawa. Holographic interferometry using a reference wave with a sinusoidally modulated amplitude. *Opt. Laser Technol.*, **8**, 21–23, 1976.

[408] D. B. Neumann, C. F. Jacobsen, and G. M. Brown. Holographic technique for determining the phase of vibrating objects. *Appl. Opt.*, **9**, 1357–1368, 1970.

[409] C. C. Aleksoff. Time average holography extended. *Appl. Phys. Lett.*, **14**, 23–24, 1969.

[410] J. A. Levitt and K. A. Stetson. Mechanical vibrations: mapping their phase with hologram interferometry. *Appl. Opt.*, **15**, 195–199, 1976.

[411] C. S. Vikram. Mechanical vibrations: mapping their phase with hologram interferometry. *Appl. Opt.*, **16**, 1140–1141, 1977.

[412] J. Politch. Spectroscopic holography: a new method for mode analysis of randomly excited structures. *Journ. Opt. Soc. Amer. A*, **7**, 1355–1361, 1990.

[413] P. Hariharan and B. F. Oreb. Stroboscopic holographic interferometry: application of digital techniques. *Opt. Comm.*, **59**, 83–86, 1986.

[414] S. Nakadate. Vibration measurement using phase-shifting time-average holographic interferometry. *Appl. Opt.*, **25**, 4155–4161, 1986.

[415] K. A. Stetson. Fringe-shifting technique for numerical analysis of time-average holograms of vibrating objects. *Journ. Opt. Soc. Amer. A*, **5**, 1472–1476, 1988.

[416] H. E. Cline, A. S. Holik, and W. E. Lorensen. Computer-aided surface reconstruction of interference contours. *Appl. Opt.*, **21**, 4481–4488, 1982.

[417] H. E. Cline, W. E. Lorensen, and A. S. Holik. Automatic moire contouring. *Appl. Opt.*, **23**, 1454–1459, 1984.

[418] A. A. Friesem and U. Levy. Fringe-formation in two-wavelength contour holography. *Appl. Opt.*, **15**, 3009–3020, 1976.

[419] K. A. Haines and B. P. Hildebrand. Contour generation by wavefront reconstruction. *Phys. Lett.*, **19**, 10–11, 1965.

[420] L. O. Heflinger and R. F. Wuerker. Holographic contouring via multifrequency lasers. *Appl. Phys. Lett.*, **15**, 28–30, 1969.

[421] B. P. Hildebrand and K. A. Haines. The generation of three-dimensional contour maps by wavefront reconstruction. *Phys. Lett.*, **21**, 422–423, 1966.

[422] B. P. Hildebrand and K. A. Haines. Multiple-wavelength and multiple-source holography applied to contour generation. *Journ. Opt. Soc. Amer.*, **57**, 155–162, 1967.

[423] R. P. Tatam, J. C. Davies, C. H. Buckberry, and J. D. C. Jones. Holographic surface contouring using wavelength modulation of laser diodes. *Opt. Laser Technol.*, **22**, 317–321, 1990.

[424] J. R. Varner. Simplified multiple-frequency holographic contouring. *Appl. Opt.*, **10**, 212–213, 1971.

[425] J. S. Zelenka and J. R. Varner. New method for generating depth contours holographically. *Appl. Opt.*, **7**, 2107–2110, 1968.

[426] T. Tsuruta, N. Shiotake, J. Tsujiuchi, and K. Matsuda. Holographic generation of contour map of diffusely reflecting surface by using immersion method. *Jap. J. Appl. Phys.*, **6**, 661, 1967.

[427] J. S. Zelenka and J. R. Varner. Multiple-index holographic contouring. *Appl. Opt.*, **8**, 1431–1434, 1969.

[428] N. Abramson. Holographic contouring by translation. *Appl. Opt.*, **15**, 1018–1022, 1976.

[429] P. K. Rastogi and L. Pflug. A holographic technique featuring broad range sensitivity to contour diffuse objects. *Journal of Modern Optics*, **38**, 1673–1683, 1991.

[430] P. Carelli, D. Paoletti, G. Schirripa Spagnolo, and A. D'Altorio. Holographic contouring method: application to automatic measurements of surface defects in artwork. *Opt. Eng.*, **30**, 1294–1298, 1991.

[431] P. DeMattia and V. Fossati-Bellani. Holographic contouring by displacing the object and the illumination beam. *Opt. Comm.*, **26**, 17–21, 1978.

[432] M. Yonemura. Holographic contour generation by spatial frequency modulation. *Appl. Opt.*, **21**, 3652–3658, 1982.

[433] N. Abramson. Sandwich hologram interferometry 3: Contouring. *Appl. Opt.*, **15**, 200–205, 1976.

[434] N. Abramson. Light-in-flight recording: High-speed holographic motion pictures of ultrafast phenomena. *Appl. Opt.*, **22**, 215–232, 1983.

[435] N. Abramson. Light-in-flight recording. 2: Compensation for the limited speed of the light used for observation. *Appl. Opt.*, **23**, 1481–1492, 1984.

[436] N. Abramson. Light-in-flight recording. 3: Compensation for optical relativistic effects. *Appl. Opt.*, **23**, 4007–4014, 1984.

[437] N. Abramson. Light-in-flight recording. 4: Visualizing optical relativistic phenomena. *Appl. Opt.*, **24**, 3323–3329, 1985.

[438] N. Abramson. Time reconstructions in light-in-flight recording by holography. *Appl. Opt.*, **30**, 1242–1252, 1991.

[439] N. Abramson and K. G. Spears. Single pulse light-in-flight recording by holography. *Appl. Opt.*, **28**, 1834–1841, 1989.

[440] G. Tribillon and R. Salazar. Holographic reconstruction of a synthetised subpicosecond pulse. In W. F. Fagan, ed., *Industrial Applications of Laser Technology, Proc. of Soc. Photo-Opt. Instr. Eng.*, 181–184, **398**, 1983.

[441] T. E. Carlsson. Measurement of three-dimensional shapes using light-in-flight recording by holography. *Opt. Eng.*, **32**, 2587–2592, 1993.

[442] J. C. Wyant. Testing aspherics using two-wavelength holography. *Appl. Opt.*, **10**, 2113–2118, 1971.

[443] C. Polhemus. Two-wavelength interferometry. *Appl. Opt.*, **12**, 2071–2074, 1973.

[444] R. Dändliker, R. Thalmann, and D. Prongue. Two-wavelength laser interferometry using superheterodyne detection. *Opt. Lett.*, **13**, 339–341, 1988.

[445] P. Lam, J. D. Gaskill, and J. C. Wyant. Two-wavelength holographic interferometer. *Appl. Opt.*, **23**, 3079–3081, 1984.

[446] N. Ninane and M. P. Georges. Holographic interferometry using two-wavelength holography for the measurement of large deformations. *Appl. Opt.*, **34**, 1923–1928, 1995.

[447] W. Jüptner, Th. Kreis, and J. Geldmacher. Determination of absolute fringe order in hologram interferometry with wavelength controlled lasers. In M. Y. Y. Hung and R. J. Pryputniewicz, eds., *Industrial Laser Interferometry II, Proc. of Soc. Photo-Opt. Instr. Eng.*, **955**, 143–146, 1988.

[448] D. C. Holloway, A. M. Patacca, and W. L. Fourney. Direction-sensitive displacement analysis by multiple frequency holographic interferometry. *Appl. Opt.*, **17**, 1213–1219, 1978.

[449] C. A. Sciammarella. Holographic moire, an optical tool for the determination of displacements, strains, contours, and slopes of surfaces. *Opt. Eng.*, **21**, 447–457, 1982.

[450] C. A. Sciammarella and J. A. Gilbert. A holographic-moire technique to obtain separate patterns for components of displacement. *Exp. Mech.*, **16**, 215–220, 1976.

[451] C. A. Sciammarella and J. A. Gilbert. Holographic interferometry applied to measurement of displacements of interior points of transparent bodies. *Appl. Opt.*, **15**, 2176–2182, 1976.

[452] P. K. Rastogi, M Spajer, and J. Monneret. Inplane deformation meassurement using holographic moire. *Opt. Lasers in Eng.*, **2**, 79–103, 1981.

[453] P. K. Rastogi. Comparative holographic interferometry: A nondestructive inspection system for detection of flaws. *Exp. Mech.*, **25**, 325–337, 1985.

[454] C. A. Sciammarella and S. K. Chawla. Holographic moire technique to obtain displacement components and derivatives. *Exp. Mech.*, **18**, 373–381, 1978.

[455] S. K. Chawla and C. A. Sciammarella. Localization of fringes produced by rotation of the recording plate in focused-image holography. *Exp. Mech.*, **20**, 240–244, 1980.

[456] E. S. Simova and K. N. Stoev. Automated Fourier transform fringe-pattern analysis in holographic moire. *Opt. Eng.*, **32**, 2286–2294, 1993.

[457] M.-A. Beeck. Pulsed holographic vibration analysis on high-speed rotating objects: fringe formation, recording techniques, and practical applications. In W. Jüptner, ed., *Holography Techniques and Applications, Proc. of Soc. Photo-Opt. Instr. Eng.*, **1026**, 186–195, 1988.

[458] M.-A. Beeck. Pulsed holographic vibration analysis on high-speed rotating objects: fringe formation, recording techniques, and practical applications. *Opt. Eng.*, **31**, 553–561, 1992.

[459] J. P. Sikora and F. T. Mendenhall. Holographic vibration study of a rotating propeller blade. *Exp. Mech.*, **14**, 230–232, 1974.

[460] T. Tsuruta and Y. Itoh. Holographic interferometry for rotating subject. *Appl. Phys. Lett.*, **17**, 85–87, 1970.

[461] T. E. Carlsson, B. Nilsson, J. Gustafsson, and N. Abramson. Practical system for time-resolved holographic interferometry. *Opt. Eng.*, **30**, 1017–1022, 1991.

[462] J. Geldmacher, H. Kreitlow, P. Steinlein, and G. Sepold. Comparison of vibration mode measurements on rotating objects by different holographic methods. In W. F. Fagan, ed., *Industrial Applications of Laser Technology, Proc. of Soc. Photo-Opt. Instr. Eng.*, **398**, 101–110, 1983.

[463] E. Vogt, J. Geldmacher, B. Dirr, and H. Kreitlow. Hybrid-vibration mode analysis of rotating turbine-blade models. *Exp. Mech.*, **25**, 161–170, 1985.

[464] K. A. Stetson. The use of an image derotator in hologram interferometry and speckle photography of rotating objects. *Exp. Mech.*, **18**, 67–73, 1978.

[465] H. J. Tiziani. Holographic interferometry and speckle metrology: a review of the present state. In W. F. Fagan, ed., *Industrial Applications of Laser Technology, Proc. of Soc. Photo-Opt. Instr. Eng.*, **398**, 2–10, 1983.

[466] P. Waddell. Stopping rotary motion with a prism. *Machine Design*, **45**, 151–152, 1973.

[467] Lasermet LTD. The Lasermet Rotoviewer. Technical information sheet.

[468] U. Haupt and M. Rautenberg. Investigation of blade vibration of radial impellers by means of telemetry and holographic interferometry. *Trans. ASME: J. Eng. Power*, **104**, 838–843, 1982.

[469] D. B. Neumann. Comparative holography. In *Technical Digest of Topical Meeting on Hologram Interferometry and Speckle Metrology*, MB2-1–MB2-4. Opt. Soc. Amer., 1980.

[470] D. B. Neumann. Comparative holography: a technique for eliminating background fringes in holographic interferometry. *Opt. Eng.*, **24**, 625–627, 1985.

[471] Z. Füzessy and F. Gyimesi. Difference holographic interferometry. In W. F. Fagan, ed., *Industrial Applications of Laser Technology, Proc. of Soc. Photo-Opt. Instr. Eng.*, **398**, 240–243, 1983.

[472] Z. Füzessy, F. Gyimesi, and I. Banyasz. Difference holographic interferometry (DHI): single reference beam technique. *Opt. Comm.*, **68**, 404–407, 1984.

[473] F. Gyimesi and Z. Füzessy. Difference holographic interferometry: theory. *Journal of Modern Optics*, **35**, 1699–1716, 1988.

[474] Z. Füzessy, F. Gyimesi, and J. Kornis. Comparison of two filament lamps by difference hologram interferometry. *Opt. Laser Technol.*, **18**, 318–320, 1986.

[475] Z. Füzessy. Measurement of 3D-displacement by regulated path length interferometry. In W. F. Fagan, ed., *Industrial Applications of Laser Technology, Proc. of Soc. Photo-Opt. Instr. Eng.*, **398**, 17–21, 1983.

[476] F. Gyimesi and Z. Füzessy. Difference holographic interferometry (DHI): two-refractive-index contouring. *Opt. Comm.*, **53**, 17–22, 1985.

[477] Z. Füzessy and F. Gyimesi. Difference holographic interferometry (DHI): phase object measurement. *Opt. Comm.*, **57**, 31–33, 1986.

[478] P. K. Rastogi. Comparative holographic moire interferometry in real time. *Appl. Opt.*, **23**, 924–927, 1984.

[479] P. K. Rastogi. Interferometric comparison of diffuse objects using comparative holography. *Opt. Eng.*, **34**, 1923–1929, 1995.

[480] E. Simova and V. Sainov. Comparative holographic moire interferometry for nondestructive testing: comparison with conventional holographic interferometry. *Opt. Eng.*, **28**, 261–266, 1989.

[481] G. H. Kaufmann and P. K. Rastogi. Comparative holographic interferometry using a phase shifting technique. In K. Stetson and R. Pryputniewicz, eds., *International Conference on Hologram Interferometry and Speckle Metrology*, 180–183, 1990.

[482] P. K. Rastogi, M. Barillot, and G. H. Kaufmann. Comparative phase shifting holographic interferometry. *Appl. Opt.*, **30**, 722–728, 1991.

[483] P. M. Boone and P. Jacquot. Use of an all-holographic desensitized white-light interferometer for visualization of plastic zones. In K. Stetson and R. Pryputniewicz, eds., *International Conference on Hologram Interferometry and Speckle Metrology*, 494–499, 1990.

[484] S. S. Cha and C. M. Vest. Tomographic reconstruction of strongly refracting fields and its applications to interferometric measurements of boundary layers. *Appl. Opt.*, **20**, 2787–2794, 1981.

[485] W. Hauf and U. Grigull. *Optical Methods in Heat Transfer*, *Advances in Heat Transfer*, **6**. Academic Press, 1970.

[486] D. Vukicevic, T. Neger, H. Jäger, J. Woisetschläger, and H. Philipp. Optical tomography by heterodyne holographic interferometry. In P. Greguss and T. H. Jeong, eds., *Holography*, *SPIE Institute Series*, **IS 8**, 160–193, 1990.

[487] C. M. Vest. Interferometry of strongly refracting axisymmetric phase objects. *Appl. Opt.*, **14**, 1601–1606, 1975.

[488] R. Ladenburg, J. Winkler, and C.C.Van Voorhis. Interferometric study of faster than sound phenomena, part i. *Phys. Rev.*, **73**, 1359–1377, 1948.

[489] R. Barakat. Solution of an abel integral equation for band-limited functions by means of sampling theorems. *J. Math. Phys.*, **43**, 325–331, 1964.

[490] D. W. Sweeney. A comparison of Abel integral inversion schemes for interferometric applications. *Journ. Opt. Soc. Amer.*, **64**, 559, 1974.

[491] M. Kalal and K. A. Nugent. Abel inversion using fast Fourier transforms. *Appl. Opt.*, **27**, 1956–1959, 1988.

[492] C. M. Vest. Holographic interferometry of transparent media. In VDI, ed., *VDI Symposium on Holographic Interferometry*, 1983.

[493] I. H. Lira and Ch. M. Vest. Refraction correction in holographic interferometry and tomography of transparent objects. *Appl. Opt.*, **26**, 3919–3928, 1987.

[494] D. W. Sweeney and C. M. Vest. Reconstruction of three dimensional refractive index fields from multidirectonal interferometric data. *Appl. Opt.*, **12**, 2649–2663, 1973.

[495] D. D. Verhoeven. Application of holographic interferometry and computer tomography to full-field measurement of asymmetric refractive index distributions. In W. Jüptner, ed., *Holography Techniques and Applications*, *Proc. of Soc. Photo-Opt. Instr. Eng.*, **1026**, 111–122, 1988.

[496] M. V. Berry and D. F. Gibbs. The interpretation of optical projections. *Proc. Roy. Soc. London*, **A314**, 143–152, 1970.

[497] H. G. Junginger and W. van Haeringen. Calculation of three-dimensional refractive-index field using phase integrals. *Opt. Comm.*, **5**, 1–4, 1972.

[498] C. M. Vest. Tomography for properties of materials that bend rays: a tutorial. *Appl. Opt.*, **24**, 4089–4094, 1985.

[499] R. Snyder and L. Hesselink. Optical tomography for flow visualization of the density around a revolving helicopter rotor blade. *Appl. Opt.*, **23**, 3650–3656, 1984.

[500] S. J. Norton and M. Linzer. Correcting for ray refraction in velocity and attenuation tomography. *Ultrason. Imaging*, **4**, 201–233, 1982.

[501] K. Iwata and R. Nagata. Calculation of refractive index distribution from interferograms using the Born and Rytov's approximation. *Jpn. J. Appl. Phys.*, **14**, 379–383, 1975.

[502] S. K. Kenue and J. F. Greenleaf. Limited angle multifrequency diffraction tomography. *IEEE Trans. Son. Ultrason.*, **29**, 213–217, 1982.

[503] S. S. Cha and H. Sun. Tomography for reconstructing continuous fields from ill-posed multidirectional interferometric data. *Appl. Opt.*, **29**, 251–258, 1990.

[504] I. Prikryl and C. M. Vest. Holographic interferometry of transparent media using light scattered by embedded test objects. *Appl. Opt.*, **21**, 2554–2557, 1982.

[505] J. D. Trolinger and J. C. Hsu. Flowfield diagnostics by holographic interferometry and tomography. In W. Jüptner and W. Osten, eds., *Fringe '93*, 423–439, 1993.

[506] B. A. Tozer, R. Glanville, A. L. Gordon, M. J. Little, J. M. Webster, and D. G. Wright. Holography applied to inspection and measurement in an industrial environment. *Opt. Eng.*, **24**, 746–753, 1985.

[507] W. Osten, W. Jüptner, and U. Mieth. Knowledge assisted evaluation of fringe patterns for automatic fault detection. In R. J. Pryputniewicz, G. M. Brown, and W. Jüptner, eds., *Interferometry VI: Applications*, *Proc. of Soc. Photo-Opt. Instr. Eng.*, **2004**, 256–268, 1993.

[508] W. Jüptner, Th. Kreis, U. Mieth, and W. Osten. Application of neural networks and knowledge based systems for automatic identification of fault indicating fringe patterns. In R. J. Pryputniewicz and J. Stupnicki, eds., *Interferometry 94: Photomechanics*, *Proc. of Soc. Photo-Opt. Instr. Eng.*, **2342**, 16–26, 1994.

[509] U. Mieth, W. Osten, and W. Jüptner. Knowledge assisted fault detection based on line features of skeletons. In W. Jüptner and W. Osten, eds., *Fringe '93*, Physical Research, 367–373. Akademie-Verlag Berlin, 1993.

[510] D. A. Tichenor and V. P. Madsen. Computer analysis of holographic interferograms for nondestructive testing. *Opt. Eng.*, **18**, 469–472, 1979.

[511] Th. Kreis. *Evaluation of holographic interference patterns using methods of spatial frequency analysis.* (In German). Progress Reports VDI. VDI-Verlag, 1986.

[512] D. W. Robinson. Automatic fringe analysis with a computer image-processing system. *Appl. Opt.*, **22**, 2169–2176, 1983.

[513] N. A. Halliwell and C. J. Pickering. Analysis methods in laser speckle photography and particle image velocimetry. In D. W. Robinson and G. T. Reid, eds., *Interferogram Analysis: Digital Fringe Pattern Measurement Techniques*, 230–261, Institute of Physics Publ., Bristol and Philadelphia, 1993.

[514] B. Ineichen, P. Eglin, and R. Dändliker. Hybrid optical and electronic image processing for strain measurements by speckle photography. *Appl. Opt.*, **17**, 2191–2195, 1980.

[515] H. Kreitlow and Th. Kreis. Automatic evaluation of Young's fringes related to the study of in-plane-deformations by speckle techniques. In P. Meyrueis and M. Grosmann, eds., *2nd European Congress on Optics applied to Metrology*, *Proc. of Soc. Photo-Opt. Instr. Eng.*, **210**, 18–24, 1979.

[516] J. M. Huntley. Speckle photography fringe analysis by the Walsh transform. *Appl. Opt.*, **25**, 382–386, 1986.

[517] H. Kreitlow, Th. Kreis, and W. Jüptner. Optimierung der automatisierten Auswertung von Specklegrammen beim Einsatz eines schnellen Fouriertransformators. In W. Waidelich, ed., *Laser 83 - Optoelektronik in der Technik*, 159–164. Springer-Verlag, 1983.

[518] J. N. Butters and J. A. Leendertz. Holographic and video techniques applied to engineering measurement. *J. Meas. Control*, **4**, 349–354, 1971.

[519] A. Macovski, D. Ramsey, and L. F. Schaefer. *Appl. Opt.*, **10**, 2722–2727, 1971.

[520] O. Schwomma. Austrian Patent No. 298830, 1972.

[521] Y. Y. Hung. Electronic Shearography versus ESPI for Nondestructive Evaluation. In F.-P. Chiang, ed., *Moire Techniques, Holographic Interferometry, Optical NDT, and Applications to Fluid Mechanics*, *Proc. of Soc. Photo-Opt. Instr. Eng.*, **1554B**, 692–700, 1991.

[522] Y. Y. Hung and C. Y. Liang. Image-shearing camera for direct measurement of surface strain. *Appl. Opt.*, **18**, 1046–1051, 1979.

[523] A. R. Ganesan, D. K. Sharma, and M. P. Kothiyal. Universal digital speckle shearing interferometer. *Appl. Opt.*, **27**, 4731–4734, 1988.

[524] , K. Creath. Phase-shifting speckle interferometry. *Appl. Opt.*, **24**, 3053–3058, 1985.

[525] K. A. Stetson and W. R. Brohinsky. Electro-optic holography and its application to hologram interferometry. *Appl. Opt.*, **24**, 3631–3637, 1985.

[526] S. Leidenbach. The direct phase measurement - a new method for determination of a phase image from a single intensity image (in German). In W. Waidelich, ed., *Laser 91 - Optoelektronik Mikrowellen*, 68–72, 1991.

[527] K. R. Castleman. *Digital Image Processing*. Prentice-Hall, Inc., 1979.

[528] R. C. Gonzales. *Digital Image Processing*. Addison-Wesley, 1977.

[529] C. D. McGillem and G. R. Cooper. *Continuous and Discrete Signal and System Analysis*. Holt, Rinehart and Winston, 1974.

[530] A. V. Oppenheim and V. R. Schafer. *Digital Signal Processing*. Prentice-Hall, 1975.

[531] W. K. Pratt. *Digital Image Processing*. J. Wiley, 1978.

[532] L. R. Rabiner and B. Gold. *Theory and Applications of Digital Signal Processing.* Prentice-Hall, 1975.

[533] A. Rosenfeld and A. C. Kak. *Digital Picture Processing.* Academic Press, 2nd edition, 1982.

[534] S. D. Stearns. *Digital Signal Analysis.* Hayden Book Comp., Inc., 1975.

[535] M. J. Lighthill. *Introduction to Fourier Analysis and Generalized Functions.* Cambridge Univ. Press, 1960.

[536] A. C. Kak and M. Slaney. *Principles of Computerized Tomographic Imaging.* IEEE Press, 1988.

[537] E. O. Brigham. *The Fast Fourier Transform.* Prentice-Hall, Inc., 1974.

[538] J. Radon. Über die Bestimmung von Funktionen durch ihre Integralwerte längs gewisser Mannigfaltigkeiten. *Ber. der Sächs. Akad. der Wissensch.*, **29**, 262–277, 1917.

[539] R. H. T. Bates, K. L. Garden, and T. M. Peters. Overview of computerized tomography with emphasis on future developments. *Proc. IEEE,* **71**, 356–372, 1983.

[540] R. M. Lewitt. Reconstruction algorithms: transform methods. *Proc. IEEE,* **71**, 390–408, 1983.

[541] Y. Censor. Finite series-expansion reconstruction methods. *Proc. IEEE,* **71**, 409–419, 1983.

[542] A. L. Collins, M. W. Collins, and J. Hunter. The application of tomographic reconstruction techniques to the extraction of data from holographic interferograms. In R. J. Pryputniewicz, G. M. Brown, and W. Jüptner, eds., *Interferometry VI: Applications, Proc. of Soc. Photo-Opt. Instr. Eng.*, **2004**, 234–243, 1993.

[543] H. Tan and D. Modarress. Algebraic reconstruction technique code for tomographic interferometry. *Opt. Eng.*, **24**, 435–440, 1985.

[544] D. D. Verhoeven. MART-type CT algorithms for the reconstruction of multidirectional interferometric data. In R. J. Pryputniewicz, ed., *Laser Interferometry IV: Computer-Aided Interferometry, Proc. of Soc. Photo-Opt. Instr. Eng.*, **1553**, 376–387, 1991.

Author Index

Index